Construction Dispute Research

Sai On Cheung
Editor

Construction Dispute Research

Conceptualisation, Avoidance and Resolution

Springer

Editor
Sai On Cheung
Department of Civil and Architectural Engineering
City University of Hong Kong
Hong Kong
People's Republic of China

ISBN 978-3-319-34728-8 ISBN 978-3-319-04429-3 (eBook)
DOI 10.1007/978-3-319-04429-3
Springer Cham Heidelberg New York Dordrecht London

Printed on acid-free paper

Springer is part of Springer Science+Business Media (www.springer.com)

To May, Veronica and Bryan
With Love

To Members of the Construction Dispute Resolution Research Unit
With Undisputed Gratitude

Foreword I

In a perfect world there would not be construction disputes. There would be perfect communication, perfect understanding and perfect harmony between all peoples engaged in a decision-making process. The future would be known, the way to address uncertainty, risk and engagement would be established and trust would pervade the whole construction process. Of course in the real world all these factors are susceptible to disagreement and some method of resolution is required to achieve a solution. Consequently, processes and procedures together with legal frameworks have been developed to achieve the outcomes desired by all stakeholders.

Part of the problem is the need in management to break down the processes into smaller more bounded units which can be handled successfully by the participants. This reductionism has developed over the past century and has been accepted as an established way forward. Unfortunately, this has also led to a growing number of interfaces between individuals and groups within the construction process, each of which has the potential to create a problem. For example two individuals, A and B, need to work together to create an aspect of a building. They realise that they are interdependent and therefore they need to negotiate how they will work together. Once they have negotiated they need to clarify the process and express their agreement in some form which both understand. If this becomes complex then they both need a written contract which formally identifies what their respective duties are and where their responsibilities lie. It may also determine who bears the risk and the obligations placed upon them both. Subsequently, individual A realises that he needs help to understand and fulfil his or her obligations and appoints a manager to oversee this interface. Individual B now realises he cannot be left out so he appoints a manager too and the two managers try to resolve the situation should it go wrong. Then there is the possibility that the managers disagree and a Project Manager is appointed to resolve their dispute! The system seems bound to lead to a more adversarial approach.

This is of course an over-simplified caricature of what happens in conventional contracts. Reductionism leads to more interfaces which potentially leads to more disputes which almost inevitably leads to more time and expense. There is a strong argument that management research should be directed towards reducing the number of interfaces!

However, another approach is to make sure that all parties are on the same side and are seeking a win–win both for themselves and their client. Inevitably someone somewhere has to take the risks involved. In the USA there have been striking examples on very complex technical projects, where the building design and construction method are at the leading edge of knowledge, where the client has stepped in and taken the whole risk, and the design/construction team have found themselves in a less adversarial position and prepared to work together for the common good. They resolve to solve the new and difficult technical problems through a 'no blame' culture. Even in this situation the result is not perfect, at least as viewed by the client, but the disputes are less open and are contained within the design/construction team for internal resolution. Unfortunately, it is much more difficult to ascertain whether the client is getting better value as the element of competition is reduced.

Within this spectrum of dispute resolution, from the traditional approach to the 'client takes all the risk', there have been a number of attempts to discover an improved approach. These keep the essentials of a shared risk, together with competition and fair process which allows for a less adversarial position between the parties resulting in a quicker and more reasonable method of agreement. At their root is recognition that the establishment of 'trust' between all the parties is the key element.

This extremely interesting book contains a review of the concepts and practice within dispute resolution and provides a comprehensive understanding of all the factors involved. More than this it explores the various alternatives using useful exploratory models to evaluate their effectiveness. It is one of the most comprehensive volumes in the field that it has been my pleasure to read. It demonstrates the underlying variables in success and failure in dispute resolution and shows how trust can be enhanced within negotiation, mediation and culture. It explores modern methods of alternative dispute resolution and suggests paths for the future. For all those involved in dispute resolution it is a 'must read' and I believe it will provide a reference point for years to come.

Salford, UK, November 2013

Professor Peter Brandon
OBE, DSc, DEng, DUniv

Foreword II

I was both honoured and pleased to have been invited to write a Foreword to *Construction Dispute Research: Conceptualisation, Avoidance and Resolution.* This new book is the product of many years of assiduous work by Professor Sai On Cheung of the City University of Hong Kong and members of the Construction Dispute Resolution Research Unit under his leadership and guidance. They are to be warmly congratulated. To carry out research into the origins of construction disputes, how they are handled and how they might best be resolved are in themselves difficult and daunting tasks, and, to add to it, investigations into the research that has already been conducted makes the result even more impressive. Happily, considering the subjects under the four parts of Conceptualisation, Avoidance, Negotiation and Mediation, and then further sub-dividing each part with essays or chapters on relevant topics, makes the whole more manageable and intelligible. Students of the subjects should find the work indispensable but those in the construction industry who work at the sharp end (and their advisers) will also find much of interest. Not all the chapters call for the intellectual ability, persistence and fortitude that so evidently characterise Professor Sai On Cheung and the authors of the individual sections.

The authors expect the reader to keep a foot firmly on the ground, as they do. Many, if not most, construction disputes owe their origins to mistakes made by a person, or a group of people, or to a disagreement about the consequences of such mistakes—again a disagreement between people. These mistakes and disagreements all too frequently become disputes, either because those responsible do not promptly accept that a mistake has been made and agree to put it right or because others do not recognise the mistake or do not intervene to prevent the disagreement going further. Sometimes the reasons for inaction (or action) are said to be commercial, e.g. the risk or responsibility is to be borne by another or the effect in terms of cost is thought to be unacceptable. Points of principle may be said to have arisen—but it has been well said that men of principle put more money in the pockets of lawyers than any other class of persons.

I was very interested, and reassured, to see that throughout the work emphasis is placed on the pivotal role of trust (or lack of trust). Construction, like other enterprises, is essentially the product of a team. Unfortunately, and in certain areas perhaps even increasingly, relationships within the team are now defined by contracts which do little or nothing to deflect adversarialism. Some seem to be premised on the basis of 'them and us'. The lead given in the United Kingdom over some decades to the production and development of what is commonly known as the New Engineering Contract (the Engineering Construction Contract) has shown that it is possible to carry out major and complex works of construction with great success, and with the absence of or marked reduction in disputes, on the basis of core concepts of mutual trust and confidence, given right attitudes and proper training on the part of all the participants. That form of contract was used extensively for the work for the 2012 Olympic Games that was held in London. Sir John Armitt, Chairman of the Olympic Delivery Authority, asked by a BBC interviewer whether the Olympic Park would have been delivered as successfully if a different contract form had been used, replied: 'No, I don't think it would—you couldn't do it.' The interviewer concluded: 'so it was a document that won it', and said that the contract had turned round 'how infrastructure is delivered not just at the Olympics but within the UK'.

Alliancing and other means of collaborative working are also emerging and have their part to play. The authors of this work have necessarily, and commendably, concentrated on those parts of the world where such forms of contracting are not prevalent, or not as prevalent as they should be. Changes may not occur as rapidly as they should. In its studies and in its meticulous detail the book shows why change is required, not just as a matter of form but in approach—in averting the reasons for disagreement, in preventing them developing into disputes, and, if all else fails, in improving methods of resolving them. It should be essential reading for those concerned to do better. Professor Sai On Cheung and all his team are to be thanked and warmly commended for their immense work.

Atkin Chambers, London, November 2013

His Honour Humphrey LLoyd
QC, MA, LL.B, LL.D (Hons)

Preface

This book is intended to provide a comprehensive study on construction dispute. Most of the chapters are findings of research projects undertaken by the Construction Dispute Resolution Research Unit (CDRRU) at the City University of Hong Kong. Construction dispute has been a topical research subject in the construction community; moreover, most of the studies either take a legal perspective or have chosen to focus on a particular aspect. The studies presented in this book are mostly framed from a management perspective drawing on methods and concepts in contract law, economics, psychology and management science. There are three specific purposes of *Construction Dispute Research*. First, this volume aims to summarise studies on construction dispute. Second, apart from the theoretical constructs, appropriate empirical tests are employed. This approach serves to go beyond the commonly used anecdotal approach. Third, it is the sincere hope of the authors that this book will help in shaping the agenda of construction disputes research.

The studies present a holistic approach to construction dispute research. Each chapter can be read as a study on its own. The studies have not taken a legal approach, as others would have performed this task far better than we could. *Construction Dispute Research* will be useful to construction professionals involved in contract management and administration. Practitioners will find the book a handy reference in dispute management and resolution. Students will find the book useful in explaining in detail the causes of disputes and the processes to resolve them. The research design and empirical approaches will be particularly useful to students in construction management, architecture, surveying and civil engineering programmes both as textbook as well as reference readings.

The book has 20 chapters that are arranged in four parts covering conceptualization, avoidance, negotiation and mediation. Part I is devoted to dispute conceptualization. A building is only as strong as its foundation. Thus it is no better start to study construction dispute by its conceptualisation. Typically, construction dispute has been identified by the subject matter. Moreover, this approach offers little to the understanding of the underlying causes. An anatomy of construction disputes is offered followed by an empirical check on whether construction disputes are inevitable. Aggressive behavior is often found in protracted disputes. It is found that the change in contracting behavior from cooperative to aggressive is bimodal. The absence of protracted dispute can be an indicator of project satisfaction. Analytical tools are employed to identify key predictors of this form of project satisfaction.

The theme of Part II is dispute avoidance. The conventional wisdom of 'prevention is better than cure' can be applied to all problems. As far as construction dispute is concerned, equitable risk allocation and trust are the two most commonly accepted avoidance strategies. To this end, a risk allocation model based on widely recognised allocation principle is proposed. The use of the tool for risk allocation and evaluation of risk pattern is demonstrated. Trust is a controversial proposition in construction contracting. Its existence is often questioned. Nonetheless, trust is believed to be the most significant contributor to project success in general and dispute avoidance in particular; four chapters are used to discuss trusting observations, trust bases and building mechanisms and trust measurement.

Parts III and IV are linked as both deal with dispute resolution through negotiation. Part III focuses on negotiation—the gateway to resolution as almost all disputes are negotiated first before the service of other mechanisms. Negotiation is sometimes described as an art because settlement may not be reached solely from legal and rational approaches. Part III discusses the behavioural dimensions of construction dispute negotiation. When a negotiator loses interest to continue, the negotiation is doomed. This situation is identified as withdrawal and is considered as a form of negotiation failure. The symptoms and triggering factors of withdrawal are discussed in detail.

When a negotiation fails, what is the best alternative? Facilitated settlements are considered commercially more worthy than seeking award and judgement from arbitration and the court. In particular, voluntary contractual use of mediation as an alternative to arbitration and litigation to resolve construction disputes has gained popularity. Part IV deals with Mediation—A Form of Assisted Negotiation. The skill of the mediators in facilitating settlement, the interrelationships among dispute sources, mediator tactics and mediation outcomes are explored.

The authors are most grateful to Professor Brandon and His Honour Humphrey Lloyd for their enlightening Foreword. This book could not be completed without the contributions of the members of the Construction Dispute Resolution Research Unit. The authors are also grateful to Miss Tina Chan for the production support.

Hong Kong, November 2013

Professor Sai On Cheung
LL.B, LL.M, MBA, MSc, Ph.D

Contents

Part I
Conceptualisation: A Building is only as Strong as its Foundation

Chapter 1
The Roles of Dispute Resolution in Construction Contracts

Sai On Cheung

Abstract Dispute resolution clause sets out the procedure to settle disagreements that arise out of the contract. It also provides a gap fulfilling function to deal with unanticipated happenings. This chapter first provides a functional analysis of construction contract. The analysis explains the purposes and the inter-relationship among contract clauses. Whilst a number of dispute resolution mechanisms are available, it is advocated that the choice of mechanism should take into account the characteristics of the transaction. A mapping framework is proposed for this purpose. The use of the framework is illustrated by mapping dispute resolution mechanisms with four types of construction contract: main contract, nominated sub-contract, domestic sub-contract and direct labour contract.

1.1 The Primal Roots of Contract

According to Macneil (1974), there are four primal roots of contract. These are (i) specialisation of labour and exchange, (ii) sense of choice, (iii) conscious awareness of past, present and future and (iv) the social matrix. Macneil (1974) further suggested that the board principles of contract law are norms growing out of the four primal roots. The broad principles are characterised by (1) reciprocity; (2) role effectuation; (3) limited freedom of choice; (4) effectuation of planning; and (5) harmonising of contracts with their internal and external social matrices. Reciprocity is the fundamental underpinning of economic transactions and is manifested by exchange of mutual benefit. The norm of role effectuation reflects the need to enable the parties to perform their respective intended roles. Limited freedom of choice is the tension inherent with having a formal contract. Whilst freedom of contract means the choice by freewill, by entering into a contract, the parties are confined to

S. O. Cheung (✉)
Construction Dispute Resolution Research Unit, Department of Civil and Architectural Engineering, City University of Hong Kong, Hong Kong, People's Republic of China
e-mail: saion.cheung@cityu.edu.hk

S. O. Cheung (ed.), *Construction Dispute Research*,
DOI: 10.1007/978-3-319-04429-3_1,

those options allowed by the agreement. Planning embraces the provisions to deal with performance, risk allocation as well as dispute resolution, taking into account of the past, present and future. Ideally, contracts should be planned to facilitate performance, exercise of choice and meet the expectations of the social matrix. Contracting parties in construction businesses are mutually dependent. Rights are typically accompanied with obligations. Classical examples include payments for works completed, delay caused versus time extension, disruption versus loss and expense. Role effectuation is accomplished through conformance to norms and legal rulings. For example, architect, engineer and other agents are to perform their roles impartially and the client shall not intervene. Likewise, contractor is free to adopt construction methods under conventional design then build type of contract.

1.2 Functional Analysis of Construction Contract Clauses

In Modern Engineering (Bristol) Ltd v. Gilbert-Ash Northern [1974] AC 689, Lord Diplock described a building contract as "*an entire contract for the sale of goods and work and labour for a lump sum price payable by installments as the goods are delivered and the work done. Decisions have to be made from time to time about such essential matters as the making of variation orders, the expenditure of provisional and prime cost sums and extension of time for the carrying out of the work under the contract*". Thus, in its most basic form, a contract restates the intentions of the contracting parties. Moreover, in response to the uncertainty involved during construction, conditions of contract have become more and more sophisticated. For example, Turner (1994) discussed a building contract under the following headings: Intentions of the parties; Possession and completion; Control of works; Payment; Statutory obligations; Insurance; Determination and dispute resolution. It is now quite common to have highly elaborated contract documents setting out procedures to deal with potential contingencies (Hughes and Greenwood 1996). In these regards, contract clauses can be analysed in terms of the functions to be served. Figure 1.1 gives the framework proposed by Cheung and Pang (2013).

Eccentric circles are used to illustrate evolving and progressive nature of the essential provisions of typical construction contracts. The central core of Fig. 1.1 represents the most fundamental components: to stipulate the obligations of the contracting parties during the contract period. Changes are considered to be necessary and inevitable in all construction projects, to effectuate such planning, provisions for raising variations, acceleration and postponement together with the corresponding time and monetary adjustments are incorporated. Thus the layer on top of the central core is for adjustment. According to Macneil (1975), planning for performance should define the obligations, incorporate ways to facilitate accomplishment and recognise discharge of obligations. Control measures include supervision, inspection, testing, surety and insurance. Collectively, these serve to ensure performance as planned. Certificates are used to signify successful discharge of obligations by the contractor. The third layer thus deals with control and approval.

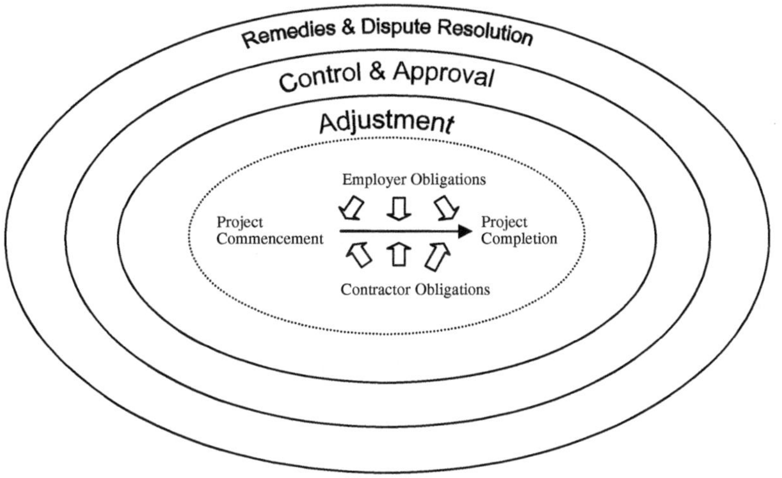

Adjustment: Variation orders; Acceleration; Postponement; Extension of Time; Loss & Expenses
Control & Approval: Testing & Inspection; Supervision; Insurance & Bond; Certificates
Remedies: Determination by Employer; Determination by Contractor; Damages for Non-completion
Dispute Resolution: Mediation, Adjudication & Arbitration etc.

Fig. 1.1 Functional analysis of construction contract clauses

The outermost layer resides the remedies available to the contracting parties for default of performance. Circumstances upon which the parties can determine the contract are typically listed together with the respective rights and obligations. Determination by either party is seldom un-contended. One common disagreement is the interpretation of the performance in terms of scope, level or both. Dispute resolution provisions are used to fill such gaps (Macneil 1975). Although dispute resolution is often regarded as stand-alone provision, its use is intimately related to the formulation and application of the provisions in the preceding layers. Where a contract cannot cater for all eventualities, a dispute resolution clause patches the holes and leaks whereby breakage of the contract is prevented.

1.3 Mapping Dispute Resolution Mechanisms with Contract Types

A dispute resolution clause set out the procedures and mechanisms to deal with disputes recognised by the contract. Macneil (1975) included dispute resolution as one of the three critical aspects of contract planning. The other two are performance and risk. Given the variety of dispute resolution mechanisms available, it is of interest to investigate how these mechanisms are to be selected with due consideration of the characteristics of the transaction.

The support for the use of alternative dispute resolution (ADR) (Hanbury 1992; Kwayke 1993; Latham 1994; Naughton 1990; Stipanowich and Henderson 1992; Tyrril 1992) is counterbalanced by the view that ADR is not a panacea to dispute epidemic (Totterdill 1991). For example, it is widely accepted that where a dispute is related to a point of law, the court should be the forum for resolution (Pengilley 1990). The choice of the dispute resolution process depends on the characteristics of the transaction. That means the choice of a dispute resolution mechanism is dependent on the characteristics of the type of contract. This chapter describes a dispute resolution mechanism mapping framework. In the following sections, the common types of dispute resolution mechanisms are first discussed. Employing the contract system classification suggested by Macneil (1974), and the transaction characteristic approach suggested by Williamson (1985), a dispute resolution mechanism—contract system mapping framework is proposed.

1.4 Dispute Resolution Mechanisms

Litigation and arbitration are well-established formal resolution mechanisms and heavily regulated by the courts or the institutions providing the service respectively. Alternative forms therefore have been promoted for use in construction with the aims of enabling a less confrontational setting that enables speedy and economical resolution. Collectively, these alternatives are called Alternative Dispute Resolution (ADR). These mechanisms are often compared with litigation and arbitration in terms of the cost and time involved. Figure 1.2 arranges the commonly used dispute resolution mechanisms in a stair-chart together with the cost, time and hostility implications.

1.5 Contract Systems and Dispute Resolution Mechanisms

The theoretical apparatus for the mapping framework draws on the work of Macneil (1974, 1975, 1978) and Williamson (1979, 1985). According to Macneil (1978), contracts can be classified into three board systems: classical, neoclassical and relational. Lyons and Mehta (1997) provide a helpful summary of the characteristics of the three contract systems (Table 1.1).

Discrete transactions typify classical contracts. A truly discrete transaction would be entirely separated not only from all other present relations but also from all past and future relations as well (Macneil 1978). Hence the identities and personal attributes of the contracting parties are irrelevant. Discrete transactions are usually of short duration, with the exchange of goods being a notable example. As these transactions are to be completed over a short duration, little change is anticipated. In the event that contingencies are to be planned, substantial efforts will

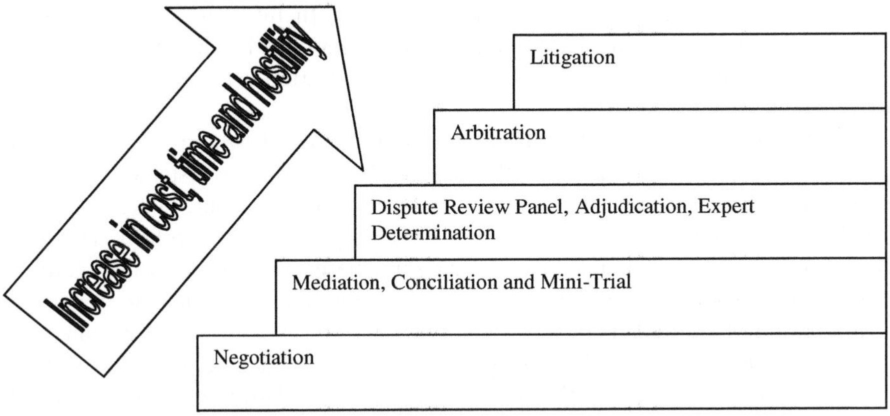

Fig. 1.2 Dispute resolution mechanisms commonly used in construction

Table 1.1 Characteristics of the three contract systems (Lyons and Mehta 1997)

Classical contract	Neoclassical contract	Relational contract
• The identities and personal attributes of parties are irrelevant • Specifies a discrete exchange (or duration) • Contingencies and penalties for non-performance are specified • Written documentation overrules any verbal agreement • Law courts adjudicate in the event of disagreement	• The identities of the parties matter • Normally specifies a fixed duration (or task to be completed) • It is accepted that not all contingencies can be specified • Written documentation provides the status quo point from which to renegotiate • Arbitration procedure for disputes	• The identities and personal attributes of parties are crucial • Normally of indeterminate duration • Norms of behaviour, or shared codes of conduct, inform responses to new developments as they unfold • Written documentation treated as a record of what has been agreed • Norms of behaviour, or shared codes of conduct, overrule written documents in settling disputes

be directed for the highest clarity. In such cases, penalties for non-performance are usually well-specified. Disputes arising out of this type of contract are best resolved in courts.

However, discrete transactions are rare. In reality, most contracts are executory and performance is affected by both internal and external factors. Hence classical contract law no longer suffices for exchanges that project into the future. In those situations, adjustment flexibility is critical. Two common characteristics of these 'projected' contracts are the existence of gaps in their planning and the presence of a range of processes and techniques used by contract planners to create flexibility. In this type of contract, it is acknowledged that eventualities cannot be exhausted.

Adjustments are necessary as the project unfolds. In this regard, written documentations shall provide the bases from which to negotiate. Furthermore, exercising flexibility will inevitably invite disagreement. Arbitration is the suggested method to fill the gaps that may arise.

A relational contract refers to a long-term contract where the contracting partners are tied not so much by the words of the contract; instead, the performance of the contract is underpinned by norms of behaviour, shared codes of conduct, and informed responses to new developments as they unfold. The identities and personal attributes of the parties in these circumstances are therefore extremely crucial, thus rendering confrontational mode of dispute resolution inappropriate. Disagreements are often negotiated for a solution, which can occur without jeopardising the relationship between the contracting parties. Resolving dispute through assisted negotiation such as Alternative Dispute Resolution (ADR) is considered appropriate. The contract system classification used by Macneil (1978) examines the characteristics of the transaction, hence more appropriately relate to the types of construction contracts instead of the procurement strategies. Nonetheless, the spirit of serial contracting and partnering type of procurement resemble those of relational contracts.

1.6 Transaction Characteristics and Contract Systems

Williamson (1985) sees contracts as 'Governance Structures'. Contracts are frameworks under which transactions are conducted in a changing world. The variations among these structures can be expressed in terms of the extent of formality and flexibility. The optimal choice therefore should cater for the key transaction characteristics. In Williamson's view, three technical characteristics are central in describing a transaction: specific investment, frequency and uncertainty. Specific investment describes expenditure on plant and machinery, time or effort that has a reduced value if used for any purpose other than to service a particular customer or supplier (Lyons and Mehta 1997). It is this latter point that inspires the inclusion of identity of the contracting party in the proposed dispute resolution mapping framework.

The second characteristic is frequency. Repeated transactions make it worthwhile to make special investments. The third characteristic is uncertainty. The greater the degree of uncertainty over future requirements, the greater the need for contracts to allow room for adapting to new conditions. To facilitate model integration and empirical study, Williamson (1979) left out uncertainty and provided the following integrated model of contract systems and transaction characteristics (Table 1.2).

Table 1.2 Transaction characteristics and contract systems (adapted from Williamson 1985)

		Transaction characteristics		
		Non-specific	Mixed	Idiosyncratic
Frequency	Occasional	Classical Contracting	Neoclassical Contracting	Neoclassical Contracting
	Recurrent	Classical Contracting	Relational Contracting	Relational Contracting

1.7 Mapping Framework

Figure 1.3 presents the interrelationships between contract systems, transaction characteristics and dispute resolution mechanisms. In addition to the three transaction characteristics used by Williamson (1979), discreteness and presentiation are also included in the framework.

For classical contracts, litigation is the dispute resolution mechanism. Transactions under this contract type are discrete and are characterised by "*sharp in by agreement and sharp out by performance*" (Macneil 1974). Litigation is employed to ensure that the parties shall keep their promises. With neoclassical contracts, the reality of incomplete presentiation is acknowledged. Planning for flexibility and hence the ability to fill gaps becomes critical. *Presentiate* is defined in Oxford English Dictionary as: "*to make or render present in place or time; to cause to be perceived or realised as present*". Arbitration has evidently been employed to effect gap-filling. The desire to continue with the relationship while disputes are being arbitrated typifies transactions under the neoclassical contracting system.

The increase in transaction cost between the parties encourages idiosyncratic investments for which vertical integration is favored over trading. The growth of relational contracting responds to this sort of situation and preservation of relationship becomes the dominant objective. The spirit of partnering is a close example of relational contracting in construction (Baxendale and Greaves 1997; Fellows 1997). Examples of idiosyncratic investment include the establishment of design office and contracting arm within a developer. A commonly observed modified form of integration occurs when a developer uses the same design consultant and contractor repeatedly. The need to minimise transaction costs in these cases has prompted the formation of stable coalitions (the client, contractor and subcontractors) across a series of transactions (Alsagoff and McDermott 1994; Lyons 1994).

In sum, five transaction characteristics are used in the mapping framework. These are discreteness, presentiation, uncertainty, frequency and identity. Accordingly, the differences among the three contract systems can be described by their respective degrees of variation in relation to the five transaction characteristics.

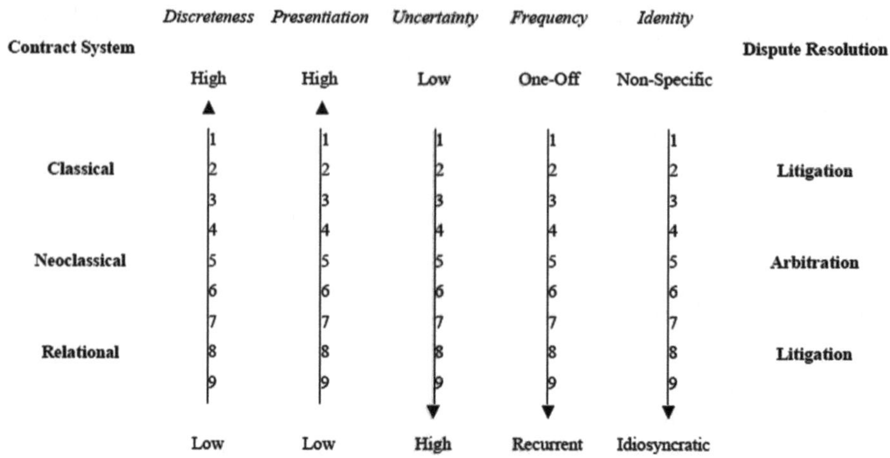

Fig. 1.3 A mapping framework for dispute resolution mechanisms and contract types

Figure 1.3 presents a mapping framework developed through the integration of transaction characteristics, contract systems and their associated dispute resolution processes. By examining their respective transaction characteristics, construction contracts can be mapped to contract systems (classical, neoclassical and relational). The choice of a dispute resolution mechanism can then be based on the mapping framework as presented in Fig. 1.3.

1.7.1 Illustration on the Use of the Mapping Framework

This section demonstrates the use of the proposed mapping framework. During the construction process, various types of contract are used. Those regularly used include main contracts, nominated subcontracts, domestic subcontracts and labour contracts. Firstly, it is suggested that these four types of construction contract vary in different degrees, in terms of the five transaction characteristics as described. Secondly, the five transaction characteristics are having different degrees of importance in relation to the selection of dispute resolution mechanism. In these connections, the mapping involves the following steps:

(1) Measurement of the transaction characteristic ratings.
(2) Establishing the relative importance weightings of the transaction characteristics.
(3) Developing the contract mapping scores.
(4) Interpretation of the contract mapping scores.

1.7.2 Measurement of Transaction Characteristic Ratings

Figure 1.4 shows the instrument used for the measurement of transaction characteristics. For each of construction contracts, the respondents were asked to assign a rating (1–9) against the five transaction characteristics. The scales are arranged as follows:

Transaction characteristics	Scale
Discreteness (high–low)	1–9
Presentiation (high–low)	1–9
Uncertainty (low–high)	1–9
Frequency (one-off–recurrent)	1–9
Identity (non-specific–idiosyncratic)	1–9

The scales are so arranged that the higher the rating, the more relational is the construction contract type. For ease of comparison, the measurement sheet is also arranged so that under each of the transaction characteristics, the four types of construction contract are compared seriatim. For illustration purpose, Table 1.3 presents the transaction characteristic ratings obtained from Respondent "A".

1.8 Establishing the Relative Importance Weightings of the Transaction Characteristics

To recognise the non-uniform impacts of the transaction characteristics toward the mapping, weightings are therefore necessary to reflect their relative importance. In this regard, the second step of the study involves the solicitation of the relative importance weightings of the transaction characteristics. The sum of the weightings must be one.

1.9 Developing Mechanisms/Contract Types Mapping Scores

For each of the four types of construction contract, with the results obtained from the previous two steps, a contract mapping score (M_c) can then be calculated by:

$$M_c = \sum_{i=1}^{5} W_i T_i \tag{1.1}$$

Transaction Characteristics of Construction Contracts

Please indicate the characteristics of the transactions as described on a scale of 1 to 9.

I. Discreteness

Contract Type	1 very high		3 high		5 medium		7 low		9 very low
Main Contract	1	2	3	4	5	6	7	8	9
Nominated Subcontract	1	2	3	4	5	6	7	8	9
Domestic Subcontract	1	2	3	4	5	6	7	8	9
Labor Contract	1	2	3	4	5	6	7	8	9

II. Presentiation

Contract Type	1 very high		3 high		5 medium		7 low		9 very low
Main Contract	1	2	3	4	5	6	7	8	9
Nominated Subcontract	1	2	3	4	5	6	7	8	9
Domestic Subcontract	1	2	3	4	5	6	7	8	9
Labor Contract	1	2	3	4	5	6	7	8	9

III. Uncertainty

Contract Type	1 *very low*		3 *low*		5 *medium*		7 *high*		9 *very high*
Main Contract	1	2	3	4	5	6	7	8	9
Nominated Subcontract	1	2	3	4	5	6	7	8	9
Domestic Subcontract	1	2	3	4	5	6	7	8	9
Labor Contract	1	2	3	4	5	6	7	8	9

IV. Frequency

Contract Type	1 *very low*		3 *low*		5 *medium*		7 *high*		9 *very high*
Main Contract	1	2	3	4	5	6	7	8	9
Nominated Subcontract	1	2	3	4	5	6	7	8	9
Domestic Subcontract	1	2	3	4	5	6	7	8	9
Labor Contract	1	2	3	4	5	6	7	8	9

V. Identity

Contract Type	1 Irrelevant		3 unimportant		5 medium		7 important		9 very important
Main Contract	1	2	3	4	5	6	7	8	9
Nominated Subcontract	1	2	3	4	5	6	7	8	9
Domestic Subcontract	1	2	3	4	5	6	7	8	9
Labor Contract	1	2	3	4	5	6	7	8	9

Fig. 1.4 Transaction characteristic ratings

where W_i is the weighting of the transaction characteristic i; T_i is the rating of transaction characteristic i.

Table 1.4 gives the contract mapping scores of Respondent "A".

Table 1.3 Summary of transaction characteristic ratings by Respondent "A"

	Respondent "A"
Discreteness	
Main contract	2
Nominated subcontract	4
Domestic subcontract	5
Labour contract	5
Presentiation	
Main contract	1
Nominated subcontract	2
Domestic subcontract	2
Labour contract	5
Uncertainty	
Main contract	8
Nominated subcontract	6
Domestic subcontract	5
Labour contract	2
Frequency	
Main contract	6
Nominated subcontract	5
Domestic subcontract	5
Labour contract	6
Identity	
Main contract	6
Nominated subcontract	4
Domestic subcontract	8
Labour contract	5

1.10 Interpretation of Contract Mapping Scores

The mapping exercise was conducted with construction professionals in Hong Kong. One hundred and forty-five sets of data had been successfully obtained. The average contract mapping scores for the four types of construction contract are given in Fig. 1.5.

The framework maps dispute resolution mechanisms to contract types. The empirical study in essence classifies the four commonly used construction contracts into the contract systems expounded by Macneil (1978). The classification was achieved through the assignment of ratings and relative importance weightings for the five transaction characteristics. The contract mapping scores as calculated by Eq. 1.1 can be interpreted as:

Mapping score	Dispute resolution mechanism suggested
1–3.33	Litigation
3.34–6.66	Arbitration
6.67–9	ADR

Table 1.4 Contract mapping scores of Respondent "A"

		Respondent "A"			
		Head Con	Nom Sub	Dom Sub	Lab Con
Discreteness	0.15	2	4	5	5
Presentiation	0.15	1	2	2	5
Uncertainty	0.40	8	6	5	2
Frequency	0.20	6	5	5	6
Identity	0.10	6	4	8	5
		5.45	4.7	4.85	3.25

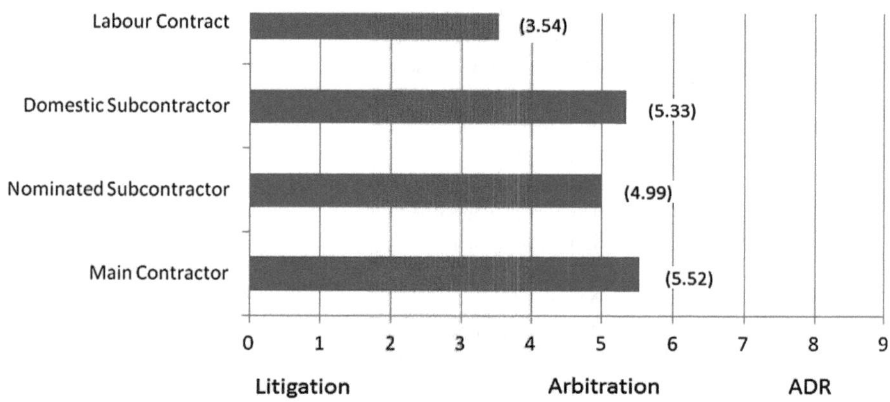

Fig. 1.5 Average contract mapping scores

In Fig. 1.4, the mapping range (1–9) is presented as a continuum of dispute resolution mechanisms. Against this continuum, the average contract mapping scores obtained from the thirty-three respondents are plotted. The four average mapping scores all fall within the band of 3.34–6.66. Strict interpretation of the selection framework would suggest the use of arbitration for all four types of construction contracts. This can be explained by the phenomenon that arbitration has been used for a long time and that the industry has somewhat accepted its use as the norm. However, a more detailed study of the relative positions of the four contract types on the continuum provides valuable insight.

The average mapping score for main contracts was 5.52, the highest among the four. The contractual arrangements between employers and main contractors have undergone tremendous changes in the past two decades, notably with heavier involvement of the main contractor in the design of the works. This requires a co-operative working relationship between the parties. In this regard, maintenance of working relationship is of prime concern if successful project delivery is to be achieved. Partnering, alliance contracting and the like have been advocated as the way forward (Alsagoff and McDermott 1994). These types of contracting arrangements resemble relational contracting and have been advocated as a model procurement strategy in the both United Kingdom (Egan 1998) and Hong Kong (CIRC 2001; HKHA 2000). The main contract is therefore viewed as the most relational type of contract among the four contract types.

Domestic subcontracts obtain an average contract mapping score of 5.33. It is a well-established principle that a main contractor is responsible for the work of his subcontractors, both domestic and nominated. Construction is a risky business and working with strangers adds further risks. This equally applies to both the main contractor and the domestic subcontractor. Domestic subcontractors tend to form alliances with several main contractors for work. The identity of the parties therefore is crucial in this type of contracting. Nevertheless, as the number of main contractors in the market far exceeds that of clients, it is easier for domestic subcontractors to make associations with a greater number of main contractors. In this respect, domestic subcontracts can be less relational than main contracts.

Nominated subcontracts obtain an average contract mapping score of 4.99. The use of nominated subcontractors is a unique form of subcontracting method under the British system. Nominated subcontractors are usually responsible for specialist works. They are selected by the employer and then forced upon the main contractor. The main contractor has no involvement in the selection process. It is perfectly possible that the main contractor has to enter a contract with a nominated subcontractor with no previous working relationship, a contracting mode analogous to neoclassical contracting. The average contract mapping score of 4.34 suggests the use of arbitration.

Labour contracts obtain an average contract mapping score of 4.33. Construction activities on site are labour intensive. Labourers are usually paid on a weekly or piece meal manner. The performance requirements are fairly clear-cut and the contract duration is relatively short. These contracts exhibit the characteristics of discrete transactions, for which litigation is the mode of dispute resolution.

In summary, the empirical study suggests that:

Contract type	Dispute resolution process
Main contract	Towards ADR
Domestic subcontract	Towards ADR
Nominated subcontract	Arbitration
Labour contract	Towards litigation

It is also prudent to note that the empirical result presented in the study is obtained in Hong Kong. The mapping framework can be used as an aid for contract planner in planning dispute resolution in construction contracts. The assessment of the relative importance weightings for the transaction characteristics can reflect situational factors.

1.11 Chapter Summary

Having a hard and fast rule for the selection of dispute resolution mechanisms is not advisable. The mapping framework suggested in this chapter is underpinned by the theoretical constructs of contract law systems (Macneil 1974, 1978) and transaction characteristics (Williamson 1979, 1985). The mapping framework is introduced through a detailed descriptive analysis and its use is illustrated by an empirical study. The results of the empirical study make good practical senses as these reflect the prevalent practices in the construction industry in Hong Kong. Notable examples include the dominant use of design-then-build as a procurement methodology and arbitration as the dispute resolution method. The findings suggest that main contracts are the most relational; the use of ADR for dispute resolution would be expected once the contracting environment becomes more co-operative, as in the case of partnering. In addition, the discrete nature of labour contracts is also spot-on. The average contract mapping score of 3.54 is indeed very close to 3.33 (the upper-range figure for classical contracts in the mapping model). Domestic subcontracts are more relational than nominated subcontracts can also be explained by the contracting practices commonly used in Hong Kong. The mapping framework can be used by contract planner as a decision aid to select dispute resolution mechanism according to the transaction characteristics.

References

Alsagoff, S. A., & McDermott, P. (1994). Relational contracting: a prognosis for the UK construction industry? In Proceedings of the CIB W92 Conference, December 1994, University of Hong Kong, pp. 11–19.

Baxendale, T., & Greaves, D. (1997). Competitive partnering—A link between contractor and subcontractor. In Proceedings of the CIB W92 Symposium on Procurement, University of Montreal, pp. 21–28.

Construction Industry Review Committee (CIRC) (2001). Construct for Excellence, Construction Industry Review Committee, the Government of the Hong Kong Special Administrative Region.

Cheung, S. O., & Pang, H. Y. (2013). Anatomy of construction disputes. *The Journal of Construction Engineering and Management, 139*(1), 15–23.

Egan, J. (1998). Rethinking Construction, Department of the Environment, Transport and the Regions, The Government of the United Kingdom.

Fellows, R. (1997). The culture of partnering. In Proceedings of the CIB W92 Symposium on Procurement, University of Montreal, pp. 193–202.

Hanbury, W. (1992). Alternative dispute resolution—The Australian model. *Solicitors Journal, April*, 334–335.

HKHA (2000). Quality housing, partnering for change. The Hong Kong Housing Authority.

Hughes, W. P., & Greenwood, D. G. (1996). The standardization of contracts for construction. *International Construction Law Review, 13*(2), 196–206.

Kwayke, A.A. (1993). Alternative dispute resolution in construction. Chartered Institute of Building, Occasional Paper no. 21.

Latham, M. (1994). Constructing the Team: Final Report of a Joint Review of Procurement and Contractual Arrangements in the United Kingdom Construction Industry, HMSO.
Lyons, B. (1994). Contracts and specific investment: An empirical test of transaction cost theory. *Journal of Economics and Management Strategy, Massachusetts Institute of Technology, 3*(2), 257–278.
Lyons, B., & Mehta J. (1997). Private sector business contracts: The text between the lines. In S. Deakin & J. Michie (Eds.), *Contracts, co-operation and competition, studies in economics, management, and law* (pp. 43–66). Oxford: Oxford University Press.
Macneil, I. R. (1974). The many futures of contracts. *Southern California Law Review, 47*, 688.
Macneil, I. R. (1975). A primer of contract planning. *Southern California Law Review, 48*, 627.
Macneil, I. R. (1978). Contracts: Adjustment of long-term economic relations under classical, neoclassical, and relational contract law. *Northwestern University Law Review, 72*, 854.
Naughton, P. (1990). Alternate forms of dispute resolution—Their strengths and weakness. *Construction Law Journal, 6*(3), 195–206.
Pengilley, W. (1990). Alternative dispute resolution: The philosophy and the need. *Australian Dispute Resolution Journal, 1*(2), 81–95.
Stipanowich, T. J., & Henderson, D. A. (1992). Settling construction disputes by mediation, mini-trials and other processes—The ABA forum survey. Construction Lawyer, pp. 6–10.
Totterdill, B. W. (1991). Does the construction industry need ADR? *Construction Law Journal, 8*(3), 189–199.
Turner, D. (1994). *Building contracts: A practical guide*. London: Longman Scientific & Technical Press.
Tyrril, J. (1992). Construction industry dispute resolution—A brief overview. *Australian Dispute Resolution Journal, 3*(3), 167–183.
Williamson, O.E. (1979). Transaction-cost economics: the governance of contractual relations. *The Journal of Law and Economics, 22*, 233–261.
Williamson, O. E. (1985). *The economic institutions of capitalism*. New York: The Free Press.

[illegible]

Chapter 2
Conceptualising Construction Disputes

Sai On Cheung and Hoi Yan Pang

Abstract Construction dispute resolution is a topical research area. These studies typically start from dispute identification and subject matter is the most commonly used approach. However, this approach does not take account of the contextual factors that may in fact the true causes. This prompts the diagnostic approach. This chapter gives an overview of these two approaches to identify construction disputes. In addition, a third approach that draws on the concepts of bounded rationality and opportunism is proposed. Minefields and manifestations of opportunism in construction contracting in relation to occurrence of construction disputes are also discussed. Accordingly, an anatomy of construction disputes is provided. It is suggested that construction disputes are mostly contractual but can also be speculative where people factor is a major trigger.

2.1 Subject Matter Approach to Identify Construction Dispute

The issues in dispute must be clearly stated in all claims. The subject matter approach employs these issues to identify the dispute. This approach is widely used for the convenience and ease of understanding. Semple et al. (1994) illustrated this observation by suggesting that site overhead, loss of productivity, loss

S. O. Cheung (✉) · H. Y. Pang
Construction Dispute Resolution Research Unit, Department of Civil and Architectural Engineering, City University of Hong Kong, Hong Kong, People's Republic of China
e-mail: saion.cheung@cityu.edu.hk

H. Y. Pang
e-mail: karen-pang@hotmail.com

S. O. Cheung (ed.), *Construction Dispute Research*,
DOI: 10.1007/978-3-319-04429-3_2,

of revenue and financing costs are the main types of construction dispute. Likewise, Yates (1998) argued that the main types of construction dispute arising from the contract document include (1) variations; (2) ambiguities in contract documents; (3) inclement weather; (4) late issue of design information/drawings; (5) delayed possession of site; (6) delay by other contractors employed by the client and (7) postponement of part of the project. Furthermore, Hewit (1991) found six principal types of construction dispute and these are change of scope, change conditions, delay, disruption, acceleration and termination. With reference to the construction disputes that reached the Supreme Courts of New South Wales and Victoria, Australia in 1989 and 1990, Watts and Scrivener (1993) assembled 59 categories of dispute with 117 sources. The 59 categories of dispute fall into the following subject matters: (1) determination of the agreement; (2) payment related; (3) the site and execution of work; (4) time related; (5) final certificate and final payment and (6) tort related. Heath (1994) also found seven main subject matters of construction dispute; (1) contract terms; (2) payments; (3) variations; (4) extensions of time; (5) nomination; (6) re-nomination and (7) availability of information. Similarly, Conlin et al. (1996a, b) summarised that payment, performance, delay, negligence, quality and administration are major issues of construction disputes. Kumaraswamy (1997) also found that construction disputes can be categorised as (1) variation due to site conditions; (2) variations due to client changes; (3) variations due to design errors; (4) unforeseen ground conditions; (5) ambiguities in contract documents; (6) variations due to external events; (7) interferences with utility lines; (8) exceptional inclement weather; (9) delayed design information and (10) delayed site possession. This categorisation is another manifestation of the subject matter approach. In fact, Totterdill (1991) pointed out that construction contract disputes must have a contractual base. Sykes (1996) further elaborated that construction disputes originate from two main interrelated sources; construction contracts and unexpected events. As construction works are subject to many uncertainties, exhaustive planning for the possible eventualities within the contract is daunting. This can be the result of outright failure to recognise the sources of uncertainties. More problematic though is having unintended contradicting contractual provisions to deal with them. With reference to Sheridan (2003) data collected by the Adjudication Reporting Centre (ARC), the typical disputes settled by adjudication in the United Kingdom include: 'valuation of variations', 'valuation of final account' and 'failure to comply with payment provisions'. Brooker (2002) examined the types of disputes where mediation had been used in U.K. and found that payment, delay, defect/quality and professional negligence as subject matters contributed 72 percent of the reported cases. A similar study on construction mediation conducted in Hong Kong also found that variation, delay in work progress, parties' expectations and intra-parties' problem were the significant types of dispute source (Yiu and Cheung 2004). Table 2.1 summarises the studies that employ the subject matter approach to identify construction disputes.

Table 2.1 Subject matter approach to construction dispute identification

Subject matters of construction disputes	References
(1) Change of scope, (2) change conditions, (3) delay, (4) disruption, (5) acceleration and (6) termination	Hewit (1991)
(1) Determination of the agreement; (2) payment related; (3) the site and execution of work; (4) time related; (5) final certificate and final payment and (6) tort related	Watts and Scrivener (1993)
(1) Contract terms; (2) payments; (3) variations; (4) extensions of time; (5) nomination; (6) re-nomination and (7) availability of information	Heath et al. (1994)
(1) Payment, (2) performance, (3) delay, (4) negligence, (5) quality and administration as headings of construction disputes	Conlin et al. (1996a, b)
(1) Variation due to site conditions; (2) variations due to client changes; (3) variations due to design errors; (4) unforeseen ground conditions; (5) ambiguities in contract documents; (6) variations due to external events; (7) interferences with utility lines; (8) exceptional inclement weather; (9) delayed design information and (10) delayed site possession	Kumaraswamy (1997)
(1) Variations; (2) ambiguities in contract documents; (3) inclement weather; (4) late issue of design information/drawings; (5) delayed possession of site; (6) delay by other contractors employed by the client (e.g. utility companies) and (7) postponement of part of the project	Yates (1998)
(1) Valuation of variations, (2) valuation of final account and (3) failure to comply with payment provisions	Sheridan (2003)
(1) Payment, (2) delay, (3) defect/quality and (4) professional negligence	Brooker (2002)
(1) Ambiguous contract documents, (2) competitive/adversarial attitude and (3) dissimilar perceptions of fairness by the participants	Spittler and Jentzen (1992)
(1) Project uncertainty; (2) contractual problems, (3) opportunistic behaviour, (4) contractors' financial position and (5) cost of conflict and culture	Mitropoulos and Howell (2001)

2.2 Diagnostic Approach to Identify Construction Dispute

Diagnostic approach to identify construction disputes aims to unveil the underlying causes. To certain extent, this approach is more informative as far as understanding construction dispute is concerned. Construction disputes are often underpinned the conflicting interests of the contracting parties. Mururu (1991) described that dispute is the formation of a position to maintain in conflict. Brown and Marriott (1999) suggested that dispute can be viewed as a class or kind of conflict that require resolution. Furthermore, according to Hellard (1987), construction dispute is the opposition of interests, values or objectives. Spittler and Jentzen (1992) showed that ambiguous contract documents, competitive/adversarial attitude and dissimilar perceptions of fairness by the participants are the main sources of construction dispute. It is further suggested that if the interests of the participants can be satisfied, disputes can be resolved by managing the time, cost and quality factors. Tillet (1991) defined construction dispute as the incompatibility of two (or more) people's (or groups') interests, needs or goals. As they seek to achieve their own interest through compromise, one party may yield to the counterpart on less important issues. When this happens, the dispute is having a better chance to be settled. This was consistent with the view of Fenn et al. (1997) who opined that dispute requires resolution and is associated with distinct justifiable issues. Similar proposition is also suggested by Burton (1990) who maintained that dispute is always negotiable. Bristow and Vasilopoulos (1995) and Sykes (1996) are also concerned with personality and suggested that disputes are due to unrealistic expectation, lack of team spirit and misunderstandings. Apparently, these studies suggested that conflict is a prime driver of dispute. Diekmann et al. (1994) suggested that people, process and product are the main sources of construction disputes. Likewise, Rhys Jones (1994) enlisted ten main sources: (1) management; (2) culture; (3) communications; (4) design; (5) economics; (6) tendering pressure; (7) law; (8) unrealistic expectations; (9) contracts and (10) workmanship.

In search for theoretical anchors, Mitropoulos and Howell (2001) applied the contracting framework of Williamson (1979) and used the effect of project uncertainty, contract, working relations and problem solving effectiveness to explain the development of disputes. It is suggested that "environmental" and "behavioural" factors play important roles in problem making. Notable examples include (1) project uncertainty; (2) contractual problems and (3) opportunistic behaviour. Table 2.2 summarises the identification of construction dispute from a diagnostic perspective.

2.3 Towards a Conceptualisation

Yates (1998) argued that complex contracts are invariably incomplete due to bounded rationality and uncertainty (Williamson 1975). As a consequence of contract incompleteness, whenever events/contingencies occur ex post which are not fully specified ex ante, one or both of the parties may behave opportunistically. Such behaviour predictably results in conflict and disputes.

2.3.1 Bounded Rationality and Opportunism

Transaction Cost Economics (TCE) has been recognised as a major theoretical underpinning in explaining governance structures of economic exchange activities (Coase 1975). In construction, TCE has been applied to analyse governance structure (Reve and Levitt 1984), project management (Winch 1995), conflict and dispute (Yates 2003), procurement systems (Ive and Chang 2007) and trust

Table 2.2 Diagnostic approach to construction dispute identification

Contributors to construction disputes	Reference
(1) People, (2) process and (3) product	Diekmann et al. (1994)
(1) Management; (2) culture; (3) communications; (4) design; (5) economics; (6) tendering pressure; (7) law; (8) unrealistic expectations; (9) contracts and (10) workmanship	Rhys Jones (1994)
(1) Technical, (2) legal and (3) managerial dispute issues must have a contractual reference	Totterdill (1991)
(1) Construction contracts and (2) unpredictable events	Sykes (1996)
Dispute is the formation of a position to maintain in conflict	Mururu (1991)
Dispute can be viewed as a class or kind of conflict that require resolution	Brown and Marriott (1999)
Construction dispute is the opposition of interests, values or objectives	Hellard (1987)
Construction dispute is linked with difference in perspectives, interests and agenda of human beings	Spittler and Jentzen (1992)
Construction dispute is the incompatibility of two (or more) people's (or groups') interests, needs or goals	Tillet (1991)
Dispute requires resolution is associated with distinct justifiable issues	Fenn et al. (1997)
Construction disputes are due to unrealistic expectation, lack of team spirit and misunderstandings	Bristow and Vasilopoulos (1995) and Sykes (1996)

(Cheung 2007). TCE aspires to describe 'man as he is' in cognitive and self-interestedness respects and works out of two key behavioural assumptions: bounded rationality and opportunism (Coase 1984). The principal ramifications of these behavioural assumptions for economic organisation (Williamson 1985) therefore include:

(i) All complex contracts are unavoidably incomplete and many incentive alignment processes cannot be complemented (because of bounded rationality), thus most contingent adjustment mechanisms would fail for unanticipated eventualities.
(ii) To rely on contract as promise is fraught with hazard (because of opportunism). Hence, ideal forms of organisation are disallowed.
(iii) Added value will be realised by organising in such a way as to economise on bounded rationality and to safeguard transactions against the hazards of opportunism. As such, transaction cost economising is implicated.

According to Simon (1997), the behaviour of decision makers is intendedly rational, but only limitedly so. Bounded rationality is used to identify rational choice that takes into account the cognitive limitations of the decision maker. Moreover, Williamson (1991), from a strategic perspective, advocates that there are two substantive approaches: strategising and economising. The latter is considered as more important and involves effective adaptation and the elimination of waste.

Within the remit of economising, Eisenberg (2001) suggested that a decision maker will only evaluate all possible options if the cost of searching and possessing information are zero and human information possessing capabilities are perfect. In reality, such searches will normally be limited because of the cost required. Furthermore, decision makers are also bounded by their limitations on computational capacity as well as the ability to calculate consequences, understand implications, make comparative judgments on complex alternatives, organise and utilise memory (Simon 1979). In these regards, rationality is bounded by both the limitation on cost of information search and the computational effort. The decision so made may not be optimal as suggested under a neo-classical economic framework.

Actual happenings thus deviate from the game of perfect information presumed in a rational model; *"Theories of bounded rationality are thus theories of decision making that assume that the decision maker wishes to attain goals, and uses his or her mind as well as possible to that end"* (Simon 1997). Under the rational model, good administration is the up keeping efficiency for which scarce resources of an organisation shall be deployed to accomplish its objective through rational behaviour. Rationality is therefore concerned with the selection of preferred behaviour alternatives in terms of some system of values whereby the consequences of behaviour can be evaluated (Simon 1957). Economic man as described by Simon (1957) is having a complete and consistent system of preferences that allows him to choose among the options open to him; he is completely aware of what these options are; there are no limits on the complexity of the computations he can perform in order to determine which option is best; probability calculations are therefore neither frightening nor mysterious. However, using the chess game as illustration, Simon (1972) suggested that instead of finding the optimal solution, choices are made when the decision maker regards an option is satisfactory. Simon (1957) introduced the concept of 'satisfying' which suggests that 'people will satisfy when they make a decision that satisfies and suffices for the purpose'.

To put these concepts in perspective, Dyner and Franco (2004) incorporated bounded rationality in modeling choice of electricity users. It was found that administrative men exhibit a kind of rational behaviour that is compatible with the access to information and the computational capacities that are actually possessed by organisations (Simon 1957). This is the difference between 'economic men' from 'administrative men' and is significant as far as decision making is concerned. An economic man is assumed to have evaluated all alternatives before making a choice. However sequential evaluation of options is the reality and the first satisfactory option is often chosen. The decision is therefore satisfying. The concept of bounded rationality is built on this characterising behaviour of 'administrative man'.

Opportunism often goes hand in hand with bounded rationality. Williamson (1993) further elaborated that contractual incompleteness (due to bounded rationality) never gives rise to contractual difficulties if parties to a contract can be relied on to self-enforce the agreement. As such, incompleteness, notwithstanding all gaps, omissions, errors etc. will be cured. A general clause such as "disclose all

Table 2.3 Contracting environment under bounded rationality and opportunism (Williamson 1985)

		Condition of bounded rationality	
		Absent	Admitted
Condition of opportunism	Absent	Bliss[a]	"General clause" contracting[b]
	Admitted	Comprehensive contracting[c]	Serious Contracting difficulties[d]

[a] An Utopia condition
[b] Example of a general clause like "I agreed candidly to disclose all relevant information and thereafter to propose and cooperate in joint profit-maximising courses of action during the contract execution interval, the benefits of which gains will be divided without dispute according to the sharing ratio herein provided"
[c] A scenario whereby perfect presentation is achieved
[d] Typical contracting environments in reality

relevant information candidly and to behave in a co-operative way during contract execution and at contract renewal intervals" may be included in the contract to formalise this desired state of cooperation. However, Hobbes (1928) was more cautious and claimed that *"Words… [are] too weak to hold men to the performance of their covenants"*. Gauss (1952) added that if opportunism is accepted to be the appropriate way to describe self-interest seeking, breaking promises to suit one's purposes can be expected. In his seminal paper about contractual man, Williamson (1995) described that opportunism is the strong form of manifestation of self-interest seeking behaviour. He further discusses the relationship between bounded rationality and opportunism. His view is presented in Table 2.3.

2.3.1.1 Opportunism in Construction Contracting

As afore-stated, opportunism refers to a lack of candor or honesty and is characterised by self-interest seeking with guile (Williamson 1975). Opportunism can be viewed as unethically calculated efforts to mislead, distort, disguise, obfuscate, or confuse (Williamson 1985). Wathne and Heide (2000) further noted that opportunists practice guileful behaviours such as lying, stealing and cheating. Opportunism is at work when an individual attempts to maximise his interest in any situation where he could gain one way or another (O'Donovan 1962). Goffman (1969) suggested that an opportunist is one who makes false or empty, self-disbelieved threats and promises. For example, an opportunist may not abide by the terms of the agreement in order to exploit the other's short-term difficulties (Parkhe 1993; Wathne and Heide 2000). A firm is behaving opportunistically if it pursues acts for unilateral gains (Brown et al. 2000). In the situation of supplier–buyer relationship, opportunism is exemplified by those conscious behaviours engaged by a dependent supplier firm to influence the decisions of the dominant buyer through deceit and guile in ways that are presumed by the supplier to enhance its position or outcomes (Provan 1993). Opportunism therefore is exemplified by *"given the opportunity, decision-makers may unscrupulously seek to serve their self-interests"* (Judge and Dooley 2006). Opportunism is at work when the following occurs: (1) misrepresenting information, activities or effects; (2) distorting results; and (3) misrepresenting intentions (Anderson 1988; Muris 1981). In construction, contracting parties may behave opportunistically by pursuing acts that will lead to cost increase/revenue reduction of the other party. For example, contracting parties may take advantage of unforeseen circumstances and exploit their counterparts (Lee et al. 2009). Other examples of opportunism in construction include speculative pricing (Winch 1989), blatant underperformance of consultants (Reve and Levitt 1984), unjustified claims for extra money and/or time by contractors (Yates and Hardcastle 2002), and unreasonable rejection of contractors' claims by clients (Yates 2003). Yates and Hardcastle (2002) added that if *"there were no gaps in the contract document and no subsequent changes in client requirements and design, there would be no requirement for ex post adjustment and consequently no room for the contractor to behave opportunistically"*. Yates (2003) added that opportunism is pre-empted by *"incomplete or distorted disclosure of information, especially to calculated efforts to mislead"*.

2.3.1.2 Opportunism at Work

John (1984) also advocated that unrestrained self-interest maximising behaviour best illustrates opportunism. Mitropoulos and Howell (2001) advocated that construction projects are transactions of long duration and exposed to high degree of uncertainty and complexity; it is impossible to foresee every contingency and indicate the respective contract provisions ex ante. Bounded rationality refers to the fact that *"decision makers have constraints on their cognitive capabilities and limits on their rationality"* (Rindfleisch and Heide 1997). Wathne and Heide (2000) summarised the idea of bounded rationality as human's physical limitation to process information. Bac (2001) stated that one or both contracting parties may behave opportunistically to seek for their own interest. Opportunism arises when the principal contracts are vague and incomplete (Williamson 1985). Muris (1981) suggested that opportunism occurs when either contracting party *"retaliates against post contractual manipulation of the terms of trade … in order to effect an unexpected transfer of wealth from the other party"*. Luo (2007) advocates that party who sustained or prolonged uncertainty of gains may behave opportunistically.

Construction works are exposed to weather conditions and unforeseen ground conditions that are beyond the control of contracting parties. The unique design coupled with complex production processes prohibits complete contracts. Under these situations construction contracts can never provide exhaustive detailing of the rights and obligations of the contracting parties. Ayres and Gertner (1992) described this type of incompleteness as *"obligationally incomplete"*. Contract is also incomplete when it does not include the necessary instruction for some eventualities. Therefore, contracts are viewed as incomplete when they are (i)

Table 2.4 Minefields of opportunism in construction contracting

Code	Descriptions	Occurrence rankings
Ambiguity		
A1	The scope of work is unclear	2
A2	The specification is unclear	4
A3	The rules to evaluate star rate is unclear	6
A4	Work activities are unclear	10
A5	Completion milestones are unclear	17
Deficiency		
B1	The rules to evaluate substantial change in quantity of works are not addressed	8
B2	There is no provision to deal with re-nomination of the nominated sub-contractor	12
B3	There is no provision to deal with the consequence of re-nomination of the nominated sub-contractor	12
B4	There is no obligation to report inconsistency of contract documents	19
B5	The specification of material is inadequate	15
B6	The performance specification is inadequate	17
B7	The drawings provide insufficient details	1
B8	The guidelines for the preparation and submission of work schedule is inadequate	23
B9	There is no statement of resources in the work schedule	19
Inconsistency		
C1	The specification of material is contradictory	22
C2	The performance specification is contradictory	19
C3	The drawings contradict with the specification	6
C4	The details in the drawings are inconsistent	3
C5	The specified design standard is different from statutory requirement	10
Defectiveness		
D1	Items in the contract bills of quantity are being omitted	5
D2	Abundant items are found in the contract bills of quantity	12
D3	Over measured items are found in the contract bills of quantities	15
D4	Some items are missing from the contract bills	8

ambiguous on the statement of obligations and responsibilities; and (ii) do not fully specify the contingencies and risk allocation in contract clauses. Embedded in the concept of bounded rationality, minefields of opportunism are arranged in four major categories: ambiguity, deficiency, inconsistency and defectiveness, are listed in Table 2.4.

While contract incompleteness cultivates the minefields, opportunism may manifest through violation of commitment, forced renegotiation, evasion of obligation and refusal to adapt (Wathne and Heide 2000). In addition, party may not honestly oblige to their agreements as in the case of inflation of Contractor's claims (violation of commitment), exploiting one's advantage and forcing renegotiation of unfavourable original contract terms (forced renegotiation). Contracting parties

Table 2.5 Manifestations of opportunism in construction contracting

Code	Descriptions	Occurrence rankings
Violation of commitments		
E1	The Contractor over-claims her cost entitlement	1
E2	The Contractor over-claims her time entitlement	2
E3	The Contractor over-claims her costs for progress acceleration	3
Forced renegotiation		
F1	The Client attempts to renegotiate the terms of signed contract	15
Evasion of obligations		
G1	The Contractor purposely fails to disclose the specifications of the materials used	17
G2	The Contractor purposely fails to notify potential implication arising from changes orders	20
G3	The Contractor purposely fails to notify over measured items in the contract BQ	12
G4	The Contractor purposely fails to notify under measured items in the contract BQ	12
G5	The Contractor purposely fails to notify omission of items in the contract BQ	7
G6	The Contractor purposely fails to notify abundant items in the contract BQ	17
G7	The Contractor purposely works below the specified standard	10
G8	The Contractor purposely fails to provide material of the required standard	12
G9	The Contractor purposely not provides invoice for the material used	21
G10	The Client orders extra without providing proper cost reimbursement	5
G11	The Client orders extra without granting justifiable extension of time	7
G12	The Client rejects Contractor's claims for variation outright without providing reasons	9
G13	The Client rejects outright extension of time claim submitted by the Contractor	6
G14	The Client rejects outright monetary claim submitted by the Contractor	4
Refusal to adapt change		
H1	The Contractor refuses to agree the valuation methods proposed/used by the Consultant QS according to the contract	11
H2	The Contractor refuses to respond to late design change requested by the Client	15
H3	The Contractor refuses to accelerate work progress requested by the Client	17

may ignore contractual requirements like late payment and late information (evasion of obligation). Furthermore, they may not be willing to respond to frequent design change (refusal to adapt change). In this regard, contracting behaviours manifesting opportunism have been long-listed and summarised in Table 2.5. These behaviours are arranged in four groups: violation of commitment, forced renegotiation, evasion of obligation and refusal to adapt change.

Table 2.6 Respondents' characteristics

Respondents' characteristics	Number of respondents	Percentage (%)
(i) Company		
Law firm	10	10.53
Client's group	56	58.95
Contractor's group	29	30.53
(ii) Profession		
Construction lawyer	12	12.63
Architect/Engineer	27	28.42
Project manger	18	18.95
Surveyor	38	40.00
(iii) Working experience		
Less than 10 years	43	45.26
10–20 years	13	13.68
More than 20 years	39	41.05

2.4 Data Collection and Analysis

A questionnaire was designed and used to collect occurrence likelihood assessments of the minefields and manifestations. Over 300 sets of questionnaire were issued to construction professionals in Hong Kong. 100 professionals returned the questionnaire representing a response rate is 33 %. 95 of them were valid and used in the data analysis. The questionnaire has two sections. The first section includes questions on demographic characteristics of the respondents that are summarised in Table 2.6. Over 58 % of the respondents are working in developer companies, and 30 % of the respondents are working in property developers. 10 % of the respondents are working in law firms. Besides, the majority of the respondents have more than 10 years of experience in administering construction contracts. In the second section, the respondents were asked to assess the occurrence likelihood of the artifacts.

Based on respondents' subjective judgment of occurrence, the relative occurrence rankings of the minefield and manifestations of opportunism are given in Tables 2.4 and 2.5 respectively.

2.5 Towards an Anatomy of Construction Disputes

The happening of the highly ranked minefields will amount to changes that trigger the variation clause. Many of these changes are either not preferred or not expected or both. Classic examples include blatant omissions that disturb planned construction activity. Other causes of change include unclear scope, inconsistency/contradiction between contract documents and missing items in bills of quantities. The relatively high occurrence likelihood suggests some hidden causes that go beyond bounded rationality. For example, contract documents are prepared in a hasty manner. Notably this has been a major concern (CIRC 2001).

Resolving construction disputes is time consuming and costly. Yates and Epstein (2006) suggested that understanding the provisions and disruptive issues at the earliest possible time could mitigate claims and disputes. Disputes would arise from a process involving conflict (Fenn et al. 1997). In addition, conflict can also stem from cognition, behaviour and emotion (Garcia-Prieto et al. 2003; Jehn 1997).

Cognitive conflict refers to the collaboration problems encountered during the construction stage. The bottlenecks so resulted negatively influence project implementation and thus project success. For instance, the consultants fail to provide adequate drawings; the client make changes frequently and the contractors delays in work. These problems may become disputes if not addressed appropriately and timely. Behavioural conflict describes the opportunistic strategies in construction claims. The contractor may bid opportunistically in non-competitive tendering (Ho and Liu 2004). The client may handle contractors' claims sinuously. Williamson (1975) described such behaviour as opportunistic. Opportunism is defined as *"self-interest seeking with guile"* or *"calculated efforts to mislead, distort, disguise, obfuscate or otherwise confuse"* (Williamson 1985). Contracting parties behave opportunistically by seeking own interests and benefits under the conditions of asymmetrical information and uncertainty. Emotional conflict delineates the personal and interpersonal affective conflict among project team members. It often escalates arguments and impedes seeking of win–win solutions.

Uncertainty and contract incompleteness are the triggering factors of construction dispute. Risk allocation of construction contract is pivotal. Construction clients generally opt to shift unanticipated risks to the contractors. While some of these risks are beyond the controllability and forseeability of the contractors, many are even beyond their manageability if they materialise. In addition, clients or consultants are boundedly rational in foreseeing the extent of uncertainty and risk involved (Simon 1961). The behaviours of decision makers are confined to their rationality boundary (Simon 1961). Rationality reaches its limits under the conditions of uncertainty and complex circumstance, as decision makers do not and are not able to possess perfect information (Masters et al. 2004). In these contexts, construction contracts are incomplete. As illustrated in previous section, most claims are somehow related to ambiguous general terms for which contracting parties are having different views on the rights and responsibilities arising therefrom (Jergeas 2001). Review of literature suggests that construction disputes could be attributed to (1) poor collaboration (Bristow and Vasilopoulos 1995; Conlin et al. 1996b), (2) opportunistic behaviours (Mitropoulos and Howell 2001; Yates 2003), (3) affective conflicts (Diekmann et al. 1994; Mitropoulos and Howell 2001), (4) high risks and uncertainties (Diekmann and Girard 1995; Kumaraswamy 1997) and (5) contract incompleteness (Heath et al. 1994; Yates 1998). Table 2.7 summarises the manifestations of construction disputes according to the above categorisation.

Incomplete contracts are springboards of construction claims and the key elements leading to both contractual and speculative disputes. Ambiguities of

Table 2.7 Manifestation of construction disputes

Category	Source	Description/attributes	References
Cognitive manifestation	Poor collaboration	Poor collaboration in construction represents any direct or indirect action, neglect or default of contracting parties influencing the cooperation of construction project	Bristow and Vasilopoulos (1995), Conlin et al. (1996b), Heath et al. (1994), Diekmann et al. (1994)
		Delays	Conlin et al. (1996b), Hewit (1991), Acharya et al. (2006)
		Incomplete information	Heath et al. (1994), Kumaraswamy (1997)
		Change	Conlin et al. (1996); Hewit (1991), Semple et al. (1994), Acharya et al. (2006)
Behavioural manifestation	Opportunistic behaviour	Opportunism delineates the self-interest seeking behaviour in construction project	Mitropoulos and Howell (2001), Rooke et al. (2003), Yates (2003)
		Violation of commitment	Beach et al. (2005), Chen and Chen (2007)
		Evasion of obligation	Chen and Chen (2007)
		Refusal to adapt change	Rameezdeen and Gunarathna (2003)
		Forced renegotiation	Acharya et al. (2006)
Emotional manifestation	Affective conflict	Affective conflict illustrates the negative emotion among construction project team members	Diekmann et al. (1994), Mitropoulos and Howell (2001)
		Personality clash	Belout and Gauvreau (2004)
		Interpersonal emotional conflict	Sommerville and Langford (1994), Acharya et al. (2006)

(continued)

Table 2.7 (continued)

Category	Source	Description/attributes	References
Structural problem	Risk and uncertainty	Risk and uncertainty in construction includes unpredictable events caused from construction environment, natural disaster and social interventions	Bristow and Vasilopoulos (1995), Diekmann and Girard (1995), Kumaraswamy (1997), Mitropoulos and Howell (2001)
		Natural cause	Semple et al. (1994), Diekmann and Girard (1995), Acharya et al. (2006)
		Intervention	Rhys Jones (1994), Diekmann and Girard (1995), Acharya et al. (2006)
	Contract incompleteness	Construction contract is incomplete when it does not fully address the contingents and provide accurate and adequate information	Bristow and Vasilopoulos (1995), Heath et al. (1994), Kumaraswamy (1997), Yates (1998), Mitropoulos and Howell (2001)
		Ambiguity	Kumaraswamy (1997), Yates (1998), Acharya et al. (2006)
		Deficiency	Kumaraswamy (1997), Yates (1998)
		Inconsistency	Acharya et al. (2006)
		Defectiveness	Kumaraswamy (1997), Yates (1998)

contractual agreements may cause differing interpretations of performance requirements among the contracting parties. Contract agreements address the monetary and time entitlements due to natural disasters, unpredictable environmental issues (i.e. risk and uncertainty), and delay and disruption (i.e. collaborative conflict). Hence, contractual disputes are triggered by task factors that include risk and uncertainty and collaborative conflict of construction project. To develop an anatomy of construction disputes, non-contractual disputes take into account of people factors that fully demonstrate the drivers of speculation. Speculative disputes are result of personal interest seeking behaviour (i.e. opportunistic behaviour) and personal emotion (i.e. affective conflict). In addition, an incomplete contract limits the contractual governance of the contracting parties and boosts the formation of speculative disputes.

Figure 2.1 gives the proposed anatomy of construction disputes. Construction disputes can be contractual or speculative. Risks and uncertainties and collaborative conflicts would evoke contractual disputes with an incomplete contract. Likewise, opportunistic behaviour of contracting parties and affective conflict of project team members would induce speculative disputes. Further elaboration of the proposed anatomy is given in Chap. 3. Furthermore, the proposed anatomy enables an empirical test on the occurrence likelihood of construction disputes.

2.6 Chapter Summary

This chapter discusses the subject matter and diagnostic approaches to identify construction disputes. The subject matter approach is the most commonly used and is useful in relating a dispute to the relevant contract provisions. However, it does not capture the contextual factors of the disputes. The diagnostic approach aims to fill this gap. An anatomy of construction dispute is proposed by integrating these two approaches. Construction disputes can be either contractual or speculative. Contract incompleteness is the common factor for the two types of construction disputes. Task and people factors are the other contributing drivers of contractual and speculative disputes respectively. This conceptualisation of construction disputes founds on the theoretical constructs of bounded rationality and opportunism offered by transaction cost economics. Bounded rationality restrains the ability to write a complete contract. As such an incomplete contract provides the window for practice of opportunism. Opportunism in construction is studied in detail. First, minefields and manifestations of opportunism are identified. Second, their respective occurrence likelihoods are assessed. Furthermore, while efforts can be directed at the pre-contract stage to minimise contract incompleteness, post-contract changes remain inevitable for various reasons. Thus adjustment mechanisms such as variation, extension of time and loss & expense provisions are included to deal with such changes while maintaining the validity of the contract. Moreover, opportunists take advantage of these changes. The management implications are two folded. To mitigate opportunism, the best strategy is to clear the minefields;

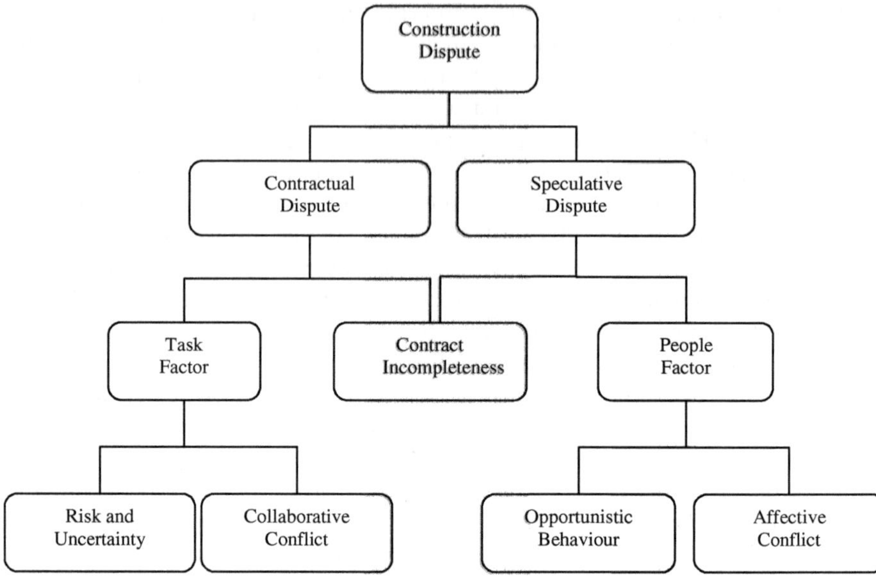

Fig. 2.1 Anatomy of construction disputes

this requires conscious effort at the pre-contract stage. For example, more time should be allowed for the preparation of contract documentation. Notwithstanding, contracts remain largely incomplete and laid the ground for opportunistic behaviours. Self-interest-seeking contracting behaviours can be opportunistic. Contractors may therefore raise inflated claim and clients may outright reject any claims from contractors. At post-contract stage, a trusting contracting environment would suppress many of the problematic opportunistic manifestations. Notwithstanding the efforts in enhancing cooperation in contracting, perhaps there is no replacement for the basics of avoiding excessive risk taking, having clear documentation and keeping changes to the minimum.

Acknowledgments Special thanks to Miss Mei Ching Choi and Miss Sin Yi Chu for collecting data for the study. Part of the contents of this chapter has been published in Volume 139(1) of the Journal of Construction Engineering and Management and is used with the permission from ASCE.

References

Acharya, N. K., Lee, Y. D., & Lm, H. M. (2006). Conflict factors in construction projects: Korean perspective, engineering. *Engineering, Construction and Architectural Management, 13*(6), 543–566.

Anderson, E. (1988). Transaction costs as determinants of opportunism in integrated and independent sales forces. *Journal of Economic Behavior & Organization, 9*(3), 247–264.

Ayres, I., & Gertner, R. (1992). Strategic contractual inefficiency and the optimal choice of legal rules. *Yale Law Journal, 101*(4), 729–773.

Bac, M. (2001). To invest or screen efficiently: A potential conflict in relationships governed by incomplete contracts. *International Journal of Industrial Organization, 19*(3), 567–588.

Beach, R., Webster, W., & Campbell, K. M. (2005). An evaluation of partnership development in the construction industry. *International Journal of Project Management, 23*, 611–621.

Belout, A., & Gauvreau, C. (2004). Factors influencing project success: The impact of human resource management. *International Journal of Project Management, 22*, 1–11.

Bristow, D., & Vasilopoulos, R. (1995). The new CCDC2: Facilitating dispute resolution of construction projects. *Construction Law Journal, 11*(2), 95–117.

Brooker, P. (2002). Construction lawyers' attitudes and experience with ADR. *Construction Law Journal, 18*(2), 97–116.

Brown, J. R., Dev, C. S., & Lee, D. J. (2000). Managing marketing channel opportunism: The efficacy of alternative governance mechanisms. *Journal of Marketing, 64*(2), 51–65.

Brown, H., & Marriott, A. (1999). *ADR principles and practice*. London: Sweet and Maxwell.

Burton, J. (1990). *Conflict: Resolution and prevention*. Basingstoke/New York: Macmillan/St. Martin's Press.

Chen, W. T., & Chen, T. T. (2007). Critical success factors for construction partnering in Taiwan. *International Journal of Project Management, 25*, 475–484.

Cheung, S. O. (2007). Bounded rationality, opportunism and trust in co-operative contracting. In Proceedings of the Fourth International Structural Engineering and Construction Conference (ISEC–4), Melbourne, pp. 1537–1542.

Coase, R. (1975). Marshall on method. *Journal of Law and Economics, 18*, 25–31.

Coase, R. (1984). The new institutional economics. *Journal of Institutional and Theoretical Economics, 140*, 229–231.

Conlin, G. J., Lanford, D. A., & Kennedy, P. (1996a). The relationship between construction procurement strategies and construction contract disputes. In R. Taylor (Eds.), GCIB W92 "North meet south" Procurement Systems Symposium Proceedings, Durban, South Africa, pp. 66–82.

Conlin, J., Lanford, D. A., & Kennedy, P. (1996b). The sources, causes, and effects of construction disputes: A research project. CIB Report 0254-4083, Construction Industry Board, London.

Construction Industry Review Committee (CIRC) (Ed.) (2001). Construct for Excellence, Report of the Construction Industry Review Committee. Hong Kong, Hong Kong SAR Government.

Diekmann, J. E., & Girard, M. J. (1995). Are contract disputes predictable? *Journal of Construction Engineering and Management, 121*(4), 335–363.

Diekmann, J., Girard, M., & Abdul-Hadi, N. (1994). Disputes potential index: A study into the predictability of contract disputes. Construction Industry Institute, Source Document 101, The University of Texas at Austin, TX.

Dyner, I., & Franco, C. J. (2004). "Consumers" bounded rationality: The case of competitive energy markets. *Systems Research and Behavioral Science, 21*, 373–389.

Eisenberg, M. A. (2001). The theory of contracts. In P. Benson (Ed.), *The theory of contract law* (pp. 213–222). Cambridge: Cambridge University Press.

Fenn, P., Lowe, D., & Speck, C. (1997). Conflict and dispute in construction. *Construction Management and Economics, 15*(6), 513–518.

Garcia-Prieto, P., Bellard, E., & Schneider, S. C. (2003). Experiencing diversity, conflict, and emotions in teams. *Applied Psychology, 52*(3), 413–440.

Gauss, C. (Ed.). (1952). *Niccolo Machiavelli The Prince*. New York: New American Library of World Literature.

Goffman, E. (Ed.). (1969). *Strategic interaction*. Philadelphia: University of Pennsylvania Press.

Heath, B. C., Hills, B., & Berry, M. (1994). The nature and origin of conflict within the construction process. In Proceedings of the CIB TG15 Conference, Kentucky, USA, CIB publication 171, pp. 35–48.

Hellard, R. B. (1987). *Managing construction conflict*. London: Longman Scientific Technical Press.
Hewit, J. (1991). *Winning construction disputes—Strategic planning for major litigation*. London: Ernst and Young.
Ho, S. P., & Liu, L. Y. (2004). Analytical model for analyzing construction claims and opportunistic bidding. *Journal of Construction Engineering and Management, 130*(1), 94–104.
Hobbes, T. (1928). *Leviathan, or, the matter, forme and power of commonwealth, ecclesiasticall and civil (1651)*. Oxford: Basil Blackwell.
Ive, G., & Chang, C. Y. (2007). The principle of inconsistent trinity in the selection of procurement system. *Construction Management and Economics, 25*(7), 677–690.
Jehn, K. A. (1997). A qualitative analysis of conflict types and dimensions in organizational groups. *Administrative Science Quarterly, 42*(3), 530–557.
Jergeas, G. F. (2001). Claims and disputes in the construction industry. *AACE International Transactions*, CDR.03.1–CDR.03.4.
John, G. (1984). An empirical investigation of some antecedents of opportunism in a marketing channel. *Journal of Marketing Research, 21*(3), 278–289.
Judge, W. Q., & Dooley, R. (2006). Strategic alliance outcomes: A transaction-cost economic perspective. *British Journal of Management, 17*(1), 23–37.
Kumaraswamy, M. H. (1997). Conflicts, claims and disputes in construction. *Engineering, Construction and Architectural Management, 4*(2), 95–111.
Lee, H. S., Seo, J. O., Park, M., Ryu, H. G., & Kwon, S. S. (2009). Transaction-cost-based selection of appropriate general contractor-subcontractor relationship type. *Journal of Construction Engineering and Management, 135*(11), 1232–1240.
Luo, Y. (2007). Are joint venture partners more opportunistic in a more volatile environment? *Strategic Management Journal, 28*(1), 39–60.
Masters, J. K., Miles, G., & Orr, J. P. (2004). Risk propensity, trust, and transaction costs in relational contracting. *Journal of Business Strategy, 21*(1), 47–66.
Mitropoulos, P., & Howell, G. (2001). Model for understanding, preventing and resolving project disputes. *Journal of Construction Engineering and Management, 127*(3), 223–231.
Muris, T. J. (1981). Opportunistic behavior and the law of contracts. *Minnesota Law Review, 65*(4), 521–590.
Mururu, N. (1991). Anatomy of a dispute. *Arbitration*, November, pp. 262–264.
O'Donovan, T. R. (1962). Opportunism and executive behavior. *Human Resource Management Journal, 1*(3), 6–9.
Parkhe, A. (1993). Strategic alliance structuring: A game theoretic and transaction cost examination of inter-firm cooperation. *Academy of Management Journal, 36*(4), 794–829.
Provan, K. G. (1993). Embeddedness, interdependence, and opportunism in organizational supplier-buyer networks. *Journal of Management, 19*(4), 841–856.
Rameezdeen, R., & Gunarathna, N. (2003). Disputes and construction industry cultures. *AACE International Transactions*, CDR.24.1–CDR.24.8.
Reve, T., & Levitt, R. E. (1984). Organization and governance in construction. *International Journal of Project Management, 2*(1), 17–25.
Rhys Jones, S. (1994). How constructive is construction law? *Construction Law Journal, 10*(1), 28–38.
Rindfleisch, A., & Heide, J. B. (1997). Transaction cost analysis: Past, present, and future applications. *Journal of Marketing, 61*(4), 30–54.
Rooke, J., Seymour, D., & Fellows, R. (2003). The claims culture a taxonomy of attitudes in the industry. *Construction Management and Economics, 21*, 167–174.
Semple, C., Hartman, F., & Jergas, G. (1994). Construction claims and disputes: cause and cost/time overruns. *Journal of Construction Engineering and Management, 120*(4), 785–795.
Sheridan, P. (2003). Claims and disputes in construction. *Construction Law Journal, 12*(1), 3–13.
Simon, H. A. (1957). *Models of man, social and rational*. New York: Wiley.
Simon, H. A. (1961). *Administrative behavior* (2nd ed.). New York: Macmillan.

Simon, H. A. (1972). Theories of Bounded Rationality. In C. B. McGuine & R. Radner (Eds.), *Decision and organization—A volume in honour of Jacob Marschak*. North-Holland, Amsterdam.

Simon, H. A. (1979). Rational decision-making in business organizations. *American Economic Review, 69*(4), 493–513.

Simon, H. A. (Ed.). (1997). *Models of bounded rationality*. Cambridge: MIT Press.

Sommerville, J., & Langford, V. (1994). Multivariate influences on the people side of projects: Stress and conflict. *International Journal of Project Management, 12*(4), 234–243.

Spittler, J. R., & Jentzen, G. H. (1992). Dispute resolution: Managing construction conflict with step negotiations. *AACE International Transactions, Accounting & Tax Periodicals, 1*(1), 9.1–9.10.

Sykes, J. (1996). Claims and disputes in construction. *Construction Law Journal, 12*(1), 3–13.

Tillet, G. (1991). *Resolving conflict: A practical approach*. Sydney: Sydney University Press.

Totterdill, B. W. (1991). Does the construction industry need alternative dispute resolution? The opinion of an engineer. *Construction Law Journal, 7*(3), 189–199.

Wathne, K. H., & Heide, J. B. (2000). Opportunism in interfirm relationships: Forms, outcomes, and solutions. *Journal of Marketing, 64*(4), 36–51.

Watts, V. M., & Scrivener, J. C. (1993). Review of Australian building disputes settled by litigation. *Building Research and Information, 21*(1), 59–63.

Williamson, O. E. (1975). *Markets and hierarchies: Analysis and antitrust implications: A study in the economics of internal organization*. New York: Free Press.

Williamson, O. E. (1979). Transaction cost economics: The governance of contractual relations. *Journal of Law and Economics, 22*(2), 233–261.

Williamson, O. E. (Ed.). (1985). *Economic institution of capitalism: Firms, markets, relational contracting*. New York: Free Press.

Williamson, O. E. (1991). Strategizing, economizing, and economic organization. *Strategic Management Journal, 12*, 75–94.

Williamson, O. E. (1993). Opportunism and its critics. *Managerial and Decision Economics, 14*(1), 97–107.

Williamson, O. E. (1995). Hierarchies, markets and power in the economy: An economic perspective. *Industrial and Corporate Change, Oxford University Press, 4*(1), 21–49.

Winch, G. (1989). The construction firm and the construction project: A transaction cost approach. *Construction Management and Economics, 7*(4), 331–349.

Winch, G. (1995). Project management in construction: Towards a transaction cost approach. Le Groupe Bagnolet Working Paper 1, University College, London.

Yates, D. J. (1998). Conflict and dispute in the development process: A transaction cost economic perspective. http://www.prres.net/proceedings/proceedings1998/Papers/Yates3Ai.PDF (March 31, 2011).

Yates, D. J. (2003). Can claims and disputes (in construction contracts) be prevented or reduced? *Building Journal Hongkong China*.

Yates, J. K., & Epstein, A. (2006). Avoiding and minimizing construction delay claim disputes in relationship contracting. *Journal of Professional Issues in Engineering Education and Practice, 132*(2), 168–179.

Yates, D. J., & Hardcastle, C. (2002). The causes of conflict and disputes in the Hong Kong construction industry: A transaction cost economics perspective. *Journal of Financial Management of Property and Construction, 7*(2), 115–126.

Yiu, T. W., & Cheung, S. O. (2004). Significant dispute sources of construction mediation. In Proceedings of the First International Conference on World of Construction Project Management. Toronto, Canada, pp. 596–604.

Chapter 3
The Occurrence Likelihood of Construction Disputes

Hoi Yan Pang and Sai On Cheung

Abstract Construction Disputes are widely considered as inevitable. The supporting evidence of this general belief is largely anecdotal. This chapter reports an empirical test of this proposition. The conceptual framework of construction disputes as presented in Chap. 2 is further developed. Furthermore, fault tree methodology is used to operationalise the framework whereby the inter-relationships among dispute artifacts are expressed in logic gates. This conceptualisation thus displays the logic relationships and further allows assessment of occurrence likelihood. In view of the imprecise nature, fuzzy sets occurrence likelihood assessments of the artifacts are used. The findings support that construction disputes are inevitable if no conscious effort is made to minimise its occurrence.

3.1 Occurrence of Construction Disputes

Construction environment and contracts have inherent characteristics that are futile for dispute. Molenaar et al. (2000) found that project complexity and inequitable risk allocation contribute to dispute occurrence. Their study suggested that people might not cause disputes directly but influence the effectiveness in resolving disputes. Competent project team would have the ability to foresee potential problems or timely address them when they materialise. In fact, most disputes are unresolved problems (Fenn 2002). Nonetheless, the frequency and severity of disputes also depend on the attitude, behaviour, knowledge and experience of the

H. Y. Pang (✉) · S. O. Cheung
Construction Dispute Resolution Research Unit, Department of Civil and Architectural Engineering, City University of Hong Kong, Hong Kong, People's Republic of China
e-mail: karen-pang@hotmail.com

S. O. Cheung
e-mail: saion.cheung@cityu.edu.hk

S. O. Cheung (ed.), *Construction Dispute Research*,
DOI: 10.1007/978-3-319-04429-3_3,

people involved. In this study, the proposed anatomy of construction disputes links the inherent characteristics of construction project and people factors together. The occurrence likelihood of construction disputes can thereby be evaluated holistically.

Diekmann and Girard (1995) and Molenaar et al. (2000) employed logistic regression and applied discrete choice modeling to develop dispute potential index. The index reflects the degree of proneness of a project having disputes. This study proposes an anatomy of construction disputes that is developed on a fault tree framework. In construction, fault tree analysis has been utilised to evaluate contractor selection (Singh and Tiong 2005), performance assessment (Pan 2006), cost estimation (Shaheen et al. 2007) and construction method selection (Pan 2008). A Fault Tree model employs a bottom-up approach to analyse failure. Fault tree analysis serves to identify the weakest part of a system. In a fault tree framework, basic events are at the bottom of the tree and need no further development; top event represents the final result from the combination of basic events linked by logic gates. "AND" and "OR" are the common logic gates used to depict the possible combination and sequences of events that contribute to the undesired top event. Therefore, with a fault tree methodology, the relationships between events can be systematically arranged with respect to hierarchical causal order. Fuzzy sets approach is used to develop a likelihood evaluation system. The resulting Fuzzy Fault Tree model (FFT) graphically presents the antecedences and consequences of dispute artifacts. The proposed FFT dispute likelihood evaluation model provides a novel approach to identify the critical contributors that trigger construction disputes.

3.2 Development of FFT Dispute Model

The operationalised anatomy of construction dispute in a fault tree framework is shown in Fig. 3.1. The right most of the fault tree lists the basic events (i.e. potential dispute artifacts) and the top event is the failure (i.e. construction disputes). Development of the FFT model can be described in three stages: (1) determination of potential dispute artifacts, (2) identification of inter-relationship of artifacts and its categories and (3) model refinement and validation. Firstly, potential dispute artifacts were long-listed from a literature review on construction disputes. The long-listed potential dispute artifacts were then arranged in a fault tree framework with logic gates giving effect to the inter-relationships of the events. Secondly, a pilot was conducted with twenty-four experienced construction professionals to enhance the authenticity of the dispute anatomy. Thirdly, the fault tree dispute framework was then refined for occurrence likelihood evaluation.

The characteristics of the panel of construction professionals participated in the pilot study are summarised in Table 3.1. All interviewees had at least 10 years working experience in the construction industry; indeed the majority of them have over 20 years working experience. The panel of professionals includes lawyers,

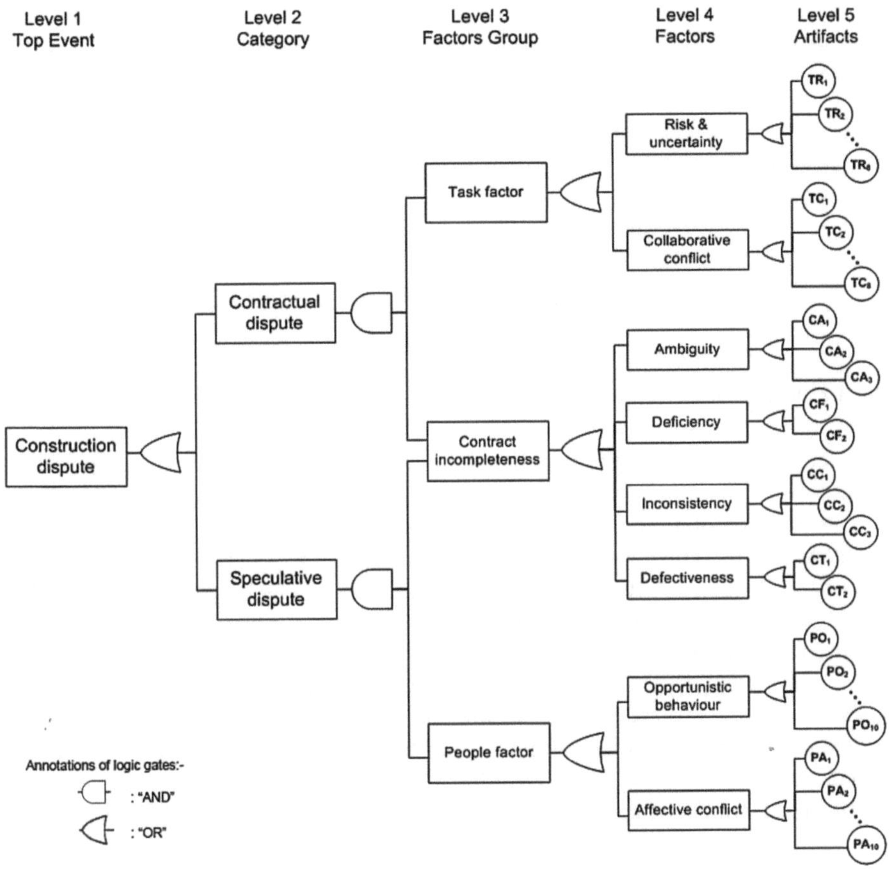

Fig. 3.1 Construction disputes: a fault tree framework

Table 3.1 Interviewee's characteristics

Company	No. of people	Working experience	No. of people
Construction law firm	6	Over 25 years	10
Claim consultant	2	20–25 years	7
Developer	3	15–20 years	3
Engineering consultant	2	10–15 years	4
Quantity surveying consultant	5		
Contractor	6		

claim consultants and senior professionals working in developers, consultant offices and contracting organisations. They all have substantial experience in claims and disputes resolution. Collectively, the panel provides insight from the perspectives of major stakeholders in construction. In these regards, the wealth of

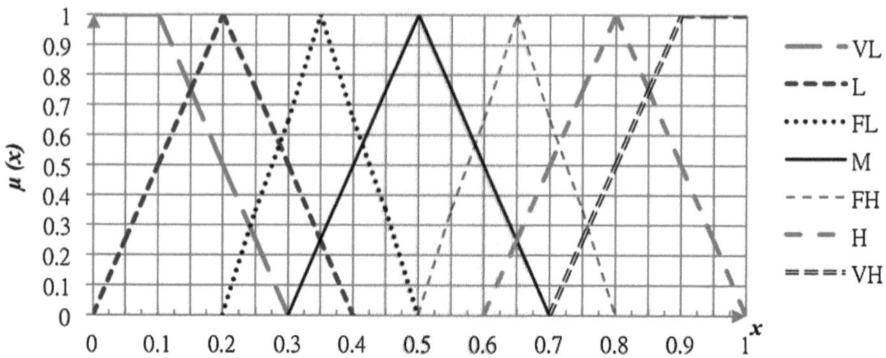

Fig. 3.2 Triangular membership functions

knowledge possessed by the panel contributes significantly to the development of the proposed anatomy.

The number of potential dispute artifacts was firstly long-listed based on a literature review. The list was then commented by the expert panel and was refined according. A total of 46-potential dispute artifacts are retained and arranged in eight factors. These eight factors are then organised into three factor-groups. The schematic of the proposed anatomy of construction dispute in the form of a fault tree is shown in Fig. 3.2. Level 1 shows the final outcome i.e. construction disputes. At Level 2, construction disputes are categorised as either contractual or speculative according to the nature of the dispute causes. Three factor-groups of dispute are shown at Level 3. Level 4 locates the eight dispute factors. Level 5 lists the dispute artifacts of each of the respective dispute factors.

3.3 Occurrence of Construction Disputes

Because of the uniqueness of construction projects, the dispute severity, dispute likelihood and dispute artifacts are different. Respondents have to provide subjective judgment on occurrence likelihood of dispute artifacts respective to the circumstances. In the following sections, application of fuzzy membership function and mathematical expression of logic gates are described. Then, the detailed evaluation of dispute likelihood is presented.

3.3.1 Fuzzy Sets and Membership Function

Fuzzy sets theory has been used to draw conclusion from subjective statements (Zadeh 1965). The fuzzy sets theory allows assessing likelihood in linguistic variables, hence is particularly useful when human judgments are involved. In fuzzy sets, the degree of belief of every fuzzy subset is represented in membership function (MF) with values in [0, 1]. MF properties comprise the number of linguistic variable, membership function shape and universe of discourse.

Triangular and trapezoidal MFs are frequently applied in construction risk analysis. For example, triangular MFs were employed to study performance improvement (Juan 2009) and project cost contingency estimation (Idrus et al. 2011); trapezoidal MFs were applied to project cost estimation (Shaheen et al. 2007) and bridge construction failure (Pan and Wang 2007). Besides, triangular MF with uniformly distributed linguistic variables is the most appropriate membership functions for the fuzzy control system (Zhao and Bose 2003). Wang and Kerre (2001) used seven-fuzzy sets assessment for system failure evaluation. Likewise, Zhao and Bose (2003) compared the performance of symmetrical membership functions for all the variables and found that seven-fuzzy sets ($N = 7$) is the optimal case of the symmetrical distribution of the triangular membership function. Singh and Tiong (2005) and Pan and Wang (2007) have applied 7 linguistic variables to the study of contractor selection and bridge failure respectively.

Linguistic variables define natural languages in words that are characterised by universe of discourse of fuzzy sets. Seven linguistic variables, namely Very Low (VL), Low (L), Fairly Low (FL), Medium (M), Fairly High (FH), High (H), and Very High (VH) are used to evaluate the occurrence likelihoods of the dispute artifacts. In the example of inclement weather, the occurrence likelihood is to be assessed by 7 linguistic variables represented by a scale of 1 to 7 respectively. The triangular MFs and linguistic variables used in the scoring system are shown as Fig. 3.2 and Table 3.2 respectively.

3.3.2 Aggregation and Defuzzification of Linguistic Variables

The linguistic variables selected by the respondents against the dispute artifacts were used to calculate the fuzzy probabilities of occurrence of construction disputes. The selected linguistic variable indicates an interval of confidence which usually be expressed by fuzzy parameters (Zadeh 1965). Aggregation on fuzzy parameters averages the respondents' preferences into group preferences (Singh and Tiong 2005). The aggregated value can be expressed as

$$\widetilde{A}'_j = \left(\frac{1}{p}\right) \otimes \left(\widetilde{a_1} \oplus \widetilde{a_2} \oplus \cdots \oplus \widetilde{a_p}\right) \quad \text{for} j = 1, 2, \ldots, p \tag{3.1}$$

where a is the first fuzzy parameter of selected linguistic variable and p is the number of respondent;

$$\widetilde{B}'_j = \left(\frac{1}{p}\right) \otimes \left(\widetilde{b_1} \oplus \widetilde{b_2} \oplus \cdots \oplus \widetilde{b_p}\right) \quad \text{for} j = 1, 2 \ldots, p \tag{3.2}$$

Table 3.2 Linguistic variables in triangular membership functions

Linguistic score	Interpretation	Fuzzy parameter (a, b, c)
Very Low (VL)	Involved event is avoidable	(0.00, 0.10, 0.30)
Low (L)	Involved event is partially likely to occur	(0.00, 0.20, 0.40)
Fairly Low (FL)	Involved event is rarely likely to occur	(0.20, 0.35, 0.50)
Medium (M)	Involved event is occasionally likely to occur	(0.30, 0.50, 0.70)
Fairly High (FH)	Involved event is somewhat frequently likely to occur	(0.50, 0.65, 0.80)
High (H)	Involved event is most frequently to occur	(0.60, 0.80, 1.00)
Very High (VH)	Involved event is almost inevitable	(0.70, 0.90, 1.00)

where b is the second fuzzy parameter of selected linguistic variable and p is the number of respondent;

$$\widetilde{C}_j' = \left(\frac{1}{p}\right) \otimes \left(\widetilde{c_1} \oplus \widetilde{c_2} \oplus \cdots \oplus \widetilde{c_p}\right) \quad \text{for} j = 1, 2 \ldots, p \tag{3.3}$$

where c is the third fuzzy parameter of selected linguistic variable and p is the number of respondent.

After the aggregation of fuzzy parameters, defuzzification of fuzzy parameters represents the expected value of selected linguistic variables (Shaheen et al. 2007) or the degree of satisfaction of the aggregated fuzzy parameters (Kales 1998). The defuzzified value equals to the mean values of triangular probability distribution (Shaheen et al. 2007) that can be expressed as

$$e = \left(\widetilde{A}_j' + \widetilde{B}_j' + \widetilde{C}_j'\right) \tag{3.4}$$

where $\widetilde{A}_j'$, $\widetilde{B}_j'$ and $\widetilde{C}_j'$ are the aggregated fuzzy parameter.

3.3.3 Evaluation of Dispute Likelihood

With reference to the logical expression illustrated in Fig. 3.1, the occurrence likelihood of top event (i.e. construction dispute) and intermediate events (i.e. two types of dispute and three dispute contributors) can be evaluated. The expression of fuzzy probability of "AND" gate and "OR" gate are as follows:

$$P_r^{\widetilde{AND}} = \prod_{i=1}^{n} P_{\mathrm{i}} \tag{3.5}$$

$$P_r^{\widetilde{OR}} = \prod_{i=1}^{n} (1 - P_{\mathrm{i}}) \tag{3.6}$$

where $P_{\mathrm{i}} = \left(\widetilde{A}_j', \widetilde{B}_j', \widetilde{C}_j'\right)$ is the aggregated fuzzy parameter of input event of the output event.

3.3.4 Web-Based Evaluation System

Figure 3.3 illustrates the system design. To analyse the occurrence probability of construction disputes, construction participants inputs their linguistic scores of each dispute artifacts to represent the fuzzy seriousness.

3.4 Likelihood Assessments

The conceptualisation of construction disputes is introduced first in the first page of the web-based evaluation system. A respondent then identifies a project that they wish to evaluate the chance of construction dispute. The project particulars, such as nature of works, procurement method, consultant performance and contractor competence, directly influence construction disputes likelihood. The respondent will then input his fuzzy-sets occurrence likelihood assessments of the 46 dispute artifacts. In the study of construction disputes likelihood, a total of 116 responses had been received from senior construction professionals including (1) the list of arbitration and accredited mediators from Hong Kong International Arbitration Centre, (2) the companies listed on The Hong Kong Institute of Architects, The Hong Kong Institution of Engineers, The Hong Kong Institute of Surveyors and Hong Kong Construction Association, and (3) registered authorised person listed in the Building Department, The Government of the Hong Kong Special Administrative Region. Furthermore, 30 % of the respondents are construction lawyers, 9 % of the respondents are claim consultant, 19 % of the respondents are architects, 10 % of the respondents are engineers, 12 % of the respondents are quantity surveyors, and 7 % of the respondents work for Clients and 13 % of the respondents are project managers.

Following the concept indicated in Eqs. 3.1, 3.2, 3.3 and 3.4, the aggregated fuzzy parameters and defuzzified values of dispute artifacts are presented in Tables 3.3, 3.4 and 3.5.

The defuzzified value illustrates the fuzzy probabilities of dispute likelihood. Firstly the likelihood values of dispute artifacts indicate that late information and instruction from consultants and unreasonably client's requirement are significant collaborative conflict in construction project; inconsistent and insufficient drawing details often reflects the incompleteness of construction contracts; and rejection of contactor's claims and over-claims by contractors are common speculative behaviours in construction contracting. The defuzzified values of these artifacts are above 0.6. Secondly, ambiguity, inconsistency and defectiveness of contract incompleteness are very likely to occur. Their dispute artifacts all have likelihood values higher than 0.5. Thus these three factors are key contributors to the happening of construction disputes. Thirdly, the likelihood evaluation of each factors group shows that contract incompleteness with defuzzified value of 0.9967 is a pernicious problem of construction contracting. Similarly, opportunistic behaviour

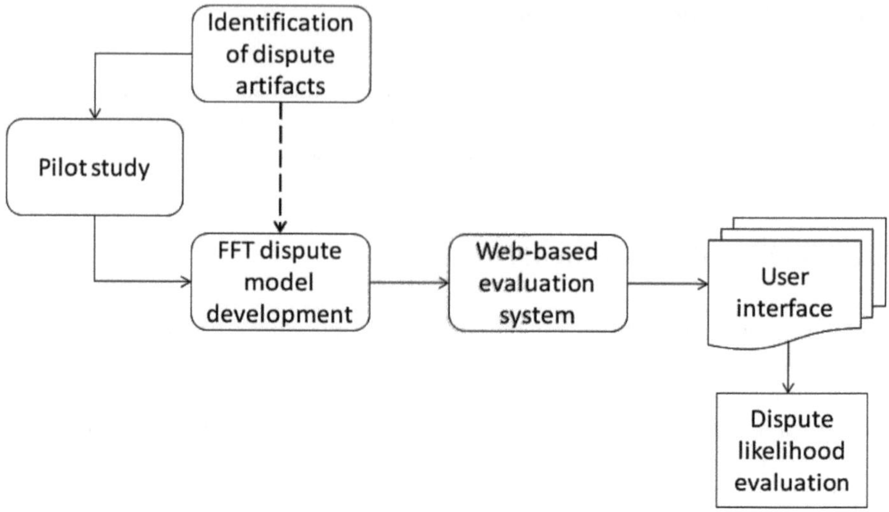

Fig. 3.3 System design

Table 3.3 Occurrence likelihood assessments of task factors

Dispute artifacts	Aggregated fuzzy parameters	Defuzzified value
Task factors	**(0.9986, 1, 1)**	**0.9995**
Risk and uncertainty	(0.9409, 0.9942, 0.9998)	0.9783
TR1 Inclement weather	(0.35, 0.53, 0.71)	0.5294*
TR2 Change of government policy	(0.21, 0.38, 0.56)	0.3833
TR3 Strike	(0.11, 0.27, 0.45)	0.2758
TR4 Fluctuations in material price	(0.41, 0.59, 0.77)	0.5905*
TR5 Fluctuations in labour cost	(0.35, 0.53, 0.70)	0.5262*
TR6 Shortage of labours	(0.29, 0.47, 0.65)	0.4700
TR7 Shortage of materials	(0.30, 0.47, 0.65)	0.4727
TR8 Uncertain ground condition	(0.32, 0.50, 0.68)	0.5002*
Collaborative conflict	(0.9758, 0.9983, 1)	0.9914
TC1 Contractors employed directly by the Client delays in works	(0.33, 0.50, 0.68)	0.5021*
TC2 Nominated Sub-Contractor delays in works	(0.31, 0.48, 0.67)	0.4871
TC3 Nominated Supplier delays in works	(0.27, 0.44, 0.62)	0.4436
TC4 Architect fails to issue instruction within time	(0.44, 0.62, 0.79)	0.6140**
TC5 Engineer fails to provide adequate site investigation details	(0.36, 0.53, 0.70)	0.5260*
TC6 Consultant fails to give information within due time	(0.46, 0.63, 0.80)	0.6307**
TC7 Client requests acceleration unreasonably	(0.37, 0.55, 0.73)	0.5522*
TC8 Client requests change unreasonably	(0.42, 0.60, 0.78)	0.6015**

Defuzzified values that above 0.5 are marked with '*', and that over 0.6 are marked with '**'

Table 3.4 Occurrence likelihood assessments of contract incompleteness

Dispute artifacts	Aggregated fuzzy parameters	Defuzzified value
Contract incompleteness	**(0.9904, 0.9996, 1)**	**0.9967**
Ambiguity	(0.6970, 0.8759, 0.9665)	0.8465
CA1 The scope of work is unclear	(0.34, 0.51, 0.69)	0.5129*
CA2 The specification is unclear	(0.37, 0.54, 0.71)	0.5373*
CA3 The rules to evaluate star rate are unclear	(0.28, 0.45, 0.63)	0.4532
Deficiency	(0.5946, 0.7904, 0.9205)	0.7694
CF1 The rules to evaluate substantial change in quantity of works are not addressed	(0.31, 0.49, 0.66)	0.4869
CF2 The drawings provide insufficient details	(0.41, 0.59, 0.76)	0.5904*
Inconsistency	(0.7684, 0.9136, 0.9801)	0.8874
CC1 The quantity of the same items in the contract bills are substantially different to the actual quantity	(0.35, 0.52, 0.69)	0.5213*
CC2 Some items are missing from the contract bills	(0.39, 0.56, 0.73)	0.5599*
CC3 The drawings contradict with the specification	(0.41, 0.59, 0.76)	0.5888*
Defectiveness	(0.6591, 0.8371, 0.9483)	0.8148
CT1 The details in the drawings are inconsistent	(0.45, 0.63, 0.80)	0.6257**
CT2 The drawings are inconsistent with the contract bills	(0.38, 0.56, 0.74)	0.5616*

Defuzzified values that above 0.5 are marked with '*', and that over 0.6 are marked with '**'

with defuzzified value of 0.9964 is popular in every construction project. Opportunism is undesirable because it would crystallise dispute. Risk and uncertainty with likelihood value of 0.9783 is relatively less significant than collaborative conflict under the task factor group. Finally, the likelihood assessments of dispute artifacts are used to calculate the fuzzy occurrence likelihood of task factors, contract incompleteness and human factors by using the concept of Eq. 3.6, the equation of fuzzy probability of task factors, contract incompleteness and people factors are expressed as Eqs. 3.7, 3.8 and 3.9 respectively:

Fuzzy probability of task factors

Task factors = Risk and uncertainty ∪ Collaborative conflict

$$
\begin{aligned}
\widetilde{P}_{\mathrm{T}} &= 1 - \prod\nolimits_{\mathrm{f}=1}^{8} \left(1 - \widetilde{P}_{TR_i}\right) \otimes \prod\nolimits_{\mathrm{f}=1}^{8} \left(1 - \widetilde{P}_{TC_i}\right) \\
&= 1 - \{1 - (0.9409.0.9942, 0.9998)\} \otimes \{1 - (0.9758, 0.9983, 1)\} \\
&= 1 - (0.0591.0.0058, 0.0002) \otimes 1 - (0.0242, 0.0017, 0) \\
&= \{1 - (0.0591 \times 0.0242), 1 - (0.0058 \times 0.0017), 1 - (0.0002 \times 0)\} \\
&= (1 - 0.0014.1, 1) \\
&= (0.9986, 1, 1)
\end{aligned} \tag{3.7}
$$

Table 3.5 Occurrence likelihood assessments of people factors

Dispute artifacts		Aggregated fuzzy parameters	Defuzzified value
People factors		**(0.9998, 1, 1)**	**0.9999**
Opportunistic behaviour		(0.9895, 0.9996, 1)	0.9964
PO1	Contractor purposely fails to notify omission of items in the contract bills of quantity	(0.33, 0.50, 0.68)	0.5004*
PO2	Contractor purposely works below the specified standard	(0.30, 0.47, 0.64)	0.4699
PO3	Contractor purposely fails to notify the substantial difference in quantity between contract bills of quantity and actual quantity	(0.30, 0.47, 0.64)	0.4695
PO4	Client rejects outright extension of time claim submitted by the contractor	(0.41, 0.59, 0.77)	0.5921*
PO5	Client rejects outright monetary claim submitted by the contractor	(0.45, 0.63, 0.80)	0.6228**
PO6	Contractor over-claims her costs for progress acceleration	(0.47, 0.65, 0.82)	0.6433**
PO7	Contractor purposely fails to disclose the specification of the materials used	(0.31, 0.48, 0.65)	0.4777
PO8	Contractor purposely not provides invoice for the materials used	(0.29, 0.46, 0.64)	0.4657
PO9	Client orders extra without providing proper cost reimbursement	(0.38, 0.56, 0.74)	0.5629*
PO10	Client orders extra without granting justified extension of time	(0.40, 0.58, 0.75)	0.5753*
Affective conflict		(0.9811, 0.9991, 1)	0.9934
PA1	Psychological distress such as fear, sadness, anger and guilt are displayed by member(s) of the project team	(0.33, 0.51, 0.69)	0.5088*
PA2	Emotions such as dominance, assertion, bullying and forcefulness are displayed by member(s) of the project team	(0.36, 0.54, 0.72)	0.5414*
PA3	Intellectually curious, behaviourally flexible, and liberal in their attitudes and values are qualities displayed by member(s) of the project team	(0.30, 0.48, 0.66)	0.4785
PA4	Hostility, callousness and cynicism are manifested by member(s) of the project team	(0.32, 0.50, 0.68)	0.4993
PA5	Excessively neat or overly exact attributes are displayed by member(s) of the project team	(0.28, 0.46, 0.64)	0.4624
PA6	Certain member(s) of the project team find it difficult to relax	(0.33, 0.51, 0.68)	0.5075*
PA7	Certain member(s) of the project team are nervous	(0.31, 0.49, 0.66)	0.4858
PA8	Certain member(s) of the project team are upset or agitated	(0.33, 0.51, 0.69)	0.5075*
PA9	Certain member(s) of the project team are irritable or over-reactive	(0.33, 0.51, 0.69)	0.5109*
PA10	Certain member(s) of the project team are impatient	(0.37, 0.54, 0.72)	0.5412*

Defuzzified values that above 0.5 are marked with '*', and that over 0.6 are marked with '**' i.e. Defuzzified value = (0.9986 + 1 + 1)/3 = 0.9995

Fuzzy probability of contract incompleteness

$$\begin{aligned}\widetilde{P}_{\mathrm{T}} &= 1-\prod\nolimits_{i=1}^{3}(1-\widetilde{P}_{CA_i})\otimes\prod\nolimits_{i=1}^{2}(1-\widetilde{P}_{CT_i})\otimes\prod\nolimits_{i=1}^{3}(1-\widetilde{P}_{CC_i})\otimes\prod\nolimits_{i=1}^{2}(1-\widetilde{P}_{CT_i})\\ &= 1-\{1-(0.6970,0.8759,0.9665)\}\otimes\{1-(0.5946,0.7904,0.9205)\}\\ &\quad\otimes\{1-(0.7684,0.9136,0.9801)\}\otimes\{1-(0.6591,0.8371,0.9483)\}\\ &= 1-(0.3030,0.1241,0.0335)\otimes(0.4025,0.2096,0.0795)\otimes(0.2316,0.0864,0.0199)\\ &\quad\otimes(0.3409,0.1629,0.0517)\\ &= \{1-(0.3030\times0.4025\times0.2316\times0.3409),1-(0.1241\times0.2096\times0.0864\times0.1629),\\ &\quad 1-(0.0335\times0.0795\times0.0199\times0.0517)\}\\ &= (1-0.0096,0.0004,0)\\ &= (0.9904,0.9996,1)\end{aligned} \tag{3.8}$$

i.e. Defuzzified value = (0.9904 + 0.9996 + 1)/3 = 0.9967

Fuzzy probability of people factors

People factors = Opportunistic behaviour ∪ Affective conflict

$$\begin{aligned}\widetilde{P}_{P} &= 1-\prod\nolimits_{f=1}^{10}(1-\widetilde{P}_{PR_i})\otimes\prod\nolimits_{f=1}^{10}(1-\widetilde{P}_{PC_i})\\ &= 1-\{1-(0.9895,0.9996,1)\}\otimes\{1-(0.9811,0.9991,1)\}\\ &= 1-(0.0105,0.0004,0)\otimes(0.0189,0.0009,0)\\ &= 1-\{1-(0.0105\times0.0189),1-(0.0004\times0.0009),1-(0\times0)\}\\ &= (1-0.0002,1,1)\\ &= (0.9998,1,1)\end{aligned} \tag{3.9}$$

i.e. Defuzzified value = (0.9998 + 1 + 1)/3 = 0.9999

The fuzzy probabilities of five categories of construction dispute range from 0.99 to 1 that suggest the inevitability of construction disputes. The results indicate that contract incompleteness is the dominant group of dispute manifestation. However, external risk and uncertainty are less detrimental in construction project in comparison to internal collaborative conflict, contract incompleteness and contracting behaviours. Construction contracts usually address compensation on force majeure, inclement weather and strikes. It could limit the implications from risk and uncertainties. The previous studies identified that claims and disputes mostly arose from people factors, such as arguments and incompatibilities (Diekmann and Girard 1995; Tsai and Chi 2009). The findings also indicated that opportunism is relatively more significant in construction projects. Contracting parties tend to seek their own interest. Besides, psychological problems are the subsequent matter of concern in construction projects. It is difficult for all construction participants to have common goals and values. As such, incompatibilities usually exist in construction collaboration. The tight bidding periods and construction periods often increases the associated stress and tension of project team members.

With reference to the FFT dispute model, the fuzzy probabilities of contractual disputes is expressed as *Contractual disputes = Task factors ∩ Contract incompleteness* as shown in Eq. 3.10, where Contractual disputes is the union of Task factors and Contract incompleteness.

Fuzzy probability of contractual disputes

$$\begin{aligned}\tilde{P_{CD}} &= \tilde{P_T} \otimes \tilde{P_C} \\ &= (0.9986, 1, 1) \otimes (0.9904, 0.9996, 1) \\ &= (0.9986 \times 0.9904, 1 \times 0.09996, 1 \times 1) \\ &= (0.9890, 0.9996, 1)\end{aligned} \tag{3.10}$$

i.e. Defuzzified value = (0.9890 + 0.9996 + 1)/3 = 0.9962

Similarly, the fuzzy probabilities of speculative disputes is expressed as *Speculative disputes = Contract incompleteness ∩ People factors* as shown in Eq. 3.11, where Speculative disputes is the union of People factors and Contract incompleteness.

Fuzzy probability of speculative disputes

$$\begin{aligned}\tilde{P_{SD}} &= \tilde{P_C} \otimes \tilde{P_P} \\ &= (0.9904, 0.9996, 1) \otimes (0.9998, 1, 1) \\ &= (0.9904 \times 0.9998, 0.9996 \times 1, 1 \times 1) \\ &= (0.9902, 0.9996, 1)\end{aligned} \tag{3.11}$$

i.e. Defuzzified value = (0.9902 + 0.9996 + 1)/3 = 0.9966

Based on the FFT dispute model, the fuzzy probabilities of both contractual and speculative disputes are relatively high, with defuzzified values of 0.99. Accordingly, both contractual and speculative disputes are unavoidable in construction industry because there is much uncertainty, cognitive issues, contract incompleteness, behavioural problems and affective issues in construction.

3.5 Chapter Summary

It is complicated and costly for contracting parties to prepare a complete contract if it is at all possible (Hart and Moore 1988). Williamson (1975) explain contract is inherently incomplete as a result of bounded rationality and asymmetric information. Ability to identify critical dispute contributors would help the installation of alleviating measures. For example, building commitment and trust are indirect ways to control the level of conflict and aggression (Bresnen and Marshall 2000). This study contributes to the research in construction disputes in a number of ways. Firstly, construction disputes are conceptualised by identifying the dispute factors and artifacts. The conceptual framework also encapsulates the contributions of dispute artifacts towards the occurrence of construction dispute. Secondly, the formulation of construction dispute is arranged in a fault-tree framework whereby the inter-relationships among dispute artifacts, factors, factor groups, category and top event are connected by logic gates. As such the framework explains how task factors, contract incompleteness and people factors influence the occurrence likelihood of construction disputes. Finally, a web-based assessment tool is developed to enable users to assessment the likelihood of dispute happening by inputting their assessments of occurrence likelihood of the dispute artifacts. The occurrence likelihoods of the factors, groups, categories and construction dispute are then calculated automatically.

Acknowledgments Part of the contents of this chapter has been published in Volume 139(1) of the Journal of Construction Engineering and Management and is used with the permission from ASCE.

References

Bresnen, M., & Marshall, N. (2000). Motivation commitment and the use of incentives in partnerships and alliances. *Construction Management and Economics, 18*, 58–98.

Diekmann, J. E., & Girard, M. J. (1995). Are contract disputes predictable? *Journal of Construction Engineering and Management, 121*(4), 335–363.

Fenn, P. (2002). Why construction contracts go wrong (or an aetiological approach to construction disputes). Society of Construction Law, http://www.scl.org.uk/node/543 (Mar. 30, 2002).

Hart, O. D., & Moore, J. (1988). Incomplete contracts and renegotiation. *Econometrica, 56*(4), 755–785.

Idrus, A., Nuruddin, M. F., & Rohman, M. A. (2011). Development of project cost contingency estimation model using risk analysis and fuzzy expert system. *Expert Systems with Applications, 38*, 1501–1508.

Juan, Y. K. (2009). A hybrid approach using data envelopment analysis and case-based reasoning for housing refurbishment contractors selection and performance improvement. *Expert Systems with Applications, 36*, 5702–5710.

Kales, P. (1998). *Reliability: For technology, engineering, and management*. Englewood Cliffs: Prentice-Hall.

Molenaar, K., Washington, S., & Diekmann, J. (2000). Structural equation model of construction contract. *Journal of Construction Engineering and Management, 126*(4), 268–277.

Pan, N. F. (2006). Evaluation of building performance using fuzzy FTA. *Construction Management and Economics, 24*, 1241–1252.

Pan, N. F. (2008). Fuzzy AHP approach for selecting the suitable bridge construction method. *Automation in Construction, 17*, 958–965.

Pan, N. F., & Wang, H. (2007). Assessing failure of bridge construction using fuzzy fault tree analysis. In: Proceedings of the Fourth International Conference on Fuzzy Systems and Knowledge Discovery, FSKD 2007 1, art. no. 4405896, pp. 96–100.

Shaheen, A. A., Fayek, A. R., & AbouRizk, S. M. (2007). Fuzzy numbers in cost range estimating. *Journal of Construction Engineering and Management, 133*(4), 325–334.

Singh, D., & Tiong, L. K. (2005). A fuzzy decision framework for contractor selection. *Journal of Construction Engineering and Management, 131*(1), 62–70.

Tsai, J. S., & Chi, C. S. F. (2009). Influences of Chinese cultural orientations and conflict management styles on construction dispute resolving strategies. *Journal of Construction Engineering and management, 135*(10), 955–964.

Wang, X. Z., & Kerre, E. E. (2001). Reasonable properties for the ordering of fuzzy quantities (I). *Fuzzy Sets and Systems, 118*(3), 375–385.
Williamson, O. E. (1975). *Markets and hierarchies: Analysis and antitrust implications*. New York: The Free Press.
Zadeh, L. A. (1965). Fuzzy sets. *Information and Control, 8*, 338–353.
Zhao, J., & Bose, B. K. (2003). Evaluation of membership functions for fuzzy logic controlled induction motor drive. *IEEE Transactions on Power Electronics*, 229–234.

Chapter 4
Catastrophic Transitions of Construction Contracting Behaviour

Sai On Cheung and Tak Wing Yiu

Abstract This chapter reports a study on the examination of construction contracting behaviour (CCB) under the influence of the competing forces of co-operation and aggression. The CCB dynamics under these forces are modeled on the Catastrophe Theory (CT) developed by Thom (1975). Mathematical treatment allows analytical examination of the dynamics among the interacting variables. A bifurcation zone within which the behaviour becomes bimodal characterises CT model. Under a CT framework, a small change in the aggression drive can produce a significant sudden change in contracting behaviour; this phenomenon is called divergence. The CCB framework is developed by the identification and establishment of indicators for the three variables; contracting behaviour, co-operation and aggression drivers. These variables are used to test the catastrophic phase transitions of CCB. It is found that if a co-operating party feels aggrieved, she remains co-operative up to a point beyond which she will suddenly attack. This jump is described as catastrophe attack. Once this happens, problems can be easily be escalated to become disputes.

4.1 Characteristics of Construction Contracting

The construction industry is infamous for its adversarial culture. The proliferation of disputes within the industry has caused acute concern over the adverse effect of protracted disputes. Furthermore, the antagonistic contracting attitude needs to be

S. O. Cheung (✉)
Construction Dispute Resolution Research Unit, Department of Civil and Architectural Engineering, City University of Hong Kong, Hong Kong, People's Republic of China
e-mail: saion.cheung@cityu.edu.hk

T. W. Yiu
Department of Civil and Environmental Engineering, Faculty of Engineering, The University of Auckland, Auckland, New Zealand
e-mail: k.yiu@auckland.ac.nz

S. O. Cheung (ed.), *Construction Dispute Research*,
DOI: 10.1007/978-3-319-04429-3_4,

overhauled (Bayliss et al. 2004; Cheung et al. (2003a, b); Cheung and Suen 2002). This view is expressed in a number of industry-wide reviews (Construction Industry Review Committee (CIRC) 2001; Egan 1998; Latham 1994). Fostering co-operation in construction contracting has been suggested to alleviate this situation. However, this is considered to be a revolutionary attitude change that can only be made possible with a transformation in culture. Co-operation fostering efforts can be broadly classified into three categories: case studies, identification of critical success factors and legal analyses. Case studies are instrumental in sharing innovations and achievements (Bayliss 2002; Bayliss et al. 2004; Black et al. 2000; Cheung et al. 2002) and are valuable learning models for the practice of co-operative contracting. Nonetheless, skeptics often comment that every construction project is unique; hence it is risky to generalise the success attained in a particular venture. Identification of success factors often goes hand in hand with case studies (Liu and Fellows 2001). The identified success factors are mostly behavioural or attitudinal, thus augmenting the common belief that contracting behaviour is in fact manifestation of the attitude of those involved. Liu and Fellows (2001) suggest that the Chinese culture appears to be more receptive to the concept of co-operative contracting. This notion is echoed by the study of Cheung (2001) which points out that the contract law regime of the People's Republic of China features many characteristics of relational contracting forwarded by Macneil (1980, 1981). Flexibility in contractual relations was succinctly advocated. His suggestion was later supported by the empirical work of Macaulay (1985) who observed that re-negotiation of contract terms is commonly practised and that adjustments should occur without resorting to court. To this end, the legal footing of co-operative contracting has to be identified. In sum, examining the compatibility of the legal system in supporting the practice of co-operation in construction contracting form the backbone of legal analyses in this area. Yet not surprisingly, the legal profession under the common law system has been swift to point out the lack of a legal basis for any contractual duty to cooperate (Newman 2000) and that such a duty is difficult to enforce due to the absence of a recognised legal concept (Colledge 2000). Furthermore, the sole reliance on contractual force in executing construction contracts already marks a clear departure from the spirit of co-operation. More importantly, commanding co-operative contracting behaviour is a management issue, and improving the performance of construction projects is one of the driving forces to promote co-operation between contracting parties. Its failure would germinate seeds for disputes, and eventually lead to programme disruption, relation deterioration, time and financial loss (Cheung 2001).

Notwithstanding the call for reforms as aforementioned, contracting behaviour remains largely adversarial in the construction industry (Construction Industry Review Committee (CIRC) 2001; Egan 1998; Latham 1994). The conventional design-bid-build approach is not conducive in enhancing co-operation (Cheung et al. (2003a, b). Contractual terms, however comprehensive, would not be able to cover all eventualities. Unanticipated happenings are testing and a co-operative contracting behaviour could curb disputes nourishing (Cheung 2002; Luo 2002). Co-operative contracting behaviour operates as a self-enforcing safeguard that enables a more effective and less costly alternative to exhaustive contractual remedies (Luo 2002). That means with a co-operative contracting attitude, a flexible approach can be adopted to deal with unanticipated eventualities (Luo 2002). In terms of implementing co-operation, Bayliss et al. (2004) suggested that *"co-operative attitude can be instilled, fostered and maintained through cogent project management, thus, commanding a co-operative contracting behaviour is a management issue, acquiring skill of managing it basically depends on the understanding of the fundamentals involved"*. Notwithstanding, the fact remains that parties to a construction contract represent the interests of their respective organisations that may not always be compatible. Cheung (2007) further demonstrated that trust is the prerequisite for co-operation in a partnering project in Hong Kong.

4.2 Construction Contracting Behaviour: Co-operation Versus Aggression Forces

According to Hill (2001), contracting behaviour is regarded as *"a means for parties to reconcile their expectations, future actions and consequent valuations to increase the size of aggregate pie"*. The view is also shared by Buckley and Casson (1988) who suggest that co-operative behaviour is a mutual forbearance in the allocation of resources such that one party is made better off and no one is worse off than it would otherwise be. In the course of an ongoing contractual relationship, disputing parties may adopt co-operative behaviour in order to retain a harmonious relationship with the other. This co-operative working environment would have allowed effective enforcement of their rights and obligations (Harmon 2003; Yiu and Cheung 2006). However, in construction, acting co-operatively is easier to be said than done, especially when conflicts are inherent in all construction projects (Fenn et al. 1997; Yiu and Cheung 2006). Opportunism is therefore common. Contracting parties would exercise opportunistic and aggressive behaviour by only taking care of one's self-interest, regardless of the detrimental consequences of their collaborators. For example, they may seek to enforce their contractual rights as much as possible on one hand, while look for means to evade their obligations on the other; they may even estimate the other party's likelihood to default. It is therefore evident that there are two co-existing conflicting forces that affect CCB: co-operation force and aggression force.

Aggression force refers to the strengths and stimuli that motivate one to make aggressive moves, whereas co-operation force is the strengths and stimuli that motivate one to make co-operative moves. These two dichotomous forces co-exist in all construction projects. As illustrated in Fig. 4.1, these forces can be framed into the classic framework of Prisoner's Dilemma (PD) (Axelrod 1984). PD refers to a two-party non-constant-sum game in which some outcomes are preferred by both parties, and the occurrence of certain outcomes depends on the behaviour of

	Contracting Party A – Co-operation	Contracting Party A - Aggression
Contracting Party B – Co-operation	Cooperate, win-win	Confront, lose much-win much
Contracting Party B - Aggression	Accommodate, win much-lose much	Attack, lose-lose

Fig. 4.1 Payoff matrix of construction contracting behaviour (CCB)

the other party. In this game, it is assumed that each individual player ("prisoner") is trying to maximise his own interest, without any concern for the well-being of the other player. The PD framework suggests that a similar payoff matrix can be applied in the area of human interaction and it has become fundamental to certain theories of human co-operation (Axelrod 1984). Hence, a similar approach as the PD framework can be applied to model CCB. A payoff matrix of CCB is constructed and displayed in Fig. 4.1.

The payoff matrix in Fig. 4.1 suggests that co-operative behaviour is not innate. Instead, practice of co-operative behaviour is characterised by reciprocal moves, i.e. if one side behaves co-operatively, he would expect a reciprocating co-operative response from the other (Cheung et al. 2003a, b; Wong et al. 2005). This implies that the contracting behaviour of one party is dynamically associated with the others. It is therefore hypothesised that a threshold exists for the transition from co-operative to aggressive contracting behaviour. When this threshold is reached, a sudden change in behaviour will occur. The theoretical explanation of such a behavioural transition can be found in Catastrophe Theory (Thom 1975).

4.3 Catastrophe Theory

Catastrophe Theory was developed by Thom (1975) and subsequently popularised by Zeeman (1976, 1977). It is a mathematical model of nonlinear systems in which discontinuous behaviour is determined by smooth changes in a small number of parameters (Wagenmakers et al. 2004). It has been applied to a wide range of areas such as physics (Tamaki et al. 2003), geology and rock mechanics (Qin et al. 2001), psychology (Ploeger et al. 2002; van der Maas et al. 2003) as well as social sciences (Holyst et al. 2000). One of the popular applications of CT is attitude-based analysis. In management, it has also been applied to study technology management (Bacck and Cullen 1992; Herbig 1991), organisational change (Gresov et al. 1993), competitive strategies (Oliva et al. 1988), customer behaviour (Oliva et al. 1992), motivation in organisations (Guastello 1987), forecasting and decision making (Wright 1983) and conflict resolution (Yiu and Cheung 2006).

4.3.1 Catastrophe Model of Construction Contracting Behaviour

Catastrophe Theory describes how small and continuous changes of independent variables can have sudden, discontinuous effect on a dependent variable. Its basic form is called 'cusp catastrophe' (Thom 1975). The cusp model involves one dependent variable and two independent variables. The independent variables take two extreme forms with different qualitative meanings: one is called the normal factor and the other is called the splitting factor (Bacck and Cullen 1992). The normal factor changes directly with the dependent variable (Gresov et al. 1993), while the splitting factor is *'a moderating variable which specifies conditions under which the normal factor will affect the dependent variable in a continuous fashion, and other circumstances under which the normal factor will produce discontinuous changes in the dependent variable…it is the splitting factor that determines the "breaking point" or threshold of change in the dependent variable…'*(Bacck and Cullen 1992). According to CT, when the intensities of the normal factor and the splitting factor reach a threshold level, the dependent variable will undergo a sudden and radical change. This unique nature is represented by the split of the contracting behaviour surface (B) of the CT model (Fig. 4.2 refers).

In this study, it is hypothesised that a party's contracting behaviour is influenced by two stimulators: co-operation force and aggression force. The CT model describes the changes in CCB, as a result of the interaction between these two forces, depicted as the contracting behaviour surface (B) in Fig. 4.2. For any combination of the co-operation and aggression forces, that means for any point on the control space (C), there is at least one likely form of corresponding behaviour indicated as a point above the corresponding point in the control space and at an appropriate height on the behaviour axis (vertical axis). The full set of such points together forms the contracting behaviour surface (B). In general, there is only one probable mode of behaviour. However, where co-operation and aggression forces are roughly equal, as shown the middle of the graph there are two sheets representing two possible forces of behaviour. They are connected by a third sheet to form a continuous pleated surface. This sheet represents the least likely behaviour, in this case, neutrality (Zeeman 1977). Towards the origin, the pleat on the contracting behaviour surface becomes increasingly narrow and eventually vanishes. The line defining the edges of the pleat is called the fold curve and its projection onto the control surface is a cusp-shaped curve.

4.3.2 Construction Contracting Behaviour as Dependent Variable

As discussed, improved performance of construction projects provides a driving force to adopt a co-operative approach, and it is necessary to better understand

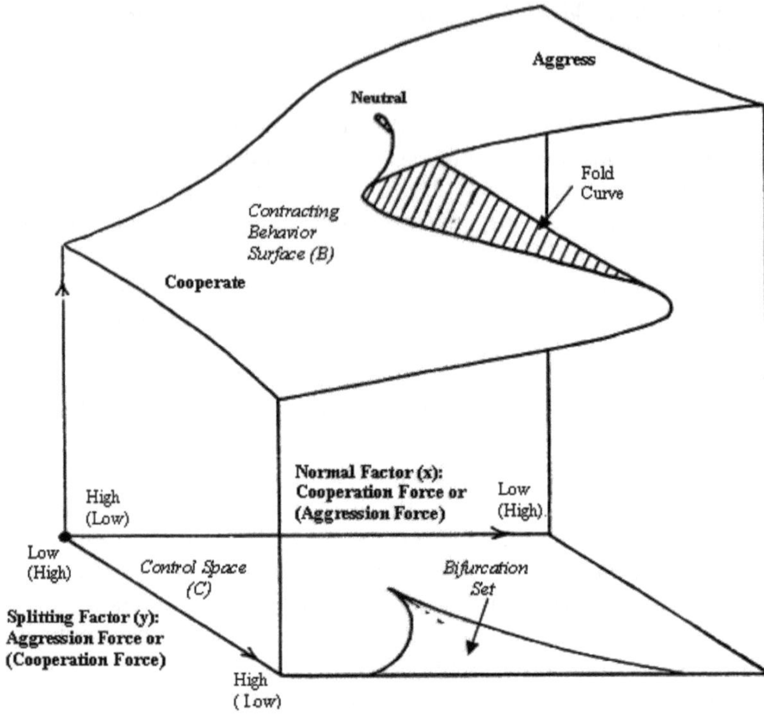

Fig. 4.2 A hypothetical catastrophe model of construction contracting behaviour

such construction contracting behaviour. As shown in Fig. 4.2, construction contracting behaviour is manifested by a combination of co-operation and aggression forces. Based on literature review, its influential variables are identified and summarised in Table 4.1.

4.3.3 Co-operation Force and Aggression Force as Independent Variables

As per the model presented in Fig. 4.2, co-operation and aggression forces are two co-existing conflicting forces that affect construction contracting behaviour. Co-operation force prompts contracting parties to focus on mutual interests and concerns. This force would generally invoke co-operative and accommodating response, which would restrain the inherent human instinct of concerning only self-interests. Aggression force, in contrast, prompts contracting parties to focus only on self-interests. These behaviours are often adversarial and invoke aggression, retaliation and defensive responses. The dichotomous nature of these two forces can be demonstrated by the framework of Prisoner's Dilemma (PD) as

Table 4.1 Influential variables of construction contracting behaviour

Variables	Definitions	References
Communication channel	The extent of effective communication affects contracting behaviour. Having a smooth and efficient communication channel among contracting parties enables them to work efficiently and effectively	Cheung et al. (2003b, 2004), Crane et al. (1999), Harmon (2003)
Possibility of goal achievement	Contracting behaviour can be influenced by the goal setting of a project team. For example, if mutual goals are likely achieved, the contracting parties would behave co-operatively	Cheung et al. (2003b), Harmon (2003), Luo (2002)
Relationship among contracting parties	The dynamic of contracting behaviour depends on the goodness of relationships among project participants	Chua et al. (1999)
Profitability	Profit-maximising is significantly affect parties' contracting behaviour. If they satisfy with their profit expectations, they would behave in a co-operative way	Swedberg (1987)
Effectiveness of problem solving	Contracting behaviour is influenced by effectiveness of problem solving. Previous studies suggested that it can be measured by the degree of mutual consultation and concerns of contracting party	Crane et al. (1999), Luo (2002)
Experience of handling similar projects	Contracting parties would unlikely behave aggressively if they have good experience on projects with similar complexity	Chua et al. (1999), Gresov et al. (1993)
Achievement of cost target	Project's financial situation affects parties' contracting behaviour. This is especially when the planned budget are probably achieved	Bacck and Cullen (1992), Cheung et al. (2004), Crane et al. (1999), Luo (2002)
Alignment of time frame	Time element of construction project affects parties' contracting behaviour. Contracting parties would behave aggressively when a project is not likely to be completed on time	
Amount of disputes	When disputes arise, no matter how specific are contractual terms, contracts alone are unable to effectively govern project operations and maintain continuity of relationship between contracting parties	Cheung (1993), Crane et al. (1999), Luo (2002)
Contract sum	The greater the contract sum of a project, the greater the defensiveness of contracting parties	Hartman (1993)

afore-described. It is therefore imperative that contracting parties shall prevent such moves so as to maintain good relationships. In summary, in modeling CCB, both co-operation and aggression forces should be considered. Their variables are presented in Tables 4.2 and 4.3 respectively.

The fitness of the model presented in Fig. 4.2 and the appropriateness of the independent variables are to be tested empirically. The steps in conducting the fit measurements are discussed here-follow.

4.4 Model Fitting

Early CT model fitting employed regression and stochastic differential equations to estimate model parameters (Gresov et al. 1993; van der Maas et al. 2003; Yiu and Cheung 2006). Cobb (1980) proved that there is a family of probability density functions, of which a stable equilibrium corresponds to a node and an unstable equilibrium corresponds to an anti-node. A stable equilibrium state is a point of high probability. The cusp surface (i.e. the contracting behaviour surface) is then viewed as a maximum probability response surface (Cobb 1981; Cobb et al. 1983). With these probability density functions, parameters can be estimated using the method of maximum likelihood estimation (Yiu and Cheung 2006; van der Maas et al. 2003; Cobb 1981; Cobb et al. 1983). In other words, the control variables can be estimated from the data with stochastic differential equations (Cobb 1978, 1980; Cobb et al. 1983; Gresov et al. 1993). Mathematically, the contracting behaviour surface can be expressed by Eq. 4.1 (Cobb 1980, 1983):

$$f(z|\alpha,\ \beta)\ \exp\left(\alpha y+\frac{1}{2}\beta y^2-\frac{1}{4}y^4\right) \tag{4.1}$$

where $y=\frac{(z-\lambda)}{\sigma}$, λ and σ scale the observed behavioural variable z to y; α and β are linear functions of the independent variables x_1 to x_n, with

$$\alpha = a_0 + a_1x_1 + a_2x_2 + \ldots + a_nx_n \tag{4.2}$$

and;

$$\beta = b_0 + b_1x_1 + b_2x_2 + \ldots + b_nx_n \tag{4.3}$$

Cobb (1980) also developed a computer program based on this model fitting technique. Although this maximum likelihood method is considered as a satisfactory method for fitting cusp catastrophe model, it is not often used (Wagenmakers et al. 2004) and unfortunately, this program often breaks down for non-apparent reasons (Ploeger et al. 2002). Hartelman (1997) later solved this problem by introducing an improved program called Cuspfit (Hartelman 1997; Ploeger et al. 2002). Hartelman (1997) and Wagenmakers et al. (2004) suggested that this program is a more robust and flexible version than Cobb's original program. It employs a more reliable optimisation routine which allows users to constrain parameter

Table 4.2 Variables of co-operation force

Variables	Definitions	References
Teamwork intensity	Effectiveness of disputes resolution by teamwork approach of a project team	Cheung et al. (2004), Crane et al. (1999), Hartman (1993)
Trust intensity	Degree of confidence and trust building in contracting parties	Luo (2002), Tallman and Shenkar (1994)
Effectiveness of communication	Satisfied previous dealings among contracting parties could facilitate effectiveness of communication	Doz (1996), Tallman and Shenkar (1994)
Goodness in relationships between project participants	A good personal and working relationship among contracting parties would intensify their co-operation forces and facilitate project progress	Chua et al. (1999), Luo (2002)
Openness level	Willingness of sharing thoughts and feelings. The extent of carrying out open communication among contracting parties	Doz (1996), Piper (2001)
Commitment maintenance	Commitments of contracting parties are enduring when they are highly involved in project issues	Luo (2002)
Goal mutuality	Establishment of common goal between contracting parties	Black et al. (2000), Luo (2002)
Availability of Information	Efficiency of information exchange among contracting parties and their experience in handling similar project(s)	Luo (2002), Zeeman (1977)
Involvement Intensity	Degree of voluntariness in project participation	Zeeman (1977)
Incentive intensity to risks and savings sharing	Degree of contractual risk allocations among contracting parties, the provision of tangible reward (s), and the degree of risk averseness of contracting parties	McKim (1992)
Effectiveness in dispute resolution	Appropriateness of incorporating contract provisions to resolve disputes, unforeseeable events and contingencies	Cheung et al. (2004), Doz (1996), Luo (2002)
Effectiveness in solving/ sharing of problem(s)	Appropriateness of incorporating contract provisions for mutual consultations among contracting parties	Cheung et al. (2003b), Doz (1996), Luo (2002), Piper (2001)
Contract completeness	Explicitness, term specificity and contingency adaptability of contract conditions	Luo (2002)
Inter-party reciprocity	Desire to maintain future business relationships among contracting parties	Black et al. (2000), Cheung et al. (2003b), Luo (2002)

Table 4.3 Variables of aggression force

Variables	Definitions	References
Quality of the past/ previous dealings	Satisfaction of previous dealings among contracting parties	Luo (2002), Tallman and Shenkar (1994)
Level of competitive pressure	Amount of pressure perceived by contracting parties would directly affect their aggressiveness	Gresov et al. (1993)
Intensity of competitive force/competitive inertia	Competitive force or competitive inertia is determined by the aggressiveness of contracting parties on comparison to the actions being taken by their competitors	Gresov et al. (1993), McKim (1992)
Likelihood of disputes	The higher the likelihood of disputes, the higher the aggression forces of contracting parties are induced	Doz (1996), Luo (2002)
Contract incompleteness	Aggression forces are likely to invoked if many ambiguous terms exist in contract conditions	Goldberg (1992), Luo (2002)

values and to employ different sets of starting values. Cobb's algorithm calculates whether the cusp model or the linear model gives the best description of the relationship between the independent and the dependent variables (Cobb 1980; Ploeger et al. 2002; Wagenmakers et al. 2004). Cuspfit, however, is equipped with additional functions and is thus capable of fitting similar models such as logistic and linear models and detect rapid changes in the dependent variable (Wagenmakers et al. 2004). It can also be used to test the three models; linear, logistic and cusp. Such comparison is useful in distinguishing an arbitrarily fast acceleration from a catastrophic change. Furthermore, Cuspfit could be used to test the presence of bifurcations by comparing the fit of the cusp model to the fit of both logistic and linear models (Hartelman 1997; Hill 2001; Ploeger et al. 2002).

In addition to the maximum likelihood method, Hartelman (1997) introduced two fit measures in Cuspfit—Akaike Information Criterion (AIC) and Bayes Information Criterion (BIC). AIC is the goodness-of-fit index that takes account of the number of parameters. Mathematically, it is defined as minus twice the log-likelihood plus twice the number of parameters, i.e. "$AIC = -2 \log L + 2k$"; the model with the smallest AIC will be the best fit (Hartelman 1997; Hill 2001; Ploeger et al. 2002). As for BIC, it is a goodness-of-fit indicator which takes into account the number of data points and implements Occam's razor (Thorburn 1915) by quantifying the trade-off and parsimony (Hill 2001; Ploeger et al. 2002; Raftery 1995; Schwarz 1978). Mathematically, BIC is calculated by the equation "$BIC = -2 \log L + k \log n$", where L is the maximum likelihood, k is the number of free parameters and n is the number of observations (Raftery 1995). Models with lower BIC values are preferred for model fitness purpose. If the AIC and BIC values of the cusp model are lower than those of the logistic and the linear models, then the cusp model shall be the best fit among the three (Hartelman 1997; Hill 2001; Ploeger et al. 2002).

Another notable feature of Cuspfit is the possibility of introducing restrictions on parameters to test specific hypotheses (Hartelman 1997). In catastrophe analysis, if one expects that one or more of the independent variables do not contribute to the normal or the splitting variable, it is possible to fix parameters at zero, so that only the non-fixed parameters are estimated. Since there are two independent variables in the cusp catastrophe model, with reference to Eqs. 4.2 and 4.3, it is possible to construct a total of 16 different cusp models by substituting the four parameters a1, a2, b1 and b2 to zero. Then, comparing the AIC and BIC values with the unrestricted catastrophe model, the appropriate independent variables—the normal and the splitting variables of the proposed model can be identified (Hartelman 1997; Ploeger et al. 2002; Van der Maas et al. 2003). The fit measures indicate which of the 16 cusp models is the most appropriate. As such the set of independent variables; i.e. the normal and the splitting variables is also identified (Schwarz 1978). A number of successful applications with this approach have been reported (Hartelman 1997; Hill 2001; Ploeger et al. 2002; Stewart and Peregoy 1983; Van der Maas et al. 2003).

4.5 Data Collection

To facilitate data collection, a questionnaire was designed to measure the perceptions of construction professionals on the dependent and independent variables. The items of this questionnaire are listed in Tables 4.1, 4.2 and 4.3. The targeted respondents were construction professionals including as project managers, architects, engineers, surveyors and mediators who had at least 5 years project management experience. With reference to their recent projects, they were asked to indicate the relative significance of the variables representing CCB, co-operation force and aggression force on a seven-point Likert scale. A total of 250 questionnaires were sent out and 91 sets were completed and returned. The overall return rate is therefore 36.40 %. The returned questionnaires were completed by construction professionals including project managers (15 %), architects (15 %), engineers (25 %), quantity surveyors (42 %), mediators (1 %) and others (2 %). Most of the respondents were holding senior positions in the industry, with 57 % having more than 10 years of experience. The profiles of the respondents assure the authenticity of this study in reflecting the industry's opinion. The profiles of the respondents according to their work experience and professional background are summarised in Fig. 4.3.

4.6 Results and Discussions

The collected data were analysed by the Cuspfit program (Cobb 1980; Hartelman 1997). The following three steps were involved:

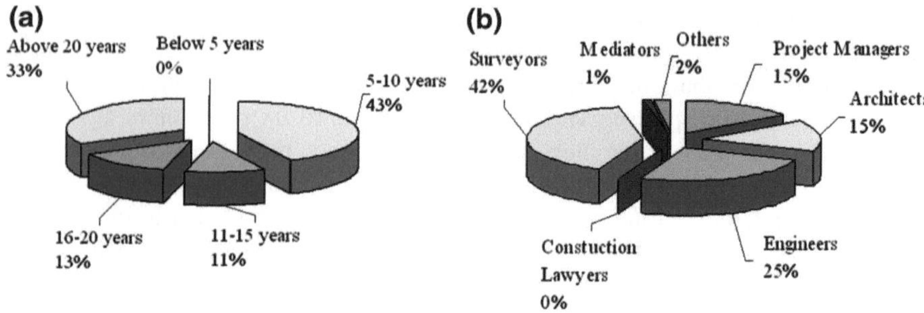

Fig. 4.3 Profiles of respondents by **a** working experience and **b** professions

(I). Step 1: Modeling and testing of the appropriateness of the control variables.
(II). Step 2: Investigating statistical fit of the models, and
(III). Step 3: Identifying the bimodal nature of CCB.

The above procedure has been successfully adopted in other studies employing the Cuspfit program (Hill 2001; Ploeger et al. 2002; Stewart and Peregoy 1983; Van der Maas et al. 2003).

(i) Step 1: Modelling and Testing of the Appropriateness of the Control Variables

Tables 4.2 and 4.3 list the influential variables of co-operation and aggression forces identified in the literature review. To examine which pair(s) of variables from these two forces is(are) appropriate to serve as the normal and the splitting factors, a total of 70 trials (devised from the combination of CCB variables, fourteen variables of co-operation force and five variables of aggression force) were analysed by the Cuspfit programme. The Cuspfit programme fits the catastrophe model with the control variables α, β, and the behaviour variable z to cross-sectional data by using the maximum likelihood method. With reference to Eqs. 4.2 and 4.3, the linear function, α (the normal factor), and β (the splitting factor), for the two control variables, (x_1: co-operation force) and (x_2: aggression force) can be written as:

$$\alpha = a_0 + a_1 x_1 + a_2 x_2 \tag{4.4}$$

$$\beta = b_0 + b_1 x_1 + b_2 x_2 \tag{4.5}$$

According to algorithm by Cobb (1980), the setting of the control variables a1 and b2 of Eq. 4.4 and a2 and b1 of Eq. 4.5 can be fixed as zero. Hence, the linear function of α (the normal factor), and β (the splitting factor) can be devised under two conditions:

Condition 1: when $a_1 = 0$, and $b_2 = 0$, then

$$\alpha = a_0 + a_2 x_2 \tag{4.6}$$

$$\beta = b_0 + b_1 x_1 \tag{4.7}$$

where x_1 = splitting factor and x_2 = normal factor

or

Condition 2: when $a_2 = 0$, and $b_1 = 0$, then

$$\alpha = a_0 + a_1 x_1 \quad (4.8)$$

$$\beta = b_0 + b_2 x_2 \quad (4.9)$$

where x_1 = normal factor and x_2 = splitting factor

To test the appropriateness of the control variables, each trial included 16 catastrophe models which were constructed by substituting the four parameters a1, a2, b1 and b2 randomly with zero. The AIC and BIC of these models were compared with those of the unrestricted model (Ploeger et al. 2002; Van der Maas et al. 2003). Significant trial(s) was (were) selected when the lowest AIC and BIC can also fulfil either Condition 1 or Condition 2. Accordingly, two significant catastrophe models (i.e. Model 10) were identified from two trials (Trials A and B) (Table 4.4 refers). Their statistical results are presented in Tables 4.5 and 4.6. These two models generally show that the degree of trust intensity (as the normal factor), contract incompleteness and competitive inertia (as the splitting factors) critically affect the sudden change of CCB.

ii) Step Two: Investigating Statistical Fit of the Models

Having confirmed the appropriateness of the normal and the splitting factors in the two identified models, the output of the Cuspfit programme also provide information on the statistical fit of the two significant models. This programme is able to test three types of models: linear, logistic and catastrophe model. The algorithm of Cobb (1980) is able to calculate whether the catastrophe or the linear model gives a better description of the relationship between the independent and dependent variables. While the work of Hartelman (1997) enables a comparison of the catastrophe model with the logistic model. The comparison is to distinguish an arbitrarily fast acceleration from a catastrophic change (Ploeger et al. 2002). When the AIC and the BIC of the catastrophe model are lower than those of the logistic and linear models, the catastrophe model then gives a better fit (Van der Maas et al. 2003). With reference to Tables 4.5 and 4.6, model 10 of both Trials A and B gave the lowest AIC and BIC values when compared with the linear and logistic models, hence, both models were statistically fit.

iii) Step 3: Identifying the Bimodal Nature of Construction Contracting Behaviour

The third step of analysis is to identify the bimodal nature of CCB. The Cuspfit programme gives a bifurcation diagram which shows how the data fit into the bifurcated region. If reasonable portion of the data points are located within the bifurcation set, the area between the bifurcation lines, the CCB is bimodal (Ploeger et al. 2002; Van der Maas et al. 2003). Figs. 4.4 and 4.5 show the plotting results and the visual displays of the bifurcation curves respectively.

Within the bimodal zone, i.e. within the area of the bifurcation line, there exists a choice of 2 points, one in the aggressive state and the other in the co-operative

Table 4.4 Findings of Catastrophe Analyses

	Model 10 from Trial A	Model 10 from Trial B
Dependent Variables	Construction Contracting Behaviour	Construction Contracting Behaviour
Normal Factor (α)	Trust Intensity[a]	Trust Intensity[a]
Splitting Factor (β)	Contract Incompleteness[b]	Competitive Inertia[c]

Surveyed variables (rated on a Likert scale from (1) strongly disagree to (7) strongly agree):-

Trust Intensity:-
1. Your project team paid due regard to the respective rights, benefits and responsibilities and the plan, polices and strategies stipulated in the Contract.
2. The previous dealing(s) between the project participants reinforced confidence of your project team in working with each other.
3. Overly detailed contractual procedures to deal with contingencies were unlikely deterred your project team's motivation to maintain commitment.

Contract Incompleteness:-
1. Guidelines and possible solutions for handling various unanticipated contingencies/future problems had been incorporated in the Contract.
2. The substantial amount (monetary) of investment in this project had led to more likely to incorporate more detailed contract conditions and contractual procedures to deal with contingencies.
3. The long project duration had led to the incorporation of more detailed contract conditions and contractual procedures to deal with contingencies.

Competitive Inertia:-
1. The actions being taken by other contracting parties were strongly aggressive.
2. The capital necessary for the project operation had been in general insufficient.
3. Low interdependency between project participants had led to your party more likely taking advantage over the others.

[a] Trust Intensity is defined as the degree of confidence and trust building in the contracting parties.
[b] Contract Incompleteness is defined as the degree of term specificity and contingency adaptability in a contract.
[c] Competitive Inertia is the degree of aggressiveness of a contracting party on comparison to the actions being taken by counterpart.

state. As a point in the bimodal zone can be in either state (co-operative or aggressive), without additional information one cannot predict the outcome of further movement from such a point. However, if prior movements (i.e. past histories) are known, one could then predict the eventual state for the next movement from that point (Herbig 1991). With reference to Fig. 4.5, in a case where the point originated from the co-operative state (point C), a change from co-operative behaviour to aggressive behaviour is looming (path CAB) if the trust intensity continues to decrease (i.e. CCB becomes aggressive, the path goes further from point A up to B because of their bimodal nature within the bimodal zone). Within a CT framework, CCB will not revert to co-operation even when trust intensity increase again. Likewise, if the CCB is in the aggressive state (point D), a significant increase in trust intensity will be required to effect a behavioural change

Table 4.5 Catastrophe analysis of significant Trial A (adopted from Ploeger et al. 2002)

Model	a_0	a_1	a_2	b_0	b_1	b_2	λ	σ	Log likelihood	Parameters	AIC	BIC
1	−0.30	0	0	−1.31	0	0	0.20	1.54	−0.1282E + 03	4	0.2645E + 03	0.2745E + 03
2	0.88	0	0	−2.01	0	−1.65	−0.56	1.59	−0.1199E + 03	5	0.2498E + 03	0.2623E + 03
3	−5.00	0	0	−2.89	−1.30	0	2.61	2.33	−0.1174E + 03	5	0.2448E + 03	0.2574E + 03
4	5.00	0	0	−4.54	1.27	−0.83	−2.10	2.32	−0.1149E + 03	6	0.2417E + 03	0.2568E + 03
5	−0.33	0	−0.51	−1.55	0	0	0.20	1.55	−0.1239E + 03	5	0.2579E + 03	0.2704E + 03
6	0.09	0	−0.35	−1.04	0	−1.17	−0.15	1.35	−0.1192E + 03	6	0.2504E + 03	0.2654E + 03
7	−5.00	0	−0.60	−3.60	−1.33	0	2.32	2.28	−0.1153E + 03	6	0.2425E + 03	0.2576E + 03
8	−4.82	0	−2.00	−5.00	−1.70	−1.96	1.75	2.21	−0.1125E + 03	7	0.2390E + 03	0.2565E + 03
9	−0.36	0.97	0	−2.08	0	0	0.18	1.55	−0.1163E + 03	5	0.2425E + 03	0.2551E + 03
10	**0.69**	**0.84**	**0**	**−2.04**	**0**	**−1.40**	**−0.40**	**1.47**	**−0.1107E + 03**	**6**	**0.2334E + 03**	**0.2485E + 03**
11	−0.43	0.96	0	−2.07	−0.08	0	0.22	1.55	−0.1163E + 03	6	0.2445E + 03	0.2596E + 03
12	0.72	0.92	0	−2.26	−0.22	−1.50	−0.38	1.50	−0.1106E + 03	7	0.2352E + 03	0.2528E + 03
13	−0.29	0.91	−0.38	−2.20	0	0	0.14	1.55	−0.1144E + 03	6	0.2407E + 03	0.2558E + 03
14	0.19	0.76	−0.23	−1.51	0	−1.14	−0.17	1.36	−0.1104E + 03	7	0.2348E + 03	0.2524E + 03
15	−0.46	0.89	−0.39	−2.18	−0.19	0	0.25	1.55	−0.1143E + 03	7	0.2426E + 03	0.2602E + 03
16	0.00	0.79	−0.32	−1.59	−0.35	−1.21	−0.04	1.36	−0.1100E + 03	8	0.2360E + 03	0.2561E + 03
Linear[a]									−0.1431E + 03	4	0.2942E + 03	0.3042E + 03
Logistic[a]									−0.1135E + 03	5	0.2370E + 03	0.2496E + 03

Note [a] Unconstrained linear and logistic models
Model 1−16: cusp models
a_0 is the constant of the normal variable
b_0 is the constant of the splitting variable
a_1 and b_2 are parameters of the normal factor
a_2 and b_1 are parameters of the splitting factor
λ is the location, σ is the scale and zeros are fixed parameters

Table 4.6 Catastrophe analysis of significant Trial B (adopted from Ploeger et al. 2002)

Model	a_0	a_1	a_2	b_0	b_1	b_2	λ	σ	Log likelihood	Parameters	AIC	BIC
1	−0.26	0	0	−2.27	0	0	0.15	1.77	−0.1287E + 03	4	0.2655E + 03	0.2755E + 03
2	0.21	0	0	−1.52	0	−0.88	−0.14	1.53	−0.1254E + 03	5	0.2608E + 03	0.2733E + 03
3	−5.00	0	0	−3.75	−1.37	0	2.37	2.37	−0.1186E + 03	5	0.2473E + 03	0.2598E + 03
4	−5.00	0	0	−3.67	−1.35	0.03	2.39	2.37	−0.1186E + 03	6	0.2492E + 03	0.2643E + 03
5	−0.26	0	−0.18	−2.30	0	0	0.14	1.77	−0.1283E + 03	5	0.2666E + 03	0.2792E + 03
6	−0.01	0	−0.09	−1.37	0	−0.82	−0.02	1.50	−0.1253E + 03	6	0.2625E + 03	0.2776E + 03
7	−5.00	0	−0.26	−3.78	−1.37	0	2.35	2.36	−0.1182E + 03	6	0.2483E + 03	0.2634E + 03
8	−5.00	0	−1.20	−5.00	−1.63	−1.43	1.98	2.35	−0.1161E + 03	7	0.2461E + 03	0.2637E + 03
9	−0.31	1.02	0	−3.17	0	0	0.14	1.78	−0.1182E + 03	5	0.2464E + 03	0.2589E + 03
10	**0.11**	**0.89**	**0**	**−2.07**	**0**	**−0.90**	**−0.07**	**1.51**	**−0.1146E + 03**	**6**	**0.2413E + 03**	**0.2563E + 03**
11	−0.69	0.97	0	−3.05	−0.30	0	0.34	1.77	−0.1181E + 03	6	0.2481E + 03	0.2632E + 03
12	0.05	0.90	0	−2.07	−0.20	−0.93	−0.02	1.51	−0.1145E + 03	7	0.2431E + 03	0.2607E + 03
13	−0.30	1.03	−0.21	−3.21	0	0	0.13	1.78	−0.1177E + 03	6	0.2474E + 03	0.2624E + 03
14	0.07	0.89	−0.02	−2.06	0	−0.89	−0.05	1.51	−0.1146E + 03	7	0.2433E + 03	0.2608E + 03
15	−0.67	0.98	−0.21	−3.09	−0.29	0	0.33	1.77	−0.1176E + 03	7	0.2491E + 03	0.2667E + 03
16	−0.11	0.88	−0.06	−2.03	−0.25	−0.89	0.06	1.50	−0.1145E + 03	8	0.2450E + 03	0.2651E + 03
Linear[a]									−0.1402E + 03	4	0.2884E + 03	0.2985E + 03
Logistic[a]									−0.1163E + 03	5	0.2426E + 03	0.2552E + 03

Note [a] Unconstrained linear and logistic models

Model 1–16: cusp models

a_0 is the constant of the normal variable

b_0 is the constant of the splitting variable

a_1 and b_2 are parameters of the normal factor

a_2 and b_1 are parameters of the splitting factor

λ is the location, σ is the scale and zeros are fixed parameters

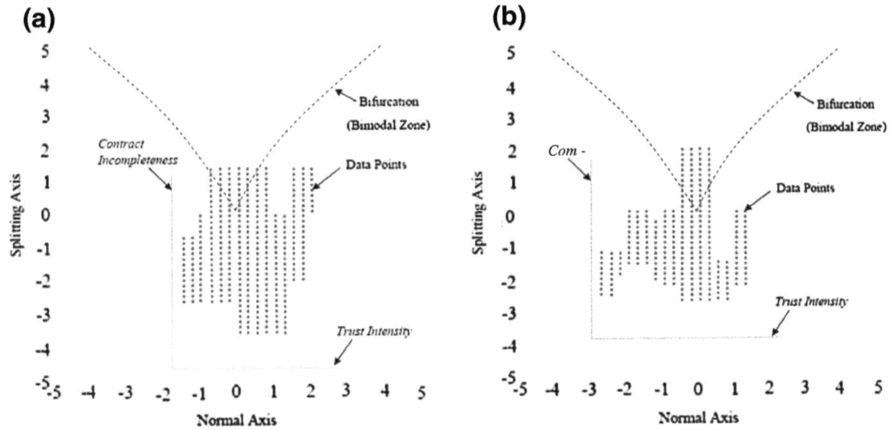

Fig. 4.4 Bifurcation diagram in the control space of the catastrophe models with **a** trust intensity as normal factor and contract incompleteness as splitting factor; **b** trust intensity as normal factor and competitive inertia as splitting factor

(called hysteresis effect) to co-operative behaviour (DEF). Hence, when the behavioural state falls within the bimodal region, it is difficult to predict the action of the contracting party. To predict which state of behaviour will occur, information of the present behavioural state on the curves and recent histories of both the control variables are needed (Herbig 1991; Hill 2001; Zeeman 1977). This highlights the importance of avoiding the building up of aggression forces. In parallel trust building is an effective way to release the tensions between the contracting parties.

4.7 Chapter Summary

Most of the industry-wide reviews recommend that construction contracting should embrace a culture of co-operation. This is considered to be one of the effective ways to reduce dispute and conflict. However, due to the fact that conflicts are inevitable in construction projects, acting co-operatively is easier to be said than done. Contracting parties often behave aggressively in order to protect and enforce their contractual rights on one hand while look for means to shun their obligations on the other. In this connection, the dichotomous pair of co-operation and aggression forces co-exists in all construction contracting environment. This chapter examines the dynamics of CCB in the light of these two co-existing forces. Modeled under a catastrophe theory (CT) based framework, three-variable Cat models were developed. In these models, CCB is the behavioural variable and co-operation and aggression forces were arranged as normal and splitting factors. A total of 70 models was analysed by the Cuspfit programme. Two catastrophe

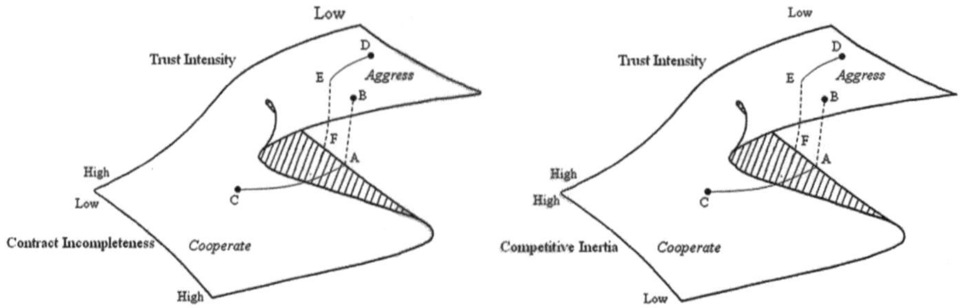

Fig. 4.5 Contracting behavioural surface of the two significant catastrophe models of construction contracting behaviour (from Trials A and B)

models were found significant. With CCB being the behavioural variable, the normal and splitting factors are trust and contract incompleteness respectively. This model affirms the positive roles that trust can play in balancing aggression. In addition, the empirical evidence fits well with the risk-based view of trust by Das and Teng (2004) who advocate that the presence of risk and uncertainty are conducive to trust development. This model suggests that if the contract is incomplete, thus unable to deal with all eventualities, the uncertainties and risks involved will be high. This type of situation is ideal platform to demonstrate the functionalities of co-operative efforts (Bhattacharya et al. 1998). It is a pragmatic approach to deal with crisis resulted from the manifestation of uncertainties and risks. In those circumstances, relying on contractual provisions or legal remedies gets the contracting parties nowhere. Instead, a flexible and co-operative problem-solving attitude is needed in order to navigate through the crisis. In this respect, trust and co-operation are indeed tightly knitted. The second significant CT CCB model is similar to the one obtained from Trial A except the splitting factor is competitive inertia (CI). CI refers to the reluctance to cooperate. This may due to the hard-line and opportunistic attitude of a self-interest seeking contracting party (Lyons and Mehta 1997). This situation is common with desperate subcontractors who have little to lose in a ruptured contractual relationship. They are not burdened by the priori capital investment or relationship building. Problems can easily be escalated to become disputes when parties are in aggressive mode.

In sum, within the CT framework (Fig. 4.5), if a contracting party is in the aggressive state, a significant increase in trust intensity is needed to install a co-operative behaviour change as suggested by the bimodal nature of CCB. In this connection, trust-building would be an important ingredient to balance aggression which dovetails the conventional wisdom of 'prevention is better than cure'.

Acknowledgments Special thanks to Professor Van der Maas for his advice on the use of the Cuspfit programme and Miss On Kiu Chiu for collecting data for the study. The content of this chapter has been published in Volume 134(12) of the Journal of Construction Engineering and Management and is used with the permission from ASCE.

References

Axelrod, R. (1984). *The evolution of cooperation*. New York: Basic Books.

Bacck, D., & Cullen, J. B. (1992). A catastrophe theory model of technological and structural change. *The Journal of High Technology Management Research, 3*(1), 125–145.

Bayliss, R. (2002). Project partnering—A case study on MTRC's Tseung Kwan O Extension. *HKIE Transactions, 9*(1), 1–6.

Bayliss, R., Cheung, S. O., Suen, C. H., & Wong, S. P. (2004). Effective partnering tools in construction: A case study on MTRC TKE Contract 604 in Hong Kong. *International Journal of Project Management, 22*, 253–263.

Bhattacharya, R., Devinney, T. M., & Pillutla, M. M. (1998). A formal model of trust based on outcomes. *Academy of Management Review, 23*(3), 459–472.

Black, C., Akintoye, A., & Fitegerald, E. (2000). An analysis of success factors and benefits of partnering in construction. *International Journal of Project Management, 18*, 423–434.

Buckley, P., & Casson, M. (1988). Cooperative strategies in international business. In F. J. Contractor & P. Lorange (Eds.), *Lexington books* (pp. 31–34) Lexington: Mass.

Cheung, S. O. (2007). *Trust in co-operative contracting in construction*. Hong Kong: City University of Hong Kong Press.

Cheung, S. O., Ng, S. T., Wong, S. P., & Suen, C. H. (2003a). Behavioral aspects of construction partnering. *The International Journal of Project Management, 21*(5), 333–343.

Cheung, S. O. (1993). A study of the application of ADR in the construction industry in Hong Kong, Research Report, Department of Building and Construction, City Polytechnic of Hong Kong.

Cheung, S. O. (2001). Relationalism: Construction contracting under the PRC contract law. *Cost Engineering, AACEI, 43*(11), 38–44.

Cheung, S.O. (2002). Mapping dispute resolution mechanism with contract types, Cost Engineering. AACE, *44*(8), 21–29.

Cheung, S. O., & Suen, C. H. (2002). A multi-attribute utility model for dispute resolution strategy selection. *Construction Management and Economics, 20*, 557–568.

Cheung, S. O., Suen, C. H., & Bayliss, R. (2002). The partnering experience of MTRC Tseung Kwan O Contract 604. *The International Construction Law Review, 19*(4), 510–520.

Cheung, S. O., Suen, C. H., & Cheung, K. W. (2003b). An automated partnering monitoring system—Partnering temperature index. *Automation in Construction, 12*, 331–345.

Cheung, S. O., Suen, C. H., & Cheung, K. W. (2004). PPMS: A web-based construction project performance monitoring system. *Automation in Construction, 13*(3), 361–376.

Chua, D. K. H., Kog, Y. C., & Loh, P. K. (1999). Critical success factors for different project objectives. *Journal of Construction Engineering and Management, ASCE, 125*(3), 142–150.

Cobb, L. (1978). Stochastic catastrophe models and multimodal distributions. *Behavioral Science, 23*(2), 360–416.

Cobb, L. (1980). *Estimation theory parameter estimation for the Cusp Catastrophe model: Proceeding of the Section on Survey Research Methods* (pp. 772–776). Washington, DC: American Statistical Association.

Cobb, L. (1981). Parameter estimation for the Cusp Catastrophe model. *Behavioral Science, 26*, 75–78.

Cobb, L., Koppstein, P., & Chen, N. H. (1983). Estimation and moment recursion Relations for multimodal distributions of the exponential family. *Journal of the American Statistical Association, 80*, 793–802.

Colledge, B. (2000). Obligations of good faith in partnering of U.K.: Construction contracts. *The International Construction Law Review, 17*(1), 174–201.

Construction Industry Review Committee (CIRC). (2001). Tang's Report on the Hong Kong Construction Industry Reform, Construction Industry Review Committee, HKSAR.

Crane, T. G., Felder, J. P., Thompson, P. J., Thompson, M. G., & Sanders, S. R. (1999). Partnering parameters. *Journal of Management in Engineering, 15*(2), 37–42.

Das, T. K., & Teng, B. S. (2004). The risk-based view of trust: a conceptual framework. *Journal of Business and Psychology, 19*(1), 85–116.

Doz, Y. L. (1996). The evolution of cooperation in strategic alliances: Initial conditions or learning processes? *Strategic Management Journal, Special Issue Summer, 17*, 55–83.

Egan, J. (1998). *Rethinking construction, department of the environment, transport and the region*. London: HMSO.

Fenn, P., Lowe, D., & Speck, C. (1997). Conflict and dispute in construction. *Construction Management and Economics, 15*(6), 513–518.

Goldberg, V. P. (1992). The past is the past—Or is it? the use of retrospective accounts as indicators of past strategy. *Academy of Management Journal, 35*, 848–860.

Gresov, C., Haveman, H., & Oliva, T. (1993). Organizational design inertia and the dynamics of competitive response. *Organization Science, 4*(2), 181–208.

Guastello, S. J. (1987). A butterfly catastrophe model of motivation in organizations: Academic Performance. *Journal of Applied Psychology, 72*, 165–182.

Harmon, K. M. (2003). Conflicts between owner and contractors: Proposed intervention process. *Journal of Management in Engineering, 19*(3), 121–125.

Hartelman, P. A. I. (1997). *Stochastic catastrophe theory*. Amsterdam: Faculteit der Psychologies.

Hartman, F.T. (1993). Construction dispute resolution through an improved contracting process in the Canadian context, PhD Thesis. Loughborough: Loughborough University of Technology.

Herbig, P. A. (1991). A Cusp Catastrophe Model of the adoption of an industrial innovation. *Journal of Production Innovation Management, 8*, 127–137.

Hill, C. A. (2001). A comment on language and norms in complex business contracting. *Chicago-Kent Law Review, 77*(29), 29–57.

Holyst, J. A., Kacperski, K., & Schweitzer, F. (2000). Phase transitions in social impact models of opinion formation. *Physica A, 285*, 199–210.

Latham, M. (1994). Constructing the Team: Final Report by Sir Michael Latham; Joint Review of Procurement and Contractual Arrangements in the United Kingdom Construction Industry. London: HMSO.

Liu, M. M., & Fellows, R. (2001). An eastern perspective on partnering. *Engineering, Construction and Architectural Management, 8*(1), 9–19.

Luo, Y. (2002). Contract, cooperation and performance in international joint ventures. *Strategic Management Journal, 23*, 903–919.

Lyons, B., & Mehta, J. (1997). Contracts, opportunism and trust: Self-interest and orientation. *Cambridge Journal of Economics, 21*, 239–257.

Macaulay, S. (1985). An empirical view of contract, Wisconsin Law Review, pp. 465–482.

Macneil, I.R. (1980). The New Social Contract: An inquiry into Modern Contractual Relations, New Haven.

Macneil, I. R. (1981). Economic analysis of contractual relations: Its shortfalls and the need for a rich classificatory apparatus. *Northwestern University Law Review, 75*, 1018–1063.

McKim, R. A. (1992). Risk behavior of contractors: a Canadian study. *Project Management Journal, 23*(3), 51–55.

Newman, P. (2000). Partnering, with particular reference to construction. *Arbitration Journal, 66*(1), 39.

Oliva, T. A., Day, D., & MacMillan, T. (1988). A generic model of competitive dynamics. *Academy of Management Review, 13*, 374–389.

Oliva, T. A., Oliva, R. L., & MacMillan, I. C. (1992). A catastrophe model for developing service satisfaction strategies. *Journal of Marketing, 56*, 83–95.

Piper, B. J. (2001). Partnering: a dream? *Newsletter of the Hong Kong Institute of Surveyors, 8*(10), 22–23.

Ploeger, A., Van der Maas, H. L. J., & Hartelman, P. A. (2002). Stochastic catastrophe analysis of switches in the perception of apparent motion. *Psychonomic Bulletin and Review, 9*(1), 26–42.

Qin, S. Q., Jiao, J. J., & Wang, S. T. (2001). A Cusp Catastrophe Model of slip-buckling slope. *Rock Mechanics and Rock Engineering, 34*(2), 119–134.

Raftery, A. E. (1995). Bayesian model selection in social research. *Sociological Methodology, 25*, 111–163.

Schwarz, G. (1978). Estimating the Dimension of a Model. *Annals of Statistics, 6*, 461–464.

Stewart, I. N., & Peregoy, P. L. (1983). Catastrophe Theory modeling in psychology. *Psychological Bulletin, 94*, 336–362.

Swedberg, R. (1987). Economic sociology: Past and present. *Current Strategic, 5*(1), 1–221.

Tallman, S., & Shenkar, O. (1994). A managerial decision model of international cooperative venture formation. *Journal of International Business Studies, 25*, 91–114.

Tamaki, T., Torii, T., & Meada, K. (2003). Stability analysis of black Holes via a Catastrophe Theory and black hole thermodynamics in generalized Theories of Gravity. *Physical Review D, 68*, 24028–1–24028-9.

Thom, R. (1975). *Structural Stability and Morphogenesis*. Benjamin: Addison-Wesley.

Thorburn, W. M. (1915). Occam's razor. *Mind, 24*, 287–288.

Van der Maas, H. L. J., Kolstein, R., & Van der Pligt, J. (2003). Sudden transitions in attitudes. *Sociological Methods and Research, 32*(2), 125–152.

Wagenmakers, E. J., Van der Mass, H. L. J. & Molenaar, P. C. M. (2004). Fitting the Cusp Catastrophe Model. Available at http://www.psycg.nwu.edu/~ej/ Encyclopediacarastrophe.pdf.

Wong, P. S. P., Cheung, S. O., & Ho, K. M. (2005). Contractor as trust initiator in construction partnering—A Prisoner's Dilemma Perspective. *Journal of Construction Engineering and Management, ASCE, 131*(10), 1045–1053.

Wright, D. J. (1983). Catastrophe Theory in management forecasting and decision making. *Journal of Operational Research Society, 37*, 935–942.

Yiu, T. W., & Cheung, S. O. (2006). A Catastrophe model of construction conflict behaviour. *Building and Environment, 40*(1), 438–447.

Zeeman, E. C. (1976). Catastrophe Theory. *Scientific American, 234*, 65–83.

Zeeman, E. C. (1977). *Catastrophe Theory: Selected papers 1972–1977*. Reading: Addison-Wesley Publishing Company.

Chapter 5
Exploring the Potential for Predicting Project Dispute Resolution Satisfaction Using Logistic Regression

Sai On Cheung and Tak Wing Yiu

Abstract The success of a construction project depends on the coordinated efforts of project team members. This is especially crucial when a project is having disputes. In fact, achieving satisfactory project dispute resolution can be a form of project success. This proposition has been empirically demonstrated a research that studied project dispute resolution satisfaction (DRS) using multivariate discriminant analysis (MDA). This chapter reports on a study that builds on that research, with the specific aim of predicting project DRS through the use of logistic regression (LR). In this study, a LR model of project DRS (Model 1) is developed, and then compared with the MDA model. The findings suggest that the LR technique provides a higher hit rate and thus a higher proportion of correct classification. With the wider acceptance of the use of alternative dispute resolution (ADR) methods, the effect, on the LR model, of changing the demarcation between adverse and favorable project DRS is also examined. For this examination, another LR model (Model 2) was developed. It is believed that Model 2 may reflect the prevailing sentiment that ADR is viewed as an amicable way to resolve disputes. Both the MDA model and LR models (Model 1 and Model 2) indicated that "design changes" are the root cause of adverse project DRS. Within the scope of the project data, these findings suggest that design changes are not just disruptive to project progress but also a critical cause of construction disputes.

S. O. Cheung (✉)
Construction Dispute Resolution Research Unit, Department of Civil and Architectural Engineering, City University of Hong Kong, Hong Kong, People's Republic of China
e-mail: saion.cheung@cityu.edu.hk

T. W. Yiu
Department of Civil and Environmental Engineering, Faculty of Engineering, The University of Auckland, Auckland, New Zealand
e-mail: k.yiu@auckland.ac.nz

S. O. Cheung (ed.), *Construction Dispute Research*,
DOI: 10.1007/978-3-319-04429-3_5,

5.1 Introduction

Achieving satisfactory dispute resolution has attracted considerable attention among construction practitioners and researchers (Cheung 1998; Cheung et al. 2000b). A number of studies have identified that unsatisfactory dispute resolution would jeopardise project success (Cheung 1999; Egan 1998; Merna and Bower 1997; Pinnel 1999; Werderitsch and Krebs 2000). Therefore, every construction project should aim at effective dispute management particularly where resources are scarce and limited. It is a well-known fact that resolving disputes using formal resolution methods such as arbitration or litigation is costly and time-consuming. These methods also disrupt the works programme (Bevan 1992; Cheung 1999; Wilsom 1992). Recent waves of dispute study have thus made tremendous efforts in investigating ways and means for preventing disputes or attaining speedy resolution (Cheeks 2003; Harmon 2003; Vallero and Vesilind 2006; Vorster 1991; Yiu and Cheung 2006). These studies generally suggest that dispute prevention could be achieved by: (i) clearly defining the project scope and (ii) establishing systems to facilitate effective communication and information exchanges amongst project team members (Gibson and Pappas 2003). For example, the use of appropriate contractual methods and equitable risk allocation has been proposed as dispute prevention measures (Allen 1993; Currie 1991; Fisher 1988; Jannadia et al. 2000; Jones 1991; Treacy 1995; Turner 1994). These studies collectively reinforce the conventional wisdom that 'prevention is better than cure'. Nonetheless, acknowledging the fact that complete dispute avoidance is not possible, a surge of studies on alternative dispute resolution was evidenced in the 90s (Brown and Marriott 1999; Cheeks 2003; Cheung 1999, 2002; Cheung et al. 2000a, b, 2002, 2004, 2005a, b, Cheung and Suen 2002; Harmon 2003; Hellard 1987; Spittler and Jentzen 1992; Tillet 1991; Vallero and Vesilind 2006; Vorster 1991; Yiu and Cheung 2006). These studies generally suggested that the implementation of alternative dispute resolution (ADR) methods could overcome the shortcomings of traditional methods such as arbitration and litigation, offering the possibility of prompt and economic resolution without deteriorating the cooperative relationship. However, negotiation is the preferred means for settling construction disputes. In terms of measuring project success, Cheung et al. (2000b) suggested the inclusion of project dispute resolution satisfaction (DRS) as a success measure in addition to time, cost and quality (Ashley et al. 1987; de Wit 1986; Tuman 1986; Wuellner 1990). This initiative was a timely response to the increasing concern over the spread of the dispute epidemic within the industry (CIRC 2001; Cheung 1998; Cheung et al. 2000a; Egan 1998; Latham 1994). In that study, Cheung et al. (2000b) employed Multivariate Discriminant Analysis (MDA) to classify projects as having either favorable or adverse project DRS. The MDA technique suggests that a number of discriminating variables are instrumental in such a classification. It appears that disputing parties are highly desirous of knowing the possibility of predicting their satisfaction level in dispute resolution. Hence, this study extends the work of Cheung et al. (2000b) by developing a DRS prediction model using logistic regression (LR). The results generated from the techniques of MDA and LR is compared. The LR model contributes to construction dispute studies by highlighting areas towards which management could direct their effort for a favorable project DRS outcome. It is predicted that this would minimise or even eliminate the uncertainty of dispute resolution, eventually lowering the antagonism during the resolution process.

5.2 Project Dispute Resolution Satisfaction (DRS)

In construction, project participants' satisfaction is one of the major factors in assessing project success (Baker et al. 1983; Cheung et al. 2000b; Murphy et al. 1974; Pinto and Slevin 1988; Sanvido et al. 1992). Cheung et al. (2000b) have addressed this point by introducing the concept of project DRS. With satisfactory dispute resolution, the contracting environment would be less confrontational and antagonistic (Cheung et al. 2000b). This is particularly important, as dispute resolution has become part of the routine of most project participants at the management level (Cheung et al. 2000b). In response to the mounting concern over the downsides of unsatisfactory construction dispute resolution, Cheung et al. (2000b) have successfully introduced an empirical approach to identifying and prioritising the variables that can be used to classify projects into favorable project DRS and adverse project DRS. In that study, project DRS was determined to be favorable if the most significant dispute was resolved through negotiation; and adverse if that dispute had to be resolved through either the alternative dispute resolution process, arbitration or litigation (Fig. 5.1).

By defining project DRS in this way MDA has been employed to statistically classify project DRS status. A dataset of 48 projects was used in that study and the list of variables used is given in Table 5.1. These variables were devised from the conceptual framework developed by Cheung et al. (2000b). This framework has captured the business environment of the industry that generically considered the social-oriented and project-related variables. Cheung et al. (2000b) hence grouped these factors into four categories (1) Environment Specific; (2) Organisation Specific; (3) Project Specific and (4) Process Specific. The following discriminant function was developed (Cheung et al. 2000b):

$$\begin{aligned} Z = & -0.167(\text{E\_TENIN}) + 0.101(\text{O\_C\_CLAM}) + 4.382(\text{P\_DESCH}) \\ & + 0.250(\text{P\_REL}) + 0.129(\text{R\_ADR}) + 0.322(\text{R\_ADV\_CL}) \\ & + 0.409(\text{R\_INC\_CL}) + 0.179(\text{R\_MAN\_CL}) - 2.996 \end{aligned} \quad (5.1)$$

The statistical fit of the MDA model is supported by (1) Eigenvalue: 1.5683; (2) Canonical Correlation: 0.7814; (3) Wilks' lambda: 0.3894 and (4) Chi square: 39.617 (df = 8, Sig. = 0.0000). Table 5.2 summarises the hit rates achieved by the MDA model in the Development and Validation stages.

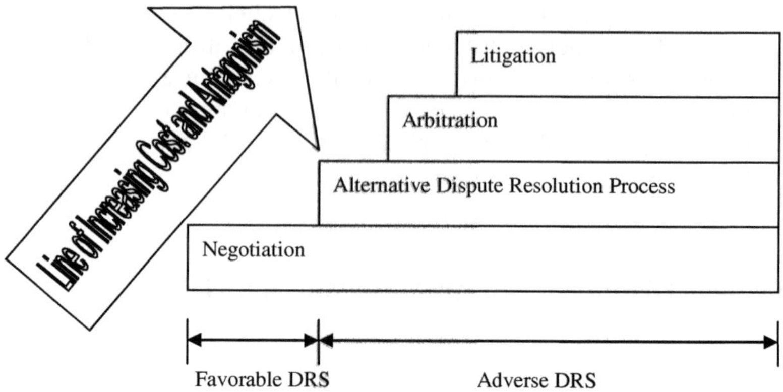

Fig. 5.1 Favorable and Adverse project DRS regions as defined by Cheung et al. (2000b)

Table 5.1 Variables suggested by Cheung et al. (2000b)

Variables	Descriptions and its measurement method
Environment specific	
Change in the tender price index (E_TENIN)[a]	*E_TENIN* = $(TPI_b—TPI_a)/n$ *Where* TPI_b *refers to the TPI of the month when the project was completed* TPI_a *refers to the TPI of the month when the project was commenced* *n refers to the project duration in months*
Organisation specific	
Claim consciousness of contractor (O_C_CLAM)[a]	*The general attitude of contractors towards claims, measured by:* *O_C_CLAM = number of claim notifications/ number of variation instructions*
Contractor's experience with similar types of project (O_COEXP)	The number of projects of similar construction type that the contractor had in the last 3 years.
Previous working relationship between parties (O_PRWKRE)	The number of projects that the two organisations had contracted together in the last 3 years.
Project specific	
Degree of design changes (P_DESCH)[a]	*The degree of design change of the project was measured by:* *(Total of the variation bill + claims)/final account figure*
Degree of involvement of client in running of project (P_INV_CL)	The mode of involvement of the clients ranged from no involvement to full participation, i.e. having in-house design and contract administration teams.

(continued)

Table 5.1 (continued)

Variables	Descriptions and its measurement method
Relationship between project personnel (P_REL)[a]	*The relationship between the project personnel was recorded on a Likert Scale of 1–6 (Good—Bad)*
Degree of reliance on price consideration in selection of contractor (P_SELCRI)	This was measured on a scale of 1–6, where 1 signifies choice solely by way of technical considerations and 6 implies selection based on price only
Process Specific	
Use of ADR to resolve contractual disputes (R_ADR)[a]	*The use of ADR to resolve contractual disputes was recorded on a Likert Scale of 1–6 (Good—bad)*
Use of an external claim advisor (R_ADV_CL)[a]	*The use of an external claim advisor was recorded on a Likert Scale of 1–6 (Low—High)*
Possibility of using the same contractor in future projects (R_FUT_CO)	The possibility of using same contractor in future projects was recorded on a Likert Scale of 1–6 (High—Low)
Incentive for client to settle (R_INC_CL)[a]	*The incentive for the client to settle was recorded on a Likert Scale of 1–6 (High-low)*
Degree of involvement of senior management in the dispute resolution process (R_MAN_CL)[a]	*The degree of involvement of senior management in the dispute resolution process was recorded on a Likert Scale of 1–6 (High—Low)*
Negotiation skill of the client's dispute resolution team (R_NEG_CL)	The negotiation skill of the client's dispute resolution team was recorded on a Likert Scale of 1–6 _(Good—Bad)

[a] The eight discriminating variables included in the discriminant function of Cheung et al. (2000b)

Table 5.2 Hit rates of MDA classification (Cheung et al. 2000b)

		Hit rates (%)
Stage of model development	Group 0 (favorable project DRS)	90.90
	Group 1 (adverse project DRS)	93.30
	Overall	91.67
Stage of model validation	Group 0 (favorable project DRS)	77.77
	Group 1 (adverse project DRS)	75.00
	Overall	76.92

In that study, eight discriminating variables were identified. These are "Change in the tender price index", "Claim consciousness of contractor", "Degree of design changes", "Relationship between project personnel", "Use of ADR to resolve contractual dispute", "Incentive for client to settle" and "Degree of involvement of senior management in the dispute resolution process". By examining the discriminant coefficient of Eq. 5.1, the variable "Degree of design changes" exhibits the highest discriminatory power. In sum, the study of Cheung et al. (2000b) provides an example of a systemic approach in identifying and prioritising variables that distinguish projects with favorable project DRS from those with adverse project DRS.

5.3 Logistic Regression (LR) Versus Multivariate Discriminant Analysis (MDA)

LR and MDA are both commonly used for analysing data on a categorical scale (Pohar et al. 2004) and are considered appropriate for developing classification models. In statistical terms, LR is considered to be an attractive alternative to MDA (Efron 1975; Hair et al. 1995; Pohar et al. 2004). This study contributes to the construction dispute resolution domain by introducing the LR technique in order to improve the prediction accuracy for project DRS. As explained by Pohar et al. (2004), LR and MDA differ in their basic ideas. The MDA technique works on the assumption of normally distributed explanatory variables, while LR does not. Consequently, when the normality assumption is fulfilled a better result can be expected with MDA. Nonetheless, the assumption of normality is often not practically achievable. The use of LR is then more appropriate when this assumption is not met (Hair et al. 1995; Pohar et al. 2004). In sum, LR analysis (a) accepts both continuous and categorical predictors, (b) produces fairly accurate results, (c) is free from any constraints of normality or equal variance/covariance assumptions and (d) relates to discriminant function analysis through Bayes theorem when compared with MDA (Flury 1997; Lei and Koehly 2000; Peng and So 1999 and Peng et al. 2001). The LR technique was used for this study as (i) it has been found to be robust even when the requirement of normality is not strictly followed (Hair et al. 1995; Kleinbaum 1994; Sharma 1996; Tung 1985), and (ii) the selection of dichotomous variables (favorable and adverse project DRS) simplifies the prediction process and facilitates a robust LR computation (Wong 2004).

5.3.1 Logistic Regression (LR)

Logistic Regression (LR) is a mathematical modeling technique which describes the occurrence or nonoccurrence of an event (Kleinbaum 1994; Tung 1985). The applications of LR have been reported in social science, psychology and organisational behaviour (Chuang 1997; DeMaris 1995; Smith 2004; Wofford et al. 1994). In construction, it has been used mainly for the development of prediction models including the prediction of contractor performance (Wong 2004), contractor failure (Russell and Jaselskis 1992), the occurrence of contract disputes (Diekmann and Girard 1995) and bid or no-bid decisions (Lowe and Parvar 2004). LR is a well-established technique for assessing the likelihood of events. It is particularly useful when the observed outcome is restricted to two values (usually coded as '1' or '0') (Hair et al. 1995). The result of LR analysis is a logistic equation that enables the calculation of the occurrence of the dependent variable as a function of the independent variables. The dichotomous dependent variable of this study is designated as '1' or '0', where '1' represents adverse project DRS and '0' represents favorable project DRS.

In essence, a LR model predicts the odds of an event occurring. It is defined as the ratio of the probability that an adverse project DRS will occur to the probability that it will not. If P is the probability of the occurrence of adverse project DRS, the odds can be written as:-

$$Odds = \frac{P}{(1 - P)} \tag{5.2}$$

The probability function can be presented as:

$$S = \text{In}\left(\frac{p}{1 - p}\right) = a_0 + a_1X_1 + a_2X_2 + \ldots + a_nX_n \tag{5.3}$$

or

$$\text{P} + \frac{1}{1 + e^{-(\text{s})}} \tag{5.4}$$

where a_0 is constant; a_n is the coefficient estimated from the data; and $X_{1..}X_n$ are the significant factors that lead to adverse project DRS (refer to Table 5.1). When considering the probabilities of the project DRS, the LR prediction model performs the function of assessing the probability of achieving a certain satisfaction level. A cut-off value of 0.5 has been advocated by previous studies for the purpose of prediction accuracy (Russel and Jaselskis 1992; Salem et al. 2004; Wong 2004).

5.4 Model Development

In this study, logistic regressions were performed by the SPSS software. LR prediction model was developed based on the 48 project data of government, quasi-government and private projects in Hong Kong used by Cheung et al. (2000b). The range of contract sums of these projects was mostly from 10 to 500 million Hong Kong dollars. These data were collected through structured interviews and a questionnaire survey. The construction of the questionnaire was based on the measurement methods of each variable as shown in Table 5.1.

As discussed in the previous section, the dichotomous occurrence of the dependent variable in LR is designated as '1' or '0', which represents adverse and favorable project DRS respectively. This designation creates a demarcation between adverse and favorable project DRS for the interpretation of the model. As suggested by Cheung et al. (2000b), whether the project DRS is adverse or favorable is defined by the type of dispute resolution method employed in resolving the most critical dispute of a construction project (refer to Fig. 5.1). Typically, project data were divided into two sets: Modeling Data (Set A) and Test Group Data (Set B) (Wong 2004). There are 48 project data sets and 33 of them with favorable project DRS (69 % of the total project data) and 15 with adverse project DRS (21 % of the total project data). 40 sets were used for the Modeling Data (Set A) and 8 sets were used for the Test Group Data (Set B). It is desirable for the Set A and Set B data to have similar composition of favorable and adverse project DRS. Accordingly, Set A involves 27 projects of favorable project DRS and 13 adverse project DRS, Set B involves in 6 projects of favorable project DRS and 2 adverse project DRS. A summary of the proposed LR prediction model and its data structure are given in Table 5.3.

5.4.1 LR Statistics

The LR analysis begins with the selection of statistically significant independent variables to be included in the analysis via a stepwise procedure. The output presented in Table 5.4 shows the stepwise LR results of the Modeling Data (Set A). It shows the stepwise selection process for the LR prediction model along with all of the selected independent variables for Steps 1–4.

As shown in Table 5.4, the LR prediction model identified four variables from the stepwise procedure, these were: "Degree of design changes", "Degree of involvement of senior management in the dispute resolution process", "Change in the tender price index" and "Incentive for client to settle". In Step 1 of the selection process, the independent variable that met the p value criterion was selected for inclusion in the model. This process was repeated until there were no further independent variables with a significant p value. The selection process was then stopped and the final LR model included four independent variables. For example, 72.50 % of 40 cases were correctly predicted in Step 1. The first selected variable was found to have the highest prediction power as compared to the second, third and fourth selected variables. The aggregate of the second, third and the fourth selected variables only contributed 15 % of the overall cases that were correctly predicted. Details of the cases that were correctly predicted were displayed in Table 5.5. Thirty-five, or 87.50 % of the total cases were correctly predicted, and hence only 5 out of the 40 cases were wrongly predicted. This indicates that this LR prediction model is reasonably good in predicting project DRS.

Table 5.3 A Summary of the proposed LR model

Dependent variable (Favorable and adverse project DRS)	Favorable project DRS	Adverse project DRS
Demarcation of Cheung et al. (2000b) (refer to Fig. 5.1)	The most critical dispute is resolved through negotiation	The most critical dispute is resolved through either an *ADR process, Arbitration or Litigation*
Data structure	Modeling data (Set A): 27 Test group data (Set B): 6	Modeling data (Set A): 13 Test group data (Set B): 2

Table 5.4 Stepwise logistic regression statistics

Step	Chi-sq.	df	Sig[a]	Class %	Variable
1	7.960	1	0.005	72.50	Degree of design changes (P_DESCH)
2	9.886	1	0.002	80.00	Degree of involvement of senior management in the dispute resolution process (R_MAN_CL)
3	6.333	1	0.012	82.50	Change in the tender price index (E_TENIN)
4	6.286	1	0.012	87.50	Incentive for client to settle (R_INC_CL)

[a] P value significant at the 0.05 level

Table 5.5 Logistic regression classification table for LR prediction

	Predicted		Hit rates
Observed	0	1	
0 (favorable project DRS)	25	2	92.60
1 (adverse project DRS)	3	10	76.90
			Overall 87.50 %

Note: Classification Table for OVERALL. The cut value was 0.50

5.4.2 Coefficients for LR Prediction Models

Column B in Table 5.6 shows the estimated coefficients and related statistics for the LR prediction model. It also gives the constant and the statistically significant variables identified in this model. Given these coefficients, the LR prediction model of project DRS can be represented by Eq. 5.5:

$$\ln\left(\frac{\mathrm{P}_1}{1-\mathrm{P}_1}\right) = -10.853 - 0.445(\mathrm{E\_TENIN}) + 28.156(\mathrm{P\_DESCH}) + 0.906(\mathrm{R\_INC\_CL}) + 1.484(\mathrm{R\_MAN\_CL}) \quad (5.5)$$

Table 5.6 Coefficients of the LR Prediction Model for Project DRS

Variable	B	S.E.	Wald	df	Sig[a]	Exp(B)
E_TENIN	–0.445	0.190	5.482	1	0.019	0.641
P_DESCH	28.156	13.560	4.312	1	0.038	1.691E + 12
R_INC_CL	0.906	0.461	3.869	1	0.049	2.474
R_MAN_CL	1.484	0.594	6.233	1	0.013	4.409
Constant	–10.853	4.145	6.857	1	0.009	0.000

[a] Significant at the 0.05 level

where P_1 is the Probability of getting an Adverse Project DRS
E_TENIN = Change in the Tender Price Index
P_DESCH = Degree of Design Changes
R_INC_CL = Incentive for Client to Settle
R_MAN_CL = Degree of Involvement of Senior Management in the Dispute Resolution Process

5.5 Model Validation

The validity of the LR prediction model was tested by how well it predicts project DRS within the test group data (Salem et al. 2004). As shown in Table 5.3, a total of eight independent sets of Testing Group Data (Set B) were designated for model validation, and the cut off value was set at 0.5 for the purpose of prediction accuracy (Russell and Jaselskis 1992 and Salem et al. 2004; Wong 2004). The results of validation for the model are presented in Table 5.7. It shows that none of the Testing Group Data were wrongly classified, and hence the model appears to predict the test group data very well. According to Pampel (2000), *"a 'perfect' model would correctly predict group membership for 100 % of the cases; a failed model would do no better than chance by correctly predicting 50 % of the cases"*. The percentage of correctly predicted cases from 50–100 provides a crude measure of predictive accuracy (Pampel 2000).

5.6 Discussion

The LR technique is often regarded as a preferred estimation technique, which provides an attractive alternative to the MDA technique (Efron 1975; Hair et al. 1995; Pohar et al. 2004). In this section, this proposition is tested by comparing the results obtained from the MDA and LR techniques. The performance of the MDA and LR models is compared by the hit rate achieved (Press and Wilson 1978). The ability of LR to cope with data sets that are not normally distributed will also be examined.

Table 5.7 Model validation

No.	E_TENIN	P_DESCH	R_INC_CL	R_MAN_CL	Original	Predicted probabilities of adverse project DRS	Predicted probabilities of favorable project DRS	Prediction[a]
1	−1.759	0.070	4.000	2.000	Favorable	18.13	81.87	Favorable
2	5.692	0.080	3.000	5.000	Favorable	26.98	73.02	Favorable
3	7.867	0.030	1.000	1.000	Favorable	0.00	100.00	Favorable
4	4.923	0.010	1.000	1.000	Favorable	0.00	100.00	Favorable
5	8.000	0.020	3.000	2.000	Favorable	0.03	99.97	Favorable
6	7.696	0.100	2.000	5.000	Favorable	9.71	90.29	Favorable
7	−1.917	0.020	6.000	5.000	Adverse	96.83	3.17	Adverse
8	1.800	0.120	5.000	5.000	Adverse	97.53	2.47	Adverse

[a] None of cases was wrongly classified, therefore 100 % of Set B data was correctly classified

Table 5.8 Summary of hit rates generated from MDA and LE techniques

Data set		MDA classification (Cheung et al. 2000b) (%)	LR classification (refer to Table 5.7) (%)
Test group data (Set B)—for model validation	0 (favorable project DRS) hit rate	77.77	100.00
	1 (adverse project DRS) hit rate	75.00	100.00
	Overall	76.92	100.00

5.6.1 Comparison of Hit Rates—Qualitative Approach

Table 5.8 gives a summary of the hit rates achieved by both the MDA and LR models. The hit rates obtained from the model validation stage are more indicative (Test Group Data—Set B). None of the cases was wrongly classified by the LR model, thus representing a 100 % hit rate. The hit rate achieved by the MDA model was 77 %.

In practice, the increase of hit rate of LR models against the MDA models implies a higher proportion of correct classification that can be recorded in the use of LR. As such in terms of classification accuracy, the LR model provides the better result.

5.6.2 Comparison of Hit Rates—Quantitative Approach

Is there a significant difference statistically in terms of the proportion correctly classified between the MDA and LR results? This can be addressed by employing McNemar's test (McNemar 1947) which has been commonly used to compare the error rates of two techniques (e.g. MDA and LR) (Fielding and Bell 1997; Chatfield 1995; Fielding and Eyden et al. 1995; Manel et al. 1999; McGurr and DeVaney 1998; Turney 1995). In order to perform McNemar's test, the predictions of both MDA and LR techniques were compared. The following classes of comparisons were employed:

1. Both MDA and LR techniques classify adverse project DRS correctly (d);
2. The MDA technique classifies favorable project DRS, and the LR technique classifies adverse project DRS (b);
3. Both MDA and LR techniques classify favorable project DRS correctly (a), and
4. The MDA technique classifies adverse project DRS, and the LR technique classifies favorable project DRS.

The above four possible classifications are tabulated in a 2 × 2 table that cross-tabulates prediction accuracy (Table 5.9).

Table 5.9 Contingency table for the McNamar test

		LR technique		Total
		0	1	
MDA technique	0 (favorable project DRS)	*a*	*b*	$a + b$
	1 (adverse project DRS)	*c*	*d*	$c + d$
Total		$a + c$	$b + d$	N

Table 5.10 Contingency table for the 48 project data

		LR technique		Total
		0	1	
MDA technique	0 (favorable project DRS)	*13*	*4*	*17*
	1 (adverse project DRS)	*1*	*30*	*31*
Total		*14*	*34*	48

In Table 5.9, the numbers of projects that both the MDA and LR techniques classify as adverse or favorable projects DRS are *a* and *d* respectively. The number of projects that the MDA technique classifies as having favorable project DRS and the LR classifies as having adverse project DRS is indicated by *b* Those projects that the MDA technique classifies as having adverse (or favorable) project DRS and LR classifies favorable (or adverse) project DRS is indicated by *c* (or *b*). The null hypothesis (H_o) stated that there is no difference between the two techniques whilst the alternative hypothesis (H_1), stated that there is a difference. Then, McNemar's test derives a χ^2 score with the following test statistic (McNemar 1947; Turney 1995):

$$\chi^2 = \frac{(b - c)^2}{(b + c)} \tag{5.6}$$

As suggested by McNemar (1947), Eyden et al. (1995) and Turney (1995), the test statistic approximately follows the Chi square distribution with one degree of freedom. At the 5 % significance level, if the magnitude obtained from the above test statistic is higher than the critical value (i.e. $\chi^2_{crit} = 3.84$), the null hypothesis (H_0) shall be rejected accordingly.

Accordingly, the predictions of the 48 project data using the MDA and LR techniques were identified and tabulated in Table 5.10. Based on Eq. 5.1, the χ^2 score was equal to 1.8, which is lower than the critical value. This result indicates that there is no significant difference in the proportions correctly classified between the MDA and LR techniques.

Table 5.11 The kolmogorov–smirnov statistics

Variables	Normal parameters		Most extreme differences			Kolmogorov-smirnov Z	Asymp. Sig. (2-tailed)
	Mean	Std. deviation	Absolute	Positive	Negative		
E_TENIN	5.137	3.911	0.195	0.080	−0.195	1.351	0.052
O_PRWKRE	2.521	3.787	0.260	0.260	−0.253	1.803	0.003[a]
O_COEXP	3.458	2.873	0.188	0.188	−0.114	1.305	0.066
O_C_CLAM	1.492	4.880	0.382	0.382	−0.380	2.644	0.000[a]
P_SELCRI	4.646	1.329	0.293	0.154	−0.293	2.027	0.001[a]
P_DESCH	0.113	0.072	0.130	0.130	−0.075	0.898	0.395
P_REL	2.583	1.302	0.256	0.256	−0.160	1.776	0.004[a]
P_INV_CL	3.833	1.326	0.290	0.190	−0.290	2.007	0.001[a]
R_ADR	3.021	1.929	0.228	0.228	−0.147	1.576	0.014[a]
R_NEG_CL	2.583	1.302	0.208	0.208	−0.119	1.440	0.032[a]
R_MAN_CL	3.583	1.648	0.159	0.144	−0.159	1.103	0.176
R_ADV_CL	1.521	1.255	0.453	0.453	−0.339	3.136	0.000[a]
R_INC_CL	2.542	1.611	0.236	0.236	−0.169	1.634	0.010[a]
R_FUT_CO	3.896	1.627	0.189	0.149	−0.189	1.308	0.065

[a] Asymp. Sig. (2-tailed) $\leq$0.05—the data is not normally distributed

5.6.3 Test of Normality

Although the result of McNemar's test indicates that there is no significant difference between the MDA and LR techniques, a more conclusive result can be obtained by testing a fundamental issue—the normality of data. This is because a feature of MDA technique is that it assumes multivariate normality, while the LR technique does not (Hair et al. 1995; Pohar et al. 2004). If this normality assumption is violated, then the data is more suitable for LR technique. Consequently, the normality assumption was tested using a statistical technique called the Kolmogorov–Smirnov (K–S) Test (Chi and Tang 2006). This test seeks to provide a quantitative determination of whether a data set is normally distributed or not. It examines the most extreme absolute deviation and determines the probability that this deviation could be explained by a normally distributed data set. Table 5.11 shows the statistics of the K–S test. It consists of a mean, standard deviation, extreme absolute deviation from normality, and the most positive and most negative deviations from normality. The value of Asymptomatic Significance (Asymp. Sig.) (2-tailed) gives this probability as a number between 0 and 1. As a rule of thumb, a value of $\leq$0.05 indicates that the data set is not normally distributed, while a value of >0.05 suggests that there is insufficient evidence to show the data is not normally distributed.

In this study, the statistics of the K–S test showed that the majority (64 %) of the variables violated the normality assumption, indicating that the data were suitable for the LR technique, and generally suggests that LR is a preferred estimation technique when compared with MDA.

5.6.4 Effect of Changing the Demarcation

Having a well-defined demarcation between adverse and favorable project DRS is one of the most important underpinnings to the development of the LR prediction model. As such, it is interesting to examine the effect of changing the demarcation between adverse and favorable project DRS in the LR model. As suggested by Cheung et al. (2000b), the demarcation between adverse and favorable project DRS was set at the interface between negotiation and the alternative dispute resolution process (refer to Fig. 5.1). With the wider acceptance of the use of ADR as a means of resolving construction disputes, the outcomes obtained from ADR process are now commonly regarded as satisfactory dispute resolutions. This is particularly the case in Hong Kong where mediation has become an integral part of the dispute resolution clause of all conditions of contract published by the Hong Kong Government in 1999. If the largest dispute gets resolved through ADR process (e.g. mediation, adjudication and conciliation), then the project DRS can be classified as favorable. However, if some formalised procedures such as arbitration and litigation are involved, then the project DRS remains adverse. This may be a better reflection of the prevailing sentiment, and with the new demarcation, a new LR prediction model of project DRS (namely Model 2), was developed. The prediction results from Model 1 and Model 2 are compared in Fig. 5.2.

Model 2 was developed following the procedures described previously for Model 1. A summary of Model 2 and its data structure is presented in Table 5.12. The stepwise procedures of LR were then carried out to select statistically significant independent variables. Table 5.13 shows the results of the stepwise selection process with all the selected independent variables. Moreover, the classification table and the coefficients of Model 2 are provided in Tables 5.14 and 5.15 respectively. Given this information, Model 2 can be represented by Eq. 5.7:

$$\ln\left(\frac{P_2}{1 - P_2}\right) = -11.796 - 0.809(E\_TENIN) + 0.246(O\_C\_CLAM) + 43.892(P\_DESCH) + 2.120(R\_NEG\_CL) \quad (5.7)$$

Where P_2 is the Probability of getting an Adverse Project DRS
E_TENIN = *Change in the Tender Price Index*
O_C_CLAM = *Claim Consciousness of Contractor*
P_DESCH = *Degree of Design Changes*
R_NEG_CL = *Negotiation Skill of Client's Dispute Resolution Team*

As shown in Table 5.14, Model 2 achieved a higher prediction power (Hit rate: 92.5 %) in predicting project DRS than Model 1 (Hit rate: 87.5 %) (Table 5.5 refers). In Model 2, the inclusion of ADR methods as signifying a favorable project DRS may thus be a reflection of the recent wave of using the ADR method as an amicable way to resolve disputes.

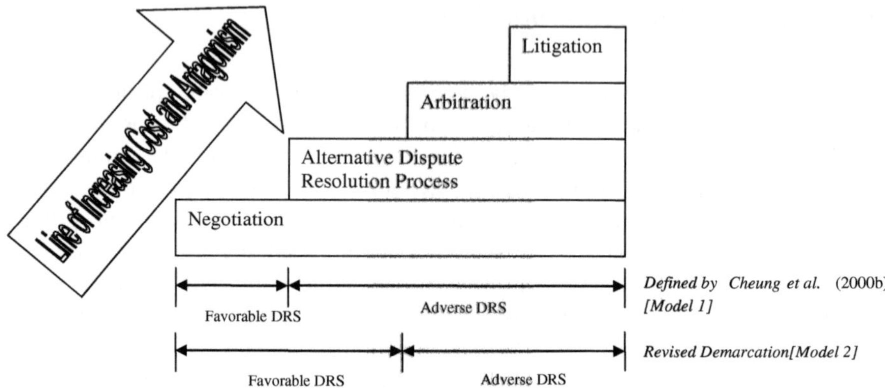

Fig. 5.2 Demarcations between Favorable and Adverse DRS regions

Table 5.12 A summary of *Model 2*

Dependent variable (Favorable and adverse project DRS)	Favorable project DRS	Adverse project DRS
Revised demarcation (refer to Fig. 5.2)	The most critical dispute is resolved through negotiation or an ADR process	The most critical dispute is resolved through either *Arbitration or Litigation*
Data structure	Modeling data (Set A): 30 Test group data (Set B): 6	Modeling data (Set A): 10 Test group data (Set B): 2

Table 5.13 Stepwise logistic regression statistics of *Model 2*

Step	Chi-sq.	df	Sig[a]	Class %	Variable
1	7.166	1	0.007	77.50	Degree of design changes (P_DESCH)
2	6.574	1	0.010	82.50	Negotiation skill of the client's dispute resolution team (R_NEG_CL)
3	8.947	1	0.003	90.00	Change in the tender price index (E_TENIN)
4	6.629	1	0.010	92.50	Claim consciousness of contractor (O_C_CLAM)

[a] P value significant at the 0.05 level

Table 5.14 Logistic regression classification table of the lr prediction model (*Model 2*)

	Predicted		Hit rate
Observed	0	1	
0 (favorable project DRS)	29	1	96.70
1 (adverse project DRS)	2	8	80.00
			Overall 92.50 %

Note: Classification Table for OVERALL. The cut value was 0.50

Table 5.15 Coefficients of LR (*Model 2*)

Variable	B	S.E.	Wald	Df	Sig[a]	Exp(B)
E_TENIN	−0.809	0.352	5.292	1	0.021	0.445
O_C_CLAM	0.246	0.173	2.016	1	0.156	1.279
P_DESCH	43.892	19.542	5.045	1	0.025	1.154E + 19
R_NEG-CL	2.120	0.972	4.760	1	0.029	8.330
Constant	−11.796	4.895	5.806	1	0.016	0.000

[a] Significant at the 0.05 level

Furthermore, both models suggested that design changes are the primal root of construction disputes and that they jeopardise dispute resolution satisfaction. Consequently, "design changes" appears to be a leading factor in increasing the probability of *adverse* project DRS. The next question concerns what the degree of design changes would need to be to lead to an *adverse* project DRS. In practice, design changes refer to the addition and/or omission of work as opposed to what has been agreed in the contract. Frequent design changes would induce project disruptions and increases construction costs. In construction, it has been a major source of disputes, especially when there is substantial revision in the design and/or work content. A substantial increase (or decrease) in the number of variations would increase the probability of causing disputes. In terms of project DRS, the chance of generating an adverse DRS may be higher with a larger extent of design changes. Even though mechanisms are available in most construction contracts for valuing variation claims, contracting parties are accustomed to shunning their responsibilities for project delay and excessive costs. As suggested by Stockenberg (2001), in the course of smoothing the dispute resolution process, difficulty is often found in the cumulative impact from a large number of changes (Stockenberg 2001). Contractors often find themselves not fairly compensated, whereas clients complain that the charges for the changes are exaggerated. Against this background, settlement is difficult and it is not unusual for the variation account not to be agreed years after practical completion. Despite the fact that the disputes may be resolved eventually, this practice may prolong the dispute resolution process and induce a higher level of hostility, putting the cooperative relationship in jeopardy.

5.7 Chapter Summary

Satisfactory project dispute resolution has been identified as one of the key indicators of project success in the construction industry. This particular phenomenon has been modeled and empirically tested with multivariate discriminant analysis (MDA) by Cheung et al. (2000b). This chapter reports on a study that builds upon this work, with the specific aim of examining the application of Logistic Regression (LR) in predicting project dispute resolution satisfaction (DRS). In this study, a LR model (Model 1) was developed, and compared with the MDA model. Model 1 gave a higher hit rate than the MDA model. This is in line with the observation derived from other reported studies that LR is a preferred alternative to the MDA technique. In response to the prevailing sentiment that resolving disputes by ADR is an acceptable method, the favorable project DRS classification included those cases where ADR was used to resolve the dispute. Another LR model (Model 2) was then developed so as to examine the effect of changing the demarcation between adverse and favorable project DRS. Finally, both the MDA model and LR models (Model 1 and Model 2) suggested that "design changes" constitute a root cause of construction disputes, and they increase the probability of generating an adverse project DRS.

Acknowledgments Special thanks to Miss Ho Wai Chan for collecting data for the study. The content of this chapter has been published in Volume 136(5) of the Journal of Construction Engineering and Management and is used with the permission from ASCE.

References

Allen, R. K. (1993). Dispute avoidance and resolution for consulting engineers. *Manuals and Reports on Engineering Practice, ASCE, 43314*, 1–81.
Ashley, D. B., Lrie, C. S., & Jaselskis, E. J. (1987). Determinants to construction project success. *Project Management Journal, 18*(2), 69–79.
Baker, B. N., Murphy, D. C., & Fisher, D. (1983). Factors affecting project success. In D. I. Cleland & W. R. King (Eds.), *Project Management Handbook*. Princeton, NJ: Van Nostrand Reinhold.
Bevan, A. (1992). *Alternative dispute resolution*. London: Sweet and Maxwell.
Brown, H., & Marriott, A. (1999). *ADR principles and practice*. London: Sweet and Maxwell.
Chatfield, C. (1995). Model uncertainty data mining and statistical inference. *Journal of the Royal Statistical Society, Series A, 158*, 419–466.
Cheeks, J. R. (2003). Multistep dispute resolution in design and construction industry. *Journal of Professional Issues in Engineering Education and Practice, 129*(2), 84–91.
Cheung, S.O. (1998). Project Dispute Resolution Satisfaction of Construction Clients in Hong Kong, PhD Thesis, UK: University of Wolverhampton.
Cheung, S. O. (1999). Critical factors affecting the use of alternative dispute resolution processes in construction. *International Journal of Project Management, 17*(3), 189–194.
Cheung, S. O., & Suen, C. H. (2002). A multi-attribute utility model for dispute resolution strategy selection. *Construction Management and Economics, 20*, 557–568.
Cheung, S. O., Yiu, T. W., & Leung, M. Y. (2005a). Third party interventions in construction dispute negotiation, the International Construction Law. *Review, 22*(4), 503–518.
Cheung, S. O., Yiu, T. W., & Chung, K. H. (2005b). The use of decision support tools in construction dispute negotiation. *Australasian Dispute Resolution Journal, 16*(3), 170–186.
Cheung, S. O. (2002). Mapping dispute resolution mechanisms with contract types. *Cost Engineering, AACEI, 44*(8), 21–29.
Cheung, S. O., Suen, C. H., & Lam, P. (2002). The fundamentals of ADR processes in construction. *Construction Engineering and Management, ASCE, 128*(5), 409–417.
Cheung, S. O., Tam, C. M., & Harris, F. C. (2000a). Project dispute resolution satisfaction classification through neural networks. *Journal of Management in Engineering, ASCE, 16*(1), 70–79.

Cheung, S. O., Tam, C. M., Ndekugri, I., & Harris, F. C. (2000b). Factors affecting clients' project dispute resolution satisfaction in Hong Kong. *Construction Management and Economics, 18*(3), 281–294.

Cheung, S. O., Yiu, T. W., & Suen, C. H. (2004). CoNegO: Construction negotiation online. *Journal of Construction Engineering and Management, ASCE, 130*(6), 844–852.

Chi, L. C., & Tang, T. C. (2006). Bankruptcy prediction: Application of logit analysis in export credit risks. *Australian Journal of Management, 31*(1), 17–28.

Chuang, H. L. (1997). High school youth dropout and re-enrolment behavior. *Economics of Education Review, 16*(2), 171–186.

Construction Industry Review Committee, CIRC. (2001). Tang's report on the Hong Kong Construction Industry Reform, HKSAR Government.

Currie, O. A. (1991). Avoiding, managing and winning construction disputes, the International construction law. *Review, 8*(3), 344–369.

de Wit, A. (1986). *Measuring project success: An illusion, Proceedings of the Project Management Institute*, 13–21.

DeMaris, A. (1995). A tutorial in logistic regression. *Journal of Marriage and the Family, 57*(4), 956–968.

Diekmann, J. E., & Girard, M. (1995). Are construction disputes predictable? *Journal of Construction Engineering and Management, 121*(4), 355–363.

Efron, B. (1975). The efficiency of logistic regression compared to normal discriminant analysis. *Journal of the American Statistical Association, 70*(352), 892–898.

Egan, J. (1998). *Rethinking construction, department of the environment, transport and the regions*. London: HMSO.

Eyden, R. J. V., de Wit, P. W. C. & Arron, J.C. (1995). *Predicting company failure—A comparison between neural networks and established statistical techniques by applying the McNermar Test, Computational Intelligence for Financial Engineering, Proceedings of the IEEE/IAFE* 9–11 April, 91–96.

Fielding, A. H., & Bell, J. F. (1997). A review of methods for the assessment of prediction errors in conservation presence/absence models. *Environmental Conservation, 24*(1), 38–49.

Fisher, J. A. (1988). Prevention is better than cure, or how to avoid costly mistakes in Hong Kong construction contracts. The International Construction Law Review, July/Oct, 342–348.

Flury, B. (1997). *A first course in multivariate statistics*. New York: Springer-Verlag.

Gibson, G. E., & Pappas, M. (2003). *Starting smart: Key practices for developing scopes of work for facility projects*. The National Academy Press, Federal Facilities Council Technical Report #146, p. 87.

Hair, A., Tatham, R. L., & Black, W. C. (1995). *Multivariate data analysis* (5th ed.). Englewood Cliffs, NJ: Prentice Hall.

Harmon, K. M. J. (2003). Effectiveness of dispute review boards. *Journal of Construction Engineering and Management, 129*(6), 674–679.

Hellard, R. B. (1987). *Managing construction conflict*. Harlow, Essex: Longman Scientific Technical Press.

Jannadia, M. O., Assaf, S., Bubshait, A. A., & Naji, A. (2000). Contractual methods for dispute avoidance and resolution (DAR). *International Journal of Project Management, 18*, 41–49.

Jones, L. S. (1991). *Zero-based contracts for dispute avoidance, proceedings of construction Congress*. Cambridge, MA: ASCE, 710–714.

Kleinbaum, D. G. (1994). *Logistic Regression: A self-learning text*. New York: Springer.

Latham, M. (1994). *Constructing the team: Joint review of procurement and contractual arrangements in the united kingdom construction industry: Final report*. London: HMSO.

Lei, P. W. & Koehly, L. M. (2000) *Linear discriminant analysis versus logistic regression: a comparison of classification errors*, Paper presented at the 2000 Annual Meeting of American Education Research Association, New Orleans, LA.

Lowe, D. J., & Parvar, J. (2004). A logistic regression approach to modeling the contractor's decision to bid. *Construction Management and Economics, 22*(6), 643–653.

Manel, S., Dias, J. M., & Ormerod, S. J. (1999). Comparing discriminant analysis, neural networks and logistic regression for predicting species distributions: A case study with a Himalayan river bird. *Ecological Modeling, 120*, 337–347.

McGurr, P. T., & DeVaney, S. A. (1998). A retail failure prediction model. The international review of retail. *Distribution and Consumer Research, 8*(3), 259–276.

McNemar, Q. (1947). Note on the sampling error of the difference between correlated proportions or percentages. *Psychometrica, 12*, 153–157.

Merna, A., & Bower, D. (1997). *Dispute resolution in construction and infrastructure projects.* Hong Kong: Asia Law and Practice Publishing.

Murphy, D. C., Baker, N. N., Fisher, D. (1974). *Determinants of project success*, Grant No. NGR 22-003-028 for the National Aeronautics and Space Administration, Management Institute, School of Management, Boston College, Boston, MA.

Pampel, F. C. (2000). *Logistic regression—A primer.* London: Sage Publications.

Peng, C. Y. & So, T. S. H. (1999). *Analysis issues and computing consideration in logistic regression*, Paper presented at the 1999 annual meeting of Am Ed Res Assoc, Montreal, Canada, April. Available from: http://php.indiana.edu/~tso/articles/aera99.pdf.

Peng, C. Y., Manz, B. D., & Keck, J. (2001). Modeling categorical variables by logistic regression. *American Journal of Health Behavior, 25*(3), 278–284.

Pinnel, S. (1999). Resolution solutions. *Civil Engineering, 69*(6), 62–64.

Pinto, J. K., & Slevin, D. P. (1988). Critical success factors in effective project implementation. In D. I. Cleland & W. R. King (Eds.), *Project Management Handbook.* Princeton, NJ: Van Nostrand Reinhold.

Pohar, M., Blas, M., & Turk, S. (2004). Comparison of logistic regression and linear discriminant analysis: A simulation study. *Metodoloski zvezki, 1*(1), 143–161.

Press, J., & Wilson, S. (1978). Choosing between logistic and discriminant analysis. *Journal of the American Statistical Association, 73*(364), 699–705.

Russell, J. S., & Jaselskis, E. J. (1992). Quantitative study of contractor evaluation programs and their impact. *Journal of Construction Engineering and Management, 118*(3), 612–624.

Salem, A. M., Rekab, K., & Whittaker, J. A. (2004). Prediction of software failures through logistic regression. *The Journal of Information and Software Technology, 46*, 781–789.

Sanvido, V., Parfitt, K., Guvenis, M., & Coyle, M. (1992). Critical success factors for construction projects. *Journal of Construction Engineering and Management ASCE, 118*(1), 94–111.

Sharma, S. C. (1996). *Applied multivariate techniques.* New York: Wiley.

Smith, B. D. (2004). Job retention in child welfare: Effects of perceived organizational support, supervisor support, and intrinsic job value. *Children and Youth Services Review, 27*, 153–169.

Spittler, J. R. & Jentzen, G. H. (1992). Dispute resolution: Managing construction conflict with step negotiations. *AACE International Transactions*, D9, 1–10.

Stockenberg, R. A. (2001). Change orders: An unhappy necessity. *Building Design and Construction, 42*(5), 41.

Tillet, G. (1991). *Resolving conflict: A practical approach.* Australia: Sydney University Press.

Treacy, T. B. (1995). Use of alternative dispute resolution in the construction industry. *Journal of Management in Engineering, 11*(1), 58–63.

Tuman, J. (1986*). Success modeling: a technique for building a winning project team, Proceedings of Project Management Institute*, 29–34.

Tung, Y. K. (1985). Channel scouring potential using logistic analysis. *Journal of Hydraulic Engineering, 111*(2), 194–205.

Turner, R. H. (1994). Avoidance and resolution of construction disputes- prior to and during the construction process. *The International Construction Law Review, 11*, 384–394.

Turney, P. (1995). Cost-sensitive classification: empirical evaluation of a hybrid genetic decision tree induction algorithm. *Journal of Artificial Intelligence Research, 2*, 369–409.

Vallero, D. A., & Vesilind, P. A. (2006). Preventing disputes with empathy. *Journal of Professional Issues in Engineering Education and Practice, 132*(3), 272–278.

Vorster, M. C. (1991). *Dispute review boards not just another alternative dispute resolution method, proceedings of Preparing for Construction in the 21st Century*, 772–776.

Werderitsch, A. J., & Krebs, J. E. (2000). Claims Avoidance—a Project Management Primer. *AACE International Transactions*, CDR1.1–CDR1.5.

Wilsom, R.L. (1992). The ultimate ADR. *AACE International Transactions, American Association of Cost Engineer, 2*, I.4.1–I.4.4.

Wofford, S., Elliott, D., & Menard, S. (1994). Continuities in marital violence. *Journal of Family Violence, 9*(3), 195–225.

Wong, C. H. (2004). Contractor performance prediction model for the United Kingdom construction contractor: A study of the logistic regression approach. *Journal of Construction Engineering and Management, 130*(5), 691–698.

Wuellner, W. W. (1990). Project performance evaluation checklist for consulting engineers. *Journal of Management in Engineering, 6*(3), 270–281.

Yiu, T. W., & Cheung, S. O. (2006). A catastrophe model of construction conflict behavior. *Building and Environment, 41*, 438–447.

Part II
Avoidance: Prevention is Better than Cure

Chapter 6
Dispute Avoidance Through Equitable Risk Allocation

Sai On Cheung

Abstract Equitable risk allocation is considered to be the gateway to dispute avoidance. To this ends, allocation of risks in construction projects should conform to accepted principles. This chapter discusses the use of equitable risk allocation to reduce claims and disputes. An allocation tool that can be used for both risk allocation and evaluation is proposed. The tool is adopted from the one developed by the Public Works Department of the Australian State of New South Wales and employs the Abrahamson allocation principles. The evaluation function of the tool is illustrated by an exercise to unveil the risk allocation pattern of the FIDIC contract. The allocation function is demonstrated by an allocation exercise conducted with construction professionals in Hong Kong.

6.1 Introduction

Equitable and efficient contracts are considered to be the gateways to dispute avoidance. This view has been advocated in a number of industry reviews. For example, the Hong Kong first-ever industry review recommended that risk allocation is one of the areas that should be improved (Construction Industry Review Committee (CIRC) 2001; Levett 2001) as fair risks allocation would reduce the happening of disputes. Similar suggestions have also been forwarded in the industry reviews conducted in the United Kingdom (Egan 1998; Latham 1994). Traditional discrete economic transaction favours 'sharp in by clear agreement, sharp out by clear performance' (Macneil 1974), but many contractual relations are not of this well-defined kind. Hence, contractual transactions and relations need more systematic planning. According to Macneil (1975), two processes are

S. O. Cheung (✉)
Construction Dispute Resolution Research Unit, Department of Civil and Architectural Engineering, City University of Hong Kong, Hong Kong, People's Republic of China
e-mail: saion.cheung@cityu.edu.hk

S. O. Cheung (ed.), *Construction Dispute Research*,
DOI: 10.1007/978-3-319-04429-3_6,

essential to contract planning, namely, defining goals and communications. A perfect contract is often referred to one that anticipates and disposes of all possible future problems and questions. The planning of such a perfect contract is difficult if not impossible, particularly with regard to risks planning.

Risk can be defined as the exposure to the probability of economic or financial loss or gain, physical damage or injury, or delay, as a consequence of the uncertainty associated with pursuing a particular course of action. The concept of risk is a recognition that human beings are in a constant state of partial ignorance about the future. Risks reduce whenever man acquires more knowledge about the occurrence or non-occurrence of future loss (Macneil 1975). Uncertainty represents situation where there is little or no empirical basis for the information of probability distribution (Chapman and Cooper 1987). The above definition indicates that risk has at least two components; risk event and potential loss or gain. Nevertheless, it is common for risk to be considered having negative impact only. The degree of riskiness varies with the complexity, size, and duration of the project. Contractual provisions distribute risks between the parties who, in turn, seek compensation, usually financial, for the risks that they assume. The risk distribution pattern thus has a major influence on contract price. The application of risk management provides explicit recognition of the risks which parties to a construction project are required to take. In extreme cases, one-sided risk allocation can result in a party withdrawing from the proposed scheme. In markets where the contractors have inferior bargaining power, they may have to take on onerous risk-laden contracts. It is not uncommon to find these projects end with major disputes. Unreasonable risk allocation therefore laid the seed of dispute.

6.2 Risk Management as a Decision Making Process

Head and Horn (1985) suggest that risk management can be implemented as a 5-stage decision-making process: (i) identifying exposures to loss; (ii) examining feasibility of alternative techniques; (iii) selecting apparent best techniques; (iv) implement the chosen techniques and (v) monitoring and improving the risk management programme. The first step involves liability exposure identification and analysis. Liability exposures can be property, income, liability or personnel. Notable methods to identify these exposures include using standardised survey, examining financial statements, searching through records and files, drawing flowcharts, personal inspection and seeking expert advice. Impact analyses shall then be performed to evaluate the degree of riskiness against organisational objectives. Furthermore, the overall impact of a certain risk can then be assessed in terms of its occurrence likelihood and loss severity. Those risks that are having high chance of happening as well as bringing significant loss top the list for special attention and treatment. Risks can either be controlled or financed. The aim of control measures is to reduce chance of happening. If complete avoidance is not feasible, the risk may have to be retained or transferred through financial or contractual arrangement as appropriate. The third step is to choose the apparent best techniques. There is certainly no obvious solution and each case will have to be examined in its own facts. It is a balanced decision of deriving the desired outcome through effective use of risk control; risk financing or a combination of both. The fourth step is to devise plan for implementation. This is often the major stumbling block. Resistance to change has been proven to be more difficult to succumb than technicalities as far as implementation is concerned. For all management processes, feedback is essential. Not just to evaluate the performance of the system, feedback can help the organisation to learn and improve (Wong et al. 2007). In the long run, an effective feedback system should enable an organisation to map out strategies to better manage the risks. For example, feedback shall provide data for more accurate assessment on occurrence likelihood of the risk events.

6.3 Risk Classification

Risks have been described as pure, speculative, static, and inherent (Doherty 1985; Greene and Serbein 1983; Grose 1987; Moore 1983). In construction, a number of classifications have also been suggested. Mason's (1973) classification emphasises the nature of obligations and losses. Erikson (1979) categorised risks in construction as being either contractual or construction oriented. The list of risks by Abrahamson (1984) highlights the potential risk areas. The classification of Casey (1979) is more systematic. Broadly, four categories of risk are proposed; physical, capabilities-related, financial & economic and political & societal. The risks identified by Casey (1979) are adapted for use in this study because these risks are characterised as being inherent in construction projects and exclude those created by the parties themselves. The left most column of Table 6.1 lists the risk events included in this study. The remainder of the table will be explained in subsequent sections of this chapter.

6.4 Risk Allocation

Risk allocation has been identified as one of the important strategies that the Australian construction industry should improve (NPWC 1990). Risk allocation determined on a sound commercial basis would reduce the occurrence of dispute (NPWC 1990). Unrealistic risk shifting is a major cause of dispute (Wall 1994). This is because construction business is highly competitive and many contractors are willing to take on projects where the risks have not been adequately priced for. If these risks did eventuate, the contractors are tempted to recoup the losses incurred by raising and protracting claims and disputes.

Allocation principles have been suggested by a number of researchers (Abrahamson 1984; Ashley 1977; Barnes 1983; Erikson and O'Connor 1979; Perry and

Table 6.1 Risk allocation pattern of studies in Hong Kong: An example of the use of the tool

Events	Risk allocation principles					Total score	Risk allocation
	P_1	P_2	P_3	P_4	P_5		
Physical							
Giving possession of site	5	5	5	5	5	25	E
Latent conditions	3	5	5	5	3	21	E
Inclement weather	3	3	3	3	3	15	S
Force majeure	3	3	3	3	3	15	S
Inadequate design	5	5	5	3	5	23	E
Errors and omissions in quantities	5	5	3	3	5	21	E
Capabilities-related							
Defective works	1	1	1	1	1	5	C
Theft and vandalism	1	1	1	1	1	5	C
Default of suppliers and subcontractors	1	2	2	1	1	7	C
Labour injuries and accidents	1	2	1	2	1	7	C
Productivity of labour and equipment	1	1	1	1	1	5	C
Financial and economic							
Inflation	3	4	3	4	3	17	S
Availability of labour and equipment	1	1	1	1	1	5	C
Political and societal							
Changes in laws and regulations	3	4	3	3	3	16	S
Public disorder	3	4	4	3	3	17	S
Labour disputes and strikes	2	2	3	2	3	12	S

Hayes 1985; Porter 1981; Thompson and Perry 1992). The five principles suggested by Abrahamson (1984) are considered as the most embracing and therefore are used in the allocation tool (NPWC 1990). Modifications in terms of using plain English instead of legal phraseology are effected for ease of understanding by the users. The five risk allocation principles used in this study are listed below.

Risk should be allocated to:

P_1 : The party who can best control the risk effectively.
P_2 : The party who can best be able to undertake the risk financially.
P_3 : The party who has the most information to forecast the risk.
P_4 : The party who can benefit most in controlling the risk.
P_5 : The party with whom the risk is inherent in its commercial role.

Risk distribution based on the above principles was implicitly approved by the House of Lords in *Photo Production v. Securior Transport 1980* (Furmston 1986) per Lord Diplock's statement, *"it is generally more economical for the person by whom the loss will be directly sustained to do so rather than be covered by the other party by liability insurance"*. These principles are also sustainable under the economic concepts of efficiency and value maximising (Harris and Veljanovski 1986).

Lloyd (1996) succinctly summarises that risk allocation in construction is a matter of assigning responsibilities in the light of ability to control, to foresee or to manage the said risk. Risk allocation principles can well be explained with this conceptualisation.

6.5 Risk Allocation Under FIDIC Conditions of Contract

FIDIC is the French acronym for the International Federation of Consulting Engineers. Founded in 1913, the FIDIC aims to promote in common the professional interests of the members associations and to disseminate information of interest to members of its component national associations. The publications of FIDIC include standard pre-qualification forms, contract documents and client/consultant agreements. The FIDIC contract is widely used in international projects. The risk allocation pattern of the FIDIC contract is examined by a desktop analysis with respect to the risks categorisation developed by Casey (1979). The FIDIC contract referred in this study is the 1999 edition of the Conditions of Contract for Construction for Building and Engineering works designed by the Employer. A brief account on the relevant clauses is given below and a tabulated summary is presented in Table 6.2.

6.5.1 Physical Risks

Under Clause 2.1, the Employer shall give the Contractor right of access, and possession of, all parts of the Site within the time (or times) stated in the Appendix to Tender. …… If no such time is stated in the Appendix to Tender, the Employer shall give the Contractor right of access to, and possession of, the Site within such times as may be required to enable the Contractor to proceed in accordance with the programme. Under Clause 4.12, "physical conditions" means natural physical conditions and man-made and other physical obstructions and pollutions, which the Contractor encounters at the Site when executing the Works. These include sub-surface and hydrological conditions but not climatic conditions. Thus latent condition is included under Clause 4.12. If the Contractor encounters physical conditions which are unforeseeable, and suffers delay and/or incurs cost due to these conditions, the Contractor shall be entitled to extension of time and such cost. Procedural requirements such as proper serving of notices shall apply. Climatic condition is not considered as physical conditions under the FIDIC contract. Moreover, exceptional adverse climatic condition is one of the causes of delay that would entitle the Contractor to an extension of time for completion under Clause 8.4. and additional payment under Clause 20.1.

Force Majeure is defined in detail under Clause 19. Force Majeure means an exceptional event or circumstance:

Table 6.2 The risk allocation pattern under FIDIC

Events	Clause No.	Risk Allocation
Physical		
Giving Possession of site	2.1	E
Latent conditions	4.12	E
Inclement weather	8.4	E
Force Majeure	19	E
Inadequate design	13.1	E
Capabilities-related		
Defective works	11.2	C
Theft and vandalism	4.22	C
Default of suppliers and subcontractors	4.4	C
Labour injuries and accidents	4.8	C
Productivity of labour and equipment	4.17	C
Financial and economic		
Inflation	13.8	E
Availability of labour and equipment	4.1	C
Political and societal		
Changes in laws and regulations	13.7	E
Public disorder	19	E
Labour disputes and strikes	19	E

E: Risk to be borne by the principal
S: Risk to be shared between the principal and the contractor
C: Risk to be borne by the contractor

(a) which is beyond a party's control,
(b) which such party could not reasonably have provided against before entering into the Contract,
(c) which having arisen, such party could not reasonable have avoided or overcome, and
(d) which is not substantially attributable to the other party.

Under those circumstances, the party can be excused from performance of such obligations for so long as such Force Majeure prevents it from performing them. The Contractor who is affected by Force Majeure is entitled to extension of time for completion and additional payment.

Instructions by the Engineer to change the quality and other characteristics of any item of work shall be treated as variations (Clause 13.1). The contractor is entitled to both time and cost reimbursement if justified.

6.5.2 Capability Related Risks

The risk events listed under this category are related to the ability of the contractor to complete the work. These include the provision of plant and labour to carry out the work in such a manner that the contractor's obligations will be honoured

(Clause 4.17). Works carried out by subcontractors and materials supplied by supplier should be properly monitored. The Contractor is responsible for any default of his subcontractors and suppliers (Clause 4.4). Any defective work shall be rectified by the Contractor at his own costs (Clause 11.2). Security of the site and protection against theft and vandalism are also the Contractor's responsibility (Clause 4.22).

6.5.3 Financial and Economic Risks

The contractor is responsible for the availability of the resources that are necessary for the proper completion of the works (Clause 4.1). Under Clause 13.8, the amounts payable to the Contractor shall be adjusted for rises or falls in the cost of labour, Goods and other inputs to the Works. The amounts are determined by applying a general formula that includes change in indices on labour, equipment and materials.

6.5.4 Political and Societal Risks

Clause 13.7 of the contract allows the contract price to be adjusted to take account of any increase or decrease in cost resulting from a change in the Laws of the Country (including the introduction of new Laws and the repeal or modification of existing Laws) or in the judicial or official governmental interpretation of such Laws, made after the Base Date, which affect the Contractor in the performance of obligations under the Contract. If delay is resulted, extension of time is also allowed. The Contractor therefore does not need to take the risk of changes in the Laws.

Public disorder is treated as Force Majeure as this is beyond the reasonable control of the contractor (Clause 19). The Contractor is entitled to time adjustment and cost recovery. Labour dispute & strikes has to be beyond the Contractors' own labour issues in order to qualify as a Force Majeure event under Clause 19.

6.6 Pattern of Equitable Risk Allocation: A Hong Kong Study

Contract planners can use the risk allocation tool to establish the allocation pattern before actual drafting of the conditions. A risk allocation tool described here follow is used to elicit the pattern of risk allocation of construction professionals in Hong Kong. Table 6.1 presents the basic structure of the tool used in this study.

Table 6.3 Comparison of risk allocation: survey data and FIDIC

Events	Risk allocation pattern under FIDIC	Pattern obtained from the Hong Kong study
Physical		
Giving Possession of site	E	(19) E
Latent conditions	E	(13) S
Inclement weather	E	(14) S
Force majeure	E	(15) S
Inadequate design	E	(22) E
Capabilities-related		
Defective works	C	(8) C
Theft and vandalism	C	(8) C
Default of subcontractors and supplier	C	(7) C
Labour injuries and accidents	C	(8) C
Productivity of labour and equipment	C	(7) C
Financial and economic		
Inflation	E	(13) S
Availability of labour, materials and plants	C	(8) C
Political and societal		
Changes in laws and regulations	E	(14) S
Public disorder	E	(15) S
Labour disputes and strikes	E	(12) S

The scores in each row against each risk event are then totaled and the interpretation of the total score is taken as:
Score of 5–10 Contractor's obligation/risk
Score of 10–20 Shared obligation/risk (score close to 10 indicates the contractor has a bigger share of the obligation/risks though the risk is shared, similarly, the principal assumes a bigger share of obligation/risk if the score is close to 20)
Score of 20–25 Employer's obligation/risk

Risk events are listed on the left most column and the risk allocation principles are listed on the top most row. Each of the respondents was requested to give scores against each risk event. A score in the range of 1–5 was to be assigned under each of the allocation principles listed. A minimum score of 1 and a maximum score of 5 was possible for each risk event examined under each allocation principle. A score of 1 indicates a presumption that the event is clearly the responsibility of the contractor; 2 that it is more than 50 % the contractor's responsibility; 3 that it is a neutral event between contractor and principal's responsibility; 4 that more than 50 % the principal's responsibility; 5 that it is clearly the responsibility of the principal. Table 6.1 gives an example of data provided by one of the respondents. The average scores obtained for the 230 respondents of the study are presented in the right most column of Table 6.3.

6.7 Discussion

The average scores for the risks are presented in Table 6.3. The score against each risk event serves as indication of the risk distribution preference. For ease of comparison, the risk distribution pattern in the FIDIC contract is also presented in the middle column of Table 6.3, together with the Hong Kong data. The "Construct for Excellence" document recommends that a contracting party should bear a risk that is under his control and unrealistic shifting of risk will increase likelihood of dispute (Construction Industry Review Committee (CIRC) 2001). Ravey (1992) is also of the same opinion that a more balanced agreement is a preferable option as a mean of avoiding conflicts. The relationship between risk allocation and incidences of dispute can be considered in the light of a life cycle model of conflict (Gardiner and Simmons 1992). Divergences of interest, value are sources of dispute (De Bono 1985). These include the difference in perceptions regarding risk ownership. In other words, the existence of a risk allocation pattern that is considered by either party to a contract as inequitable may create a conflict situation. Usually, a contractor, to whom onerous risks are shifted, will feel aggrieved. This creates tension and stress between the contractor and the employer and attitude towards problems that may arise will become more confrontational and less cooperative. Manifestation of such conflict happens when such risks eventuate. If the contractor has not made sufficient coverage or allowance for such risks, which often being the case, he will seek to redress the losses through all possible channels, typically through raising claims and disputes (Lewis and Carter 1992). Unrealistic risk shifting creates tension and hinders cooperation between the contracting parties. Disputes are, in many instances, manifestation of such conflict.

Under the category of physical risk, there is no variance between the result of the survey and the desktop analysis of the FIDIC contract for possession of site and inadequate design. The responsibilities for latent conditions, inclement weather and force majeure have always been controversial. In Hong Kong, Employers would like to shift this risk towards the contractor. However, contractors usually are not given sufficient time to carry out investigation work to enable a proper risk assessment. It is of interest that a share strategy is generally preferred in Hong Kong for both financial and political risks. The only exception is availability of labour, materials and plants. Generally where these risks materially affect the progress, time adjustment is allowed but with no loss and expense. It is quite different in the case of FIDIC where the Employer takes up responsibility. It can be said that the FIDIC approach would enable the contractor to putting in too high or too low an allowance for these risks. An inadequate allowance laid the seed for dispute should the risks materialise.

The capabilities related category shows great consistency. The risk events under this category are either under the control of the contractor or considered inherent in the contractor's commercial role. Therefore, there is no doubt that the contractor should be the risk bearer. The survey data well illustrates this.

It can be concluded that the risk allocation in the FIDIC contract acknowledges the fact that the site is owned by the employer. The employer therefore has much more time to assess the risks arising from the site conditions, be it physical or latent. Asking the tendering contractors to do the same tasks within the tender period is unrealistic and uneconomical. Furthermore, for those risks that the contractor has no control, the employer allows both time and cost adjustment if FIDIC contract is used. Although sharing is an appealing approach, its realisation by giving time redress but no monetary compensation appears to be a strategy of convenience.

6.8 Chapter Summary

The methodology used in this study is inherently subjective. However, the risk allocation tool used in this study can be used as a starting point to determine which risks to be borne by which parties. Parties to a contract can use the model as a guide in risks planning/allocation during contract negotiation. It can also be used as a risk identification and assessment tool during tender preparation. Systematically listing of risk events is useful for risks identification. The list of risk events used in this study is by no means exhaustive and can be extended or reduced to suit the particular project under scrutiny. As for risks allocation, with greater number of respondents, the risks allocation pattern so derived can be taken as a fair representation of the industry view of an acceptable 'equitable allocation'. In this study, it is found that the FIDIC contract allocates risk to the party who has the better ability either to foresee, control or manage such risks.

References

Abrahamson, M. W. (1984). Risk Management. *International Construction Law Review, 1*(3), 587–598.

Ashley, D. B. (1977). *Construction project risk sharing,* Construction Institute Technical Report 220. Stanford: Stanford University.

Barnes, M. (1983). How to allocate risks in construction contracts. *International Journal of Project Management, 1*(1), 24–28.

Casey, J. J. (1979). *Identification and nature of risks in construction project: A Contractor's perspective, Construction Risks and Liability Sharing, American Society of Civil Engineers Conference Proceedings* (Vol. 1).

Chapman, C. B., & Cooper, D. F. (1987). *Risk Analysis for Large Projects: Models, Methods and Cases.* Chichester: Wiley.

Construction Industry Review Committee (CIRC). (2001). Construct for Excellence, Construction Industry Review Committee, the Government of the Hong Kong Special Administrative Region.

De Bono, E. (1985). Conflicts: A better way to resolve them. New York: Penguin Books.

Doherty, N. A. (1985). Corporate risk management. New York: McGraw Hill.

Egan, J. (1998). *Rethinking construction, department of the environment, transport and the region*. London: HMSO.

Erikson, C. A. (1979). Risk Sharing in Construction Contracts. PhD Thesis, Civil Engineering, Graduate College of University of Illinois, Urbana-Champaign.

Erikson, C. A. & O'Connor, M. J. (1979). *Construction contract risk assignment*, U.S. army technical report P-101. Construction Engineering Research Laboratory.

Furmston, M. (1986). The Liability of Contractors: Principles of Liability in Contract and Tort. In H. Lloyd (Ed.), *The Liability of Contractors*. London: Centre for Commercial Law Studies, Queen Mary College.

Gardiner, P. D., & Simmons, J. E. L. (1992). Analysis of Conflict & Change in Construction Projects. *Construction Management and Economics, 10*(6), 459–478.

Greene, M. R., & Serbein, O. N. (1983). *Risk management, text and cases*. New Jersey: Prentice Hall.

Grose, M. R. (1987). *Managing risk, systematic loss prevention for executives*. New Jersey: Prentice Hall.

Harris, D. R. & Veljanovski, C. G. (1986). The use of economics to elucidate legal concepts: The law of contract. In T. Daintith & G. Teubner (Eds.), *Contract & organisation: Legal analysis in the light of economic and social theory*. Berlin: Walter de Gruyter & Co.

Head, G. & Horn, S. (1985). *Essentials of the risk management process* (1st ed.). Malvern: Insurance Institute of America.

Latham, M. (1994). Constructing the team: Final Report of a Joint Review of Procurement and Contractual Arrangements in the United Kingdom Construction Industry, HMSO.

Levett, D. (2001). Drafting and risk allocation in the new Hong Kong Standard Form of Building Contract. *The International Construction Law Review, 18*(4), 717–723.

Lewis, J., Cheetham D. W. & Carter, D. J. (1992). *Avoiding conflict by risk management—the role of the client's project manager: Proceedings of the first international construction management conference* (pp. 72–94). The University of Manchester Institute of Science and Technology.

Lloyd, H. (1996). Prevalent Philosophies of risk allocation- an overview. *The International Construction Law Review, 13*(4), 502–548.

Macneil, I. R. (1974). The many futures of contract. *Southern California Law Review, 47*, 691.

Macneil, I. R. (1975). A primer of contract planning. *Southern California Law Review, 48*, 627.

Mason, G. E. (1973). *A quantitative risk management approach to the selection of construction contract provisions* (pp. 26–61). Technical Report No. 173. The Construction Institute, Department of Civil Engineering, Stanford University.

Moore, P. (1983). *The business of risk*. Cambridge : Cambridge University Press.

NPWC (1990). No Dispute: Strategies for Improvement in the Australian Building and Construction Industry. *National Public Works Conference, Australia*.

Perry, J. G. & Hayes R. W. (1985). Risks and its management in construction projects. *Proceedings of Institute of Civil Engineers, Part I, 78*, 499–521

Porter, C. E. (1981). Risk Allocation in Construction Contracts. MSc Thesis, University of Manchester Institute of Science and Technology.

Ravey, S. G. (1992). *Can construction claim be avoided? In Proceedings of the First International Construction Management Conference* (pp. 202–208). The University of Manchester Institute of Science and Technology.

Thompson, P. A., & Perry, J. G. (1992). *Engineering Construction Risks*. London: Thomas Telford.

Wall, C. J. (1994). Dispute prevention and resolution for design and build contracts in Hong Kong. In *Proceedings of CIB W92 Symposium EAST MEETS WEST*, University of Hong Kong.

Wong, S. P., Cheung, S. O., & HardCastle, C. (2007). Embodying learning effect in performance prediction. *The Journal of Construction Engineering and Management, 133*(6), 474–482.

Chapter 7
Trusting Behaviours in Construction Contracting

Sai On Cheung, Pui Ting Chow and Peter Shek Pui Wong

Abstract Construction contracting is typically subjected to monitoring and control mechanisms. This policing contracting environment reflects the inherent distrust among the contracting parties. Paradoxically, construction project teams work best in a trusting environment because the members are mutually dependent on each other. Furthermore, where flexibility in performance is needed as in case of long-term or complex projects, trust is the necessary condition to suppress opportunism and avoid dispute. Whilst engendering trust is advocated in many industry reviews, trust in construction contracting is a myth remains the majority view within the construction community. This chapter explores the existence of trust in construction projects. Observations of trusting behaviours in construction contracting are solicited and analysed to unveil the bases on which trust can develop.

7.1 Introduction

Construction contracting environment is typically characterised by defensive and uncompromising behaviour (Lyons and Mehta 1997). This environment reflects the inherent distrust among the contracting parties. However, accomplishing construction tasks requires the coordinated efforts of members of project teams.

S. O. Cheung (✉) · P. T. Chow
Construction Dispute Resolution Research Unit, Department of Civil and Architectural Engineering, City University of Hong Kong, Hong Kong, People's Republic of China
e-mail: saion.cheung@cityu.edu.hk

P. T. Chow
e-mail: ronnie.chow.pt@hotmail.com

P. S. P. Wong
School of Property, Construction and Project Management, RMIT University, Melbourne, Australia
e-mail: peterspwong@rmit.edu.au

S. O. Cheung (ed.), *Construction Dispute Research*,
DOI: 10.1007/978-3-319-04429-3_7,

A trusting environment is conducive for team performance. Furthermore, where flexibility in performance is needed as in cases of long-term or complex projects, trust is a necessary condition to suppress opportunism and avoid dispute (Lyons and Mehta 1997; Williams 2001; Zand 1972). In a trustor-trustee paradigm, the employer (trustor) is willing to take risks on the ability of the contractor (trustee) in completing the project (Gambetta 1988; Mayer et al. 1995; Williams 2001; Zand 1972). In addition, trust signifies the faith between the trustor and trustee that they will not act opportunistically (Gulati 1995; Johnson-George and Swap 1982; Jones and George 1998; Lewis and Weigert 1985; Mayer et al. 1995). To what extent a trustor can trust a trustee is gauged by the trustworthiness of the trustee (Butler 1991; Gabarro 1978; McAllister 1995; Williams 2001). What are the measures of trustworthiness? Many still maintain that trust is a myth in construction contracting. This chapter explores the existence of trust in construction projects. Observations of trusting behaviours in construction contracting are solicited as a means to understand trust-building (Currall and Judge 1995; Webber and Klimosk 2004).

7.2 Observations of Trusting Behaviours in Construction

Construction contracting involves transactions cutting across different professional interfaces (Williamson 1985). Members of these interfaces like architects, surveyors and engineers are connected through various procurement arrangements within a multidisciplinary project team. Asset specificity is high in teams where relocation entails considerable switching costs (Williamson 1985). The assets, the works completed for the project, are durable and have a much higher value than the opportunity cost of the best alternative. Thus, members rarely relocate their assets for alternative usage without scarifying productive value. Moreover, fear of relocation induces distrust among project team members (Williamson 1985). The extent of surveillance and control will thereby step up. As a result, transaction cost will increase. Once distrust is rooted, members become suspicious of each others' underlying motives. Further distrusting behaviour like cautious defense against betrayal would make joint problem solving more difficult. Without trust, members will take self-interest seeking stances. As such, opportunistic acts are common in construction contracting (Cheung and Pang 2013).

Is trust a solution to guard against opportunism? Do employers and contractors value trust (Kadefors 2004)? Are they willing to build trust and cooperate for the sake of the projects? These questions collectively raise thoughts on the paradoxes of trust. Mutual trust may just be a unilateral sincere wish (Dasgupta 1988; Gambetta 1988; Luhmann 1979). Even when both parties are committed, trust takes years to develop but just needs one incidence to destroy. Once this happens, the team can rarely work together in the same trusting environment as before (Cheung et al. 2008). Project teams need stability that builds on security, reliability and certainty. With trust, project team is able to tackle challenges and uncertainties with concerted efforts. The presence of trust is important to a project team, but its development

Table 7.1 Particulars of the projects discussed for the study

P	Nature	Inter-organisation interaction history	Contract sum[a]	CFA[b]
1	An 28-sorey office building	5–6 year	1,300	70
2	A multifunction hall	More than 20 years	1	
3	An 12-storey office building	None	75	13
4	An office building	None	17	4
5	A park	None	300	30
6[c]	A residential estate	Having no previous cooperation history, but top-management members have previous interaction	380	223

Note: P: Project
Contract sum[a]: Approximate in Million RMB; CFA[b]: Construction Floor Area in Thousand m^2; Location of the project: Shanghai except project 6[c] Hangzhou

requires unwavering and enduring effort. Observations of trusting behaviour in construction contracting may unveil the underlying bases of trust-building.

Interviews were conducted to collect trust stories for this study. Interview provides flexibility for researchers to examine the various contexts and insights from the interviewees (Kaplan and Maxwell 1994; Punch 2005). The basic concepts, contributions and deterrence of trust, manifestations such as cooperation and opportunism can then be explored. A total of six dyadic pairs of employer-contractor (Contractors: C1–C6; Employers: E1–E6; Projects: P1–P6) participated in this study. Senior management of these organisations were contacted and were briefed about the research objectives. The 12 interviews were conducted in Shanghai. The information regarding the level of trust measurement mainly collected from front-line project managers who are responsible for the day-to-day operation of construction projects (Geringer and Hebert 1991; Janowicz-Panjaitan et al. 2009; Zaheer et al. 1998). Background of the interviewees and their organisations, project information and team members' portfolios were collected in each of the interviews. Furthermore, the following perceptive views of the interviewees were solicited: (1) the trust-development process and the nature of inter-organisational success, (2) the project management practices, in particular, their nature of works and interactions, and (3) the identification of risk and uncertainty faced by the project teams. Open-ended questions were used. Where possible and appropriate, examples were solicited to illustrate and confirm the views expressed. The interviewees were also asked to draw mind-maps on trust. A summary of the projects discussed in the six interviews is given in Table 7.1.

All the participating organisations are developers and contractors active in the market. Some of the interviewees expressed appreciation of their team members' effort in developing trust. Unsurprisingly, most of them talked about the risks, challenges and difficulties in rendering trust. Similar findings have been widely reported in trust literatures. While most of them are unable to describe their trust stories in full, trusting behaviours could be observed. Based on these observations, five antecedents of trust are identified. These are commitment, risk-taking, knowledge, honesty and benevolence.

7.2.1 Commitment

Commitment can be used as a trust identifier. "Can do" and "will do" are the two major manifestations of commitment. The former refers to the skills, competence and abilities that enable a party to fulfill his promises (Mayer et al. 1995). The latter refers to the willingness of the party to do so. Co-operation has also been noted by the interviewees as a form of commitment. When there is a cooperative relationship between employer and contractor, they would explain their expectations to each other clearly. If necessary, adjustments of these expectations can be made for the attainment of mutually acceptable outcomes (Bennett and Jayes 1998). The degree of interference from a trustor on a trustee is inversely related to the level of commitment. For example, a trustor who frequently reminds a trustee to perform a particular task projects an untrustworthy outlook of the trustee. Some interviewees suggested that a commanding party often distrusts his counterpart (Lau and Rowlinson 2009). The following are some relevant comments from the interviewees.

Interviewee E1: *I look for well known contractors. The contractor for this project has participated in many landmark projects, like World Expo,* etc. *His competence and abilities are well-proven.*

Interviewee E1: *I am serious about safety issues and prefer hands-on management. Solely relying the contractor on safety issues is risky to me.*

Interviewee E2: *The contractor is specialised in renovating small buildings. This specialty nicely fits the need of my kindergarten project. He is committed to the project. He has been working on this type of work for more than 20 years.*

Interviewee E2: *The contractor identified the safety hazards that may arise and proposed some useful and practical ways to deal with them.*

Interviewee C2: *Maintaining a cooperative relationship is of higher priority than making money.*

Interviewee C3: *Although the employer seldom pays late, his performance is still unsatisfactory because of the unrealistic expectations.*

7.2.2 Risk-Taking

It has been suggested that construction contracting relationship is inherently distrustful (Hanna 2007). Factors influencing initial level of distrust include risk allocation, personality, cognition and social categorisation processes, role-based behaviour and reputation of the party concerned (Lewicki et al. 2006). The most influential one is risk-taking according to Mayer et al. (1995). All interviewees indicate that as construction projects are becoming more and more complex, uncertainty is high. Thus, trust is very meaningful to handle inter-organisational and inter-disciplinary teams dynamic. Trust provides the springboard for "quantum leap" in performance. Lewis and Weigert (1985) describe this as "beyond the expectations that reason and experience alone would warrant". This is particularly true in crisis situation. Theory of bounded rationality prescribes that decision to trust depends on the potential risk (costs and benefits) and the probability of reciprocity (temptation). Moreover, there is no guarantee that cooperation will succeed, in particular when the contract is incomplete. In extreme situations, even when the circumstances appear to promise mutual gain (benefits > temptation), irrational exploitation would happen (Lorenz 1999). Trust helps members to understand such situations, make informed decisions and reinforce trusting ambience of the project team. The proffer of evidence to trust can be explicit (independent status) and/or subtle (unintentional, nonverbal or body language) (Lyons and Mehta 1997).

Interviewee C1: *The employer has sound knowledge in construction. Communication is thus easier The risks of misunderstanding is low.*

Interviewee C1: *I won't say that I totally trust the employer. No matter how well we cooperate, incompatible interests remain.*

Interviewee E3: *Having trust is academic and too abstract. I think cooperation can only be based on contract requirements. The market is too complicated. I follow the market rules instead of the theory of trust.*

Interviewee E3: *It is the practical need to cooperate rather than because of trust.*

Interviewee C4: *I am more vulnerable in the relationship since the employer is in a better position to bargain. To a greater extent, I cannot afford to trust the employer where huge money is at stake. I can rely on the contract which is the only link between the two organisations.*

Interviewee E6: *We shared the objective of effectively and efficiently complete the project with good quality. We focused on reducing risks in the project.*

7.2.3 Knowledge

Some interviewees suggested that they have to work with some opportunistic counterparts. Moreover, knowing that the counterpart is an opportunist may reduce uncertainty (Lewis and Weigert 1985). If one party can predict his counterpart's action, he would prepare for it (Gabarro 1978). However, some of the interviewees disagreed that predictability is a trusting factor. Being predicable does not necessarily trigger the willingness to take risks that underpins trust (Mayer et al. 1995). In the context of inter-organisational relationship, predictability to a large extent depends on an organisation's control mechanisms or company policy (Friedland 1990). Reputation and previous working experience are typical examples of sources of knowledge (Burt and Knez 1996). Thus, knowledge is perhaps just the top soil on which trust may be placed.

Interviewee C1: *We share similar beliefs and we have developed very strong working relationship. Previous pleasant and productive dealings engender our trusting relationship.*

Interviewee E4: *Trust is ancillary as I focus on contractual arrangement to deal with the contractor.*

Interviewee E5: *I don't trust him because I do not know him. When conflicts arise, I am not sure what he will do.*

Interviewee C6: *We have no previous interaction, but the employer's attitude towards the project impressed me a lot. He did focus on how to carry out the works. His professionalism makes him predictable, reliable and trustworthy.*

7.2.4 Honesty

One of the interviewees highlighted that the most common form of inter-organisational communication between employer and contractor is through regular meetings. However, this setting does not necessarily lead to open communication because of the "formality" and lacking of real communication. In the context of trust, the information exchanged in these meetings is of low quality (Lau and Rowlinson 2009). Truth-telling or fault-finding leads to very opposite results. Forgiveness is a considerate and appropriate response for honesty. A trustor would only share sensitive information effectively if favorable responses are reciprocated. Through effective communication, uncertainty is minimised where the parties can honestly talk and negotiate. Solving problems in this way retains the relationship and there would be fewer disputes (Lau and Rowlinson 2009).

Interviewee E1: *He is honest. It is a very important criterion in contractor selection for public projects. Face-to-face site meetings were held every week together with the supervisor. Meeting is the most preferred form of communication. In addition, we have telephone calls very frequently. We rarely use E-mail.*

Interviewee E2: *The contractor's behaviour is consistent with his words. I rarely find an honest man like him nowadays.*

Interviewee C2: *The users of the kindergarten are mostly small kids. We need to clarify every issue to avoid any tiny mistake that may cause injuries to those lovely kids. We need honest opinions.*

Interviewee C3: *His lack of knowledge in construction makes communication ineffective. He even refuses to listen to my recommendations.*

Interviewee E5: *It is impossible for me to trust him because he will not tell me what has gone wrong.*

Interviewee C5: *He often by-pass me and refers the conflicting issues to the senior management. I had to pay extra effort to communicate with him. However, he is not willing to accept my advice on technical issues despite his lack of knowledge in construction.*

7.2.5 Benevolence

Benevolence is grounded in helping acts of a trustor (Lewicki et al. 2006). Some of the interviewees gave some examples but also mentioned that it was never easy for the parties not to take advantage of one another. The employers and contractors have their own motives and objectives that are often incompatible. Viewing parties' interests as mutually opposite inhibits trust-building (Mayer et al. 1995).

Interviewee E1: *We have a culture of emphasising relationship. We are flexible to offer the contractor some help if necessary. For example, we made advanced payment to the contractor before Chinese New Year. We also offered financial support to the contractor when he was in financial difficulty.*

Interviewee E1: *The development project faced opposition from some neighbours. The contractor took over the site and dealt with the residents, even he did not have to do so. We would be in trouble if the complaints were not resolved.*

Interviewee E2: *The contractor cares about the kindergarten project. He is always available to offer some valuable advices whenever it is needed. He is very patient to evaluate and explain each advice. He even helped me to renovate the old building of the kindergarten, including replacement of those old pipes, which was outside the scope of the original contract works.*

Interviewee E2: *I called the contractor 1 day at midnight, requesting for help to fix the emergency water leakage of the old building. He did me a favor without asking for return. We care the well-being of each other.*

Interviewee C2: *The employer is so nice that he provided us with lunch, dessert and spaces for rest during the project. We care the well-being of each other.*

Interviewee E2: *The contractor always wants to earn more and I expect a building with higher quality and lower cost. That's our relationship. Conflict is inevitable.*

Interviewee E2: *The contractor was willing to finish the works which were not detailed in the contract or the drawings. He also helped a lot in the site handover.*

Interviewee C4: *The employer was too self-centered and unwilling to listen to our views. He thinks that he is always right.*

7.3 Discussion

Trust reduces uncertainty over future outcomes and its presence facilitates swift decision-making (McAllister 1995). Trusting behaviours collected from the interviewees are informative. The five trust antecedents can be explained in the light of the three trust bases suggested by McAllister (1995). These are system-based, cognition-based and affect-based.

7.3.1 System-Based

The development of system-based trust focuses on formalising the procedural arrangements to reduce the reliance of emotional and personal influences (Lewis and Weigert 1985). System-based trust emphasises trust on the integrity of system rather than trust in a particular person. Its influence on contracting relationship is found on the creation of a trusting platform. As an illustration, this form of trust assimilates the faith one puts in the legal system. For example, the use of formal contract can trigger trust on the belief that the contracting parties will honor their contractual commitments in view of the safeguards and remedies allowed therein. In an organisational context, accounting and approval systems are warranty of performance. The parties' behaviours are expected to conform to these systems. In this regard, commitment is a manifestation of system-based trust. This form of trust is of particular importance at the commencement stage of a project, as the contracting parties agree to assure their performance by installing systems to monitor one another behaviours.

7.3.2 Cognition-Based

Cognition-based trust is grounded in reliability, dependability as well as competence (Lewis and Weigert 1985). A trustor believes that a trustee is able to provide quality product/service in a timely fashion. In a construction project, an employer always wishes to award contract to a trustworthy contractor (McAllister 1995, Morrow et al. 2004). When determining whether a person, a group or an organisation is trustworthy, a trustor will evaluate the prospective trustee (Morrow et al. 2004). At the early stage of a relationship, members typically assess how the other members are likely to behave in a given situation. Over time, cognition-based trust is then built on the members' reputation, behaviour stability and consistency. In other words, the member cognitively evaluates the competence of other project team members with the information available (Morrow et al. 2004; Rousseau et al. 1998). Any mechanism that enriches such information supports the development of cognition-based trust. Cognition based trust therefore grows overtime as the project progress as both contracting parties gain better knowledge of each other.

7.3.3 Affect-Based

Affect-based trust manifests as reciprocal interpersonal care, concern and emotional bonds (Lewis and Weigert 1985; McAllister 1995). A trustor is making emotional investments in trust relationships when genuine care and concern for the welfare of the trustee is expressed. The trustor is instilling intrinsic virtue of such relationships and believes that these sentiments will be reciprocated (Bachmann 2001; McAllister 1995). Emotional bond will then develop if the trusting cycle does materialise. This bond will link members together and provide the platform for trust development (McAllister 1995). Jones and George (1998) highlighted that trust fosters when a party believes that the counterpart is trustworthy. On the other hand, distrust germinates due to lack of information to initiate cognition-based trust. In such situation, the parties may rely on "affective response" such as instincts, intuitions and feelings to gauge the other party's trustworthiness (Morrow et al. 2004).

7.4 Chapter Summary

Trust is a controversial topic in construction contracting. Whilst trust has been identified as the most important ingredient for efficient business exchanges, skeptics have maintained that trust is not possible in construction. This study posits to provide empirical evidence on the existence of trust by soliciting trust stories. Through case studies conducted in Shanghai, observations of trusting behaviours are collected and analyses. Antecedents of trust are identified and these are commitment, risk-taking, knowledge, honesty and benevolence. These are fur-ther discussed under the examined under the three trust bases advanced by McAllister (1995). The three trust bases are system, cognition and affect. It is advocated that system based trust is most relevant at the commencement stage of a project. As the contracting parties get to know each other better, their trusting relationship will shift to cognition-based. More enduring trust status is attained when affect-based trust develops with fruitful trusting exchanges. Trust building mechanisms in construction contracting are further deliberated in Chap. 8.

Acknowledgments Special thanks to Miss Chen Ling for collecting data for the study.

References

Bachmann, R. (2001). Trust, power and control in trans-organizational relations. *Organization Studies, 22*(2), 337–365.

Bennett, E., & Jayes, S. (Eds.). (1998). *Seven pillars of partnering—A guide to second generation partnering*. London: Thomas Telford.

Burt, R., & Knez, M. (1996). Trust and third-party gossip. In R. M. Kramer & T. R. Tyler (Eds.), *Trust in organizations: Frontiers of theory and research* (pp. 68–89). Thousand Oaks, CA: Sage Publications.

Butler, J. K. (1991). Toward understanding and measuring conditions of trust: Evolution of a conditions of trust inventory. *Journal of Management, 17*(3), 643–663.

Cheung, S. O., Yiu, T. W., Leung, Y. T., & Chiu, O. K. (2008). Catastrophic transition of construction contracting behaviour. *Journal of Construction Engineering and Management, ASCE, 134*(2), 942–952.

Cheung, S. O., & Pang, H. Y. (2013). Anatomy of construction disputes. *The Journal of Construction Engineering and Management, ASCE, 139*(1), 15–23.
Currall, S., & Judge, T. (1995). Measuring trust between organizational boundary role persons. *Organizational Behaviour and Human Decision Processes, 64*(2), 151–170.
Dasgupta, P. (ed.) (1988). Trust as a commodity, In Gambetta, D. (eds.), *Trust: Making and breaking cooperative relations*. Oxford: Basil Blackwell.
Friedland, N. (1990). Attribution of control as a determinant of cooperation in exchange interactions. *Journal of Applied Social Psychology, 20*(4), 303–320.
Gabarro, J. J. (1978). The development of trust influence and expectations. In A. G. Athos & J. J. Gabarro (Eds.), *Interpersonal behavior: Communication and understanding in relationships* (pp. 290–303). Englewood Cliffs, NJ: Prentice Hall.
Gambetta, D. (Ed.). (1988). *Trust: Making and breaking cooperative relations, B*. New York: Blackwell.
Geringer, J. M., & Hebert, L. (1991). Measuring the performance of international joint ventures. *Journal of International Business Studies, 22*(2), 249–264.
Gulati, R. (1995). Does familiarity breed trust? The implications of repeated ties for contractual choice in alliances. *Academy of Management Journal, 38*(1), 85–112.
Hanna, A. S. (2007). Risk allocation and increased claims in the construction industry. *Journal of Professional Issues in Engineering Education and Practice, 133*(1), 43–52.
Janowicz-Panjaitan, M. K., Bakker, R. M., & Kenis, P. N. (2009). Research on temporary organizations: The state of the art and distinct approaches toward temporariness. In P. N. Kenis, M. K. Janowicz-Panjaitan, & B. Cambré (Eds.), *Temporary Organizations: Prevalence, Logic and Effectiveness* (pp. 56–85). Cheltenham: Edward Elgar Publishing.
Johnson-George, C., & Swap, W. C. (1982). Measurement of specific interpersonal trust: Construction and validation of a scale to assess trust in a specific other. *Journal of Personality and Social Psychology, 43*(6), 1306–1317.
Jones, G. R., & George, J. M. (1998). The experience and evolution of trust: Implications for cooperation and teamwork. *Academy of Management Review, 23*(3), 531–546.
Kadefors, A. (2004). Trust in project relationships—Inside the black box. *Journal of Project Management, 22*(3), 175–182.
Kaplan, B., & Maxwell, J. A. (1994). Qualitative research methods for evaluating computer information systems. In J. G. Anderson, C. E. Aydin, & S. J. Jay (Eds.), *Evaluating health care information systems: Methods and applications*. CA: Sage Publications, Thousand Oaks.
Lau, E., & Rowlinson, S. (2009). Interpersonal trust and inter-firm trust in construction projects. *Construction Management and Economics, 27*(6), 539–554.
Lewicki, R. J., Tomlinson, E. C., & Gillespie, N. (2006). Models of interpersonal trust development: Theoretical approaches, empirical evidence, and future directions. *Journal of Management, 32*, 991–1022.
Lewis, J. D., & Weigert, A. (1985). Trust as a social reality. *Social Forces, 63*(4), 967–985.
Lorenz, E. (1999). Trust, contract and economic cooperation. *Cambridge Journal of Economics, 23*(3), 301–315.
Luhmann, N. (Ed.). (1979). *Trust and power*. Great-Britain: John Wiley and Sons.
Lyons, B., & Mehta, J. (1997). Contracts, opportunism and trust: Self-interest and social orientation. *Cambridge Journal of Economics, 21*(2), 239–257.
Mayer, R. C., Davis, J. H., & Schoorman, F. D. (1995). An integrative model of organizational trust. *Academy of Management Review, 20*(3), 709–734.
McAllister, D. J. (1995). Affect and cognition-based trust as foundations for interpersonal cooperation in organizations. *Academy of Management Journal, 38*(6), 24–59.
Morrow, J. L., Hansen, M. H., & Pearson, A. W. (2004). The cognitive and affective antecedents of general trust within cooperative organizations. *Journal of Managerial Issues, 14*(1), 48–64.
Punch, K. F. (Ed.). (2005). *Introduction to social research: Quantitative and qualitative approaches* (2nd ed.). London: Sage Publications.
Rousseau, D. M., Sitkin, S. B., Burt, R. S., & Camerer, C. (1998). Not so different after all: A cross-discipline view of trust. *Academy of Management Review, 23*(3), 393–404.

Webber, S. S., & Klimoski, R. J. (2004). Client-project manager engagements, trust, and loyalty. *Journal of Organizational Behaviour, 25*(8), 997–1013.

Williams, M. (2001). In whom we trust: Group membership as an effective context for trust development. *Academy of Management Review, 26*(3), 377–396.

Williamson, O. E. (Ed.). (1985). *Economic institution of capitalism: Firms, Markets, Relational Contracting*. New York: Free Press.

Zaheer, A., McEvily, B., & Perrone, V. (1998). Does trust matter? Exploring the effects of interorganizational and interpersonal trust on performance. *Organization Science, 9*(2), 141–159.

Zand, D. E. (1972). Trust and managerial problem solving. *Administrative Science Quarterly, 17*(2), 229–239.

Chapter 8
Trust Building in Construction Contracting

Pui Ting Chow, Sai On Cheung and Ka Ying Chan

Abstract Trust is defined as the willingness of a trustor to become vulnerable to a trustee whose behaviour is beyond his control. The efficiency of a project team can be enhanced should its members trust each other. As such, there have been notable efforts in promoting trust in the construction industry through the use of a variety of trust building mechanisms. However, the reciprocating trusting behaviours that could be expected (identified as trust expectations in this study) has not been elaborated. This study aims to investigate such relationships. For this, trust building mechanisms and trust expectations are identified and then operationalised for the development of their respective measurement scales. With data collected from practitioners, four and three taxonomies of trust-building mechanisms and trust expectations are developed respectively through the use of principal component factor analysis. The former includes: networking, procedural measure, credit rating and calculativeness. The latter consists of self-awareness, responsiveness and value congruence. Their inter-relationships were then examined by structural equation modeling. Most of the trust-building mechanisms, especially Networking and calculativeness, generally relate positively relate to most types of the trust expectations. Nevertheless, trust-building mechanisms like procedural measure and credit rating are not so related to trust expectations and may even lead to trust deterioration.

P. T. Chow (✉) · S. O. Cheung · K. Y. Chan
Construction Dispute Resolution Research Unit, Department of Civil and Architectural Engineering, City University of Hong Kong, Hong Kong, People's Republic of China
e-mail: ronnie.chow.pt@hotmail.com

S. O. Cheung
e-mail: saion.cheung@cityu.edu.hk

K. Y. Chan
e-mail: tinakychan@yahoo.com

S. O. Cheung (ed.), *Construction Dispute Research*,
DOI: 10.1007/978-3-319-04429-3_8,

8.1 Introduction

Trust has been identified as one of the determining factors to bring about avoiding or reducing cost of negotiation and increasing possibility for attaining mutually beneficial dispute settlement (Egan 1998; Khalfan et al. 2007; Latham 1994). The significance of trust becomes notable when the possibilities of exit, betrayal and defection are real (Walker 2003). Given the inherent asymmetries that characterise information exchange during construction project development, the practice of opportunism is potentially high and tempting (Lau and Rowlinson 2009). Building trust has been identified as one of the most effective means to suppress opportunism (Walker 2003). Project team members should give every effort to develop and foster trust among them. However, managing contracting relationship has been described as putting the customer at the centre of the organisation whereby customer is regarded as "king" (Peters and Waterman 2004). This unidirectional dependence motive is however paradoxical in the development of trust in construction contracting (Gummesson 2002; Johnson and Selnes 2004; Sheppard and Sherman 1998). The construction industry has well-developed institutional arrangements between contracting organisations that make reciprocating exchanges under risk-laden contracts environment. These exchanges are more likely based on fear and/or power rather than trust (Egan 1998; Latham 1994; Pretty and Ward 2001). Notwithstanding, some construction firms manage to surmount these institutionalised barriers against trust and successfully establish long-term business partnerships (Wong and Cheung 2004). These trusting relations are fragile but typically having enduring foundation germinated from trust-building mechanisms that are either planned, incidental or both (Lau and Rowlinson 2009; Sitkin and Roth 1993; Walker 2003). Reciprocating trusting behaviours are expected, as responses to effective trust building mechanisms. In this study, these trusting behaviours are termed as trust expectations to reflect the reciprocating nature. This study enriches the literature of trust in construction by demonstrating the contingent nature and importance of trust-building mechanisms with reference to trust expectations. In the following sections, the definition of trust is first presented then followed by an account each for trust-building mechanisms and trust expectations. The attributes of trust-building mechanisms and trust expectations are operationalised for data collection. With the collected data, taxonomies of mechanism and expectation are then developed with the technique of principal component factor analysis (PCFA). These taxonomies are used to develop a relationship model between trust-building mechanisms and trust expectations. Structural equation modeling then empirically tests the model. Finally, the findings are discussed.

8.2 Research Framework

8.2.1 Definition of Trust in the Construction Industry

Warren Buffet's saying *"it takes twenty years to build a reputation but five minutes to ruin it"* and English proverb *"it is an equal failing to trust everybody, and to trust nobody"* aptly reflect the fragile and party-specific nature of trust. Trust refers to a trustor's willingness to become vulnerable to a trustee whose behaviour is beyond his control (Mayer et al. 1995). In construction contracting, trust is a risk-taking act as a trustee may exploit a trustor. For example, an employer (a trustor) relies on (trusts) a contractor (a trustee) to deliver a built facility. Trust itself is a complex phenomenon with multiple attributes. A number of perspectives have been used to examine trust in organisational management studies (Ross and LaCroix 1996). For example, McAllister (1995) distinguished affect-based trust from cognition-based trust. Butler (1991) classified determinants of trust as dispositional and situational. Lau and Rowlinson (2009) emphasised that the role of trust in inter-firm exchanges is better understood at two levels, i.e. inter-organisational trust and inter-personal trust. Cheung et al. (2003) has discovered that trust is a disposition or an emergent state in construction contracting relationship. These perspectives are inspiring and have provided invaluable theoretical conceptualisation of trust (Rousseau et al. 1998; Schoorman et al. 2007). This study posits trust as a temporary state that can be induced by different mechanisms (Ross and LaCroix 1996). From a behavioural point of view, an expectation of trustworthiness induces a trustor's efforts in building trust that he/she expects in return from a trustee. Trust expectation is thus a catalyst that engenders resources commitments and facilitates working among project team members (Eriksson 2008). The presence of trust improves the chance of having quality communication and effective performance (Cheung 2007; Wong et al. 2008). As a result, team members can work together as a unified whole in a trusting environment. The outcome is enhanced cooperation.

8.2.2 A Relationship Framework Between Trust-Building Mechanisms and Trust Expectations

Despite the positive impact of trust in team management (Mayer et al. 1995), developing trust in project teams has not been forthcoming as wished. This may be a lack of understanding between theory and reality. Notable examples include identification of trusting behaviour; the relationship between trust-building mechanism and trusting behaviour; differentiation between trust antecedents and outcomes; confusion in levels of analysis due to a lack of specificity of trust referents; and a failure to consider the mutual dependence between the trustor and the trustee (Kuriyan et al. 2010; Ratnasingham 1998). Trust operates differently at interpersonal, inter-group and inter-organisational levels (Currall and Inkpen 2006). When referring to "the parties" involved in a trusting relationship, it is important to designate who are the "trustor" and the "trustee". Interestingly, researches on trust suggested that the three facets of trust are somewhat correlated (Barney and Hansen 1994; Doz 1996; Doney and Cannon 1997; Jeffries and Reed 2000; Zaheer et al. 1998). For example, decision to trust interpersonally can be motivated by another party's interorganisational trust-building mechanism (Mayer et al. 1995). Aulakh et al. (1996) and Gefen (2003) identified several interperonal trust-building mechanisms that could play critical roles in effecting inter-organisation trusting relationship. Both interpersonal and interorganisational trust-building mechanisms therefore will lead to reciprocating trusting behaviours by the trustees—identified as trust expectations in this study (Bigley and Pearce 1998). This study aims to explore the relationships between trust-building mechanisms and trust expectations. For this purpose, it is necessary first to operationalise trust-building mechanisms and trust expectations. With these, a relationship framework between trust-building mechanisms and trust expectations is proposed and examined.

8.2.2.1 Trust-Building Mechanisms

Mcknight and Chervancy (2000) suggested that perceived interactivity as an interpersonal-based antecedent, dispositional to trust as a personality-based antecedent and perceived inter-organisational-based antecedent can initiate trust. This study focuses on the last one that is considered to be a more formative approach in trust development. Studies in psychology and management fields have provided the construction community with a set of potential trust-building mechanisms (Aulakh et al. 1996; Ba and Pavlou 2002; Parkhe 1998; Pavlou and Gefen 2004). However, their applications and effectiveness in construction need more empirical evidences (Doney and Cannon 1997; Moorman et al. 1993; Williamson 1993). Every interaction between a trustor and a trustee can be a "moment of trust". And what a trustee does have far greater impact than what he says. Thus, trust building has four principles (Sheppard and Sherman 1998). First, deterrence is to penalise parties who are not abiding by an understanding or are performing unreliably. Such penalties can be either tangible (e.g. liquidated damages) or intangible (e.g. reputation). The forms of penalty can be (1) increased cost of cheating which exceeds the benefit from cheating or (2) perceived benefit of future collaboration which is no less that as a potential advantage derivable from practicing opportunism. Second, obligations are evolved respective to the roles and responsibilities of the parties. Psychological contracts are also developed based on mutually perceived obligations (Rousseau 1995). A key feature of these contracts is that the parties voluntarily assent to make and accept certain delegations as per his perceived understanding. Third, delivery is to engage in active discovery through communication and research. The extent and quality of the communication determine the depth and width of the information delivered. Research suggested that the prevalence of e-channel (e.g. email, net meeting) have negative effects on trust such as reduced social presence and demotivation (Das and Teng 1998). Fourth, internalisation is to negotiate common values, shared strategies and identity that lead to the evolution of similarity in views, beliefs, and values. The process of internalisation can only be built over time.

8.2.2.2 Trust Expectations

Successful trust-building mechanisms initiate information sharing that believingly would lead to trusting behaviours (Butler 1999). Reported studies have indicated that the degree of trust between a trustor and a trustee is a function of the trustor's own propensity to trust (the way to collect the "knowledge") and the perceived "trustworthiness" of the trustee (the "knowledge" to trust) (Cheung 2007; Colquitt et al. 2007; Mayer et al. 1995). This study focuses on the latter that reflects the trustee's effort in building the amount of "knowledge" necessary for trust. The level of knowledge lies between total ignorance and complete knowledge (McAllister 1995). Given total ignorance, there is no basis upon which to trust. And with complete knowledge, there is no need to trust (Houde et al. 2004; Morrow et al. 2004). A trustworthy trustee is one who would mitigate the risks taken by the trustor in their relationship, for example refraining from cheating, abusing, or neglecting. Eleven attributes of trust expectations; discretion, reliability, competence, integrity, concern, benevolence, predictability, consistency, foresight, intuition and empathy have been identified (Butler 1991; Das and Teng 1998; Mishra 1996).

8.3 Methodology

Two lists of attributes were identified for trust-building mechanism and trust expectation respectively and were employed in a questionnaire survey designed to collect project specific data concerning trust-building mechanisms and trust expectations. A list of prospective respondents was prepared based on information obtained from (1) trade magazines (Far East Trade 2003), (2) newspaper, (3) personal network, (4) government, professional institutes and (5) companies' websites. Targeted respondents were practicing construction professionals working in Hong Kong. The prospective respondents were contacted in person, by mail, email or facsimile. The study is particularly interested in independent responses in the client-contractor relationship. Respondents were asked to provide data on one of their recently completed project with an identifiable client or contractor. They were also asked to indicate their degree of agreement (i.e. 1: strongly disagree; 7: strongly agree) on the statements of trust-building mechanism and trust expectation in the questionnaire. Cronbach's alpha reliability test is first conducted to validate the construct reliability and inter-item relationships of the attributes.

Alpha values range from 0 to 1. The higher the alpha value, the greater is the internal consistency of the attributes. Alpha values greater than 0.7 are considered as good (Sharma 1996). Principal component factor analysis (PCFA) is then conducted to categorise the related attributes by means of class inclusion. In this study, PCFA is used to estimate factors or latent variables and to reduce dimensionality of the large number of attributes to a manageable number of factors. The structures of interrelationships among attributes to define a set of common underlying factors were thereby explored. Kaiser–Meyer–Olkin (KMO) measure of sampling adequacy and Bartlett's test of sphericity are used to assess the suitability of the data for PCFA. Finally, a relationship model between trust-building mechanisms and trust expectations is organised in a format of a structural equation model (SEM) with the taxonomies developed from the PCFA. SEM is used to analyse (1) the relationship between trust-building mechanisms and its attributes, (2) the relationship between trust expectations and its attributes, and (3) the relationship between trust-building mechanisms and trust expectations (Akgün et al. 2007; Smith and Smith 2004). Measurement models are used for (1) and (2) and structural models are used for (3) The measurement models relate observed variables to their respective latent variables. The structural model specifies relations among latent variables and regressions of latent variables on observed variables. The exogenous latent variables are connected by covariance paths and confirmatory factor analysis is conducted. This methodology has also been applied by Butler (1991) and Cummings and Bromiley (1996) in measuring conditions of trust and developing trust inventory respectively in business sectors. When the number of data set is relatively small (<200), bootstrapping procedure can be used to reduce the risk of sample non-normality and increase the accuracy of the results (Kline 1998). Bootstrapping focuses on (1) resampling, iterating and replacing random selection of observations from the original data set and (2) creating a number of new data sets with the same number of observations as the original data set (Paiva et al. 2008). A comparison of the results between the original data set and the new data sets is made possible. The structural equation model is satisfactory when the following criteria are met (Molenaar et al. 2000; Wong and Cheung 2005):

1. all GOF measures (e.g. TLI, CFI) are above the threshold level;
2. multivariate kurtosis value is greater than 1.96;
3. regression weights of the variables generated from the original data set are contained within upper and lower bounds generated from the new data sets at 95 % confidence level; and
4. computed standard errors of the regression weights are low.

The data analysis was performed by the Statistical Package for the Social Sciences Version 17.0 (SPSS 17.0) and Analysis of Moment Structures 17.0 (AMOS 17.0).

Table 8.1 Means, standard deviations, internal consistency reliabilities and correlations (n = 100)

Variable		μ	SD	2	3	4	5
1	Gen.	0.19	0.39	−0.104	−0.366*	−0.191	−0.255*
2	Com.	0.58	0.50		0.174	0.282*	0.425*
3	Exp.	11.04	9.71			0.077	0.137
4	ME	4.32	0.93			0.935	0.765*
5	EP	4.61	0.98				0.945

Note: *Gen.*: Gender, 0-female, 1-male; *Com.*: Company, 0- client, 1- contractor; *Exp.*: Experience in year; *ME*: Trust-building Mechanism; *EP*: Trust expectation; *Upper-diagonal*: Correlation coefficient *r; Diagonal*: Cronbach's alpha; *Lower-diagonal*: t-value; Significant level *p*, * $p < 0.05$; ** $p < 0.01$

8.4 Results and Findings

Two hundred and ninety-three questionnaires were distributed and one hundred valid responses were collected. Forty-two questionnaires were returned by clients and fifty-eight from contractor. The average working experience of the respondents in the construction industry and their length of service in their prevailing organisations were 11.03 and 6.42 years respectively. The projects involved were procured mainly through traditional contracting (91 %). Forty-nine percentage of the respondents were surveyors, architects or engineers and 51 % were project management staff. Both the attributes of trust-building mechanisms and trust expectations achieved Cronbach's alpha values above 0.7, suggesting that the data collection statements are statistically stable and related to the respective attributes (Hair et al. 1998). Descriptive statistics and correlations of the attributes are reported in Table 8.1. Trust-building mechanisms and trust expectations were then subjected to principal component factor analysis for the development of their respective taxonomies. To shortlist factors, only factors having an eigenvalue greater than 1 were considered significant according to the eigenvalue-greater-than-1 rule (Velicer and Jackson 1990). The factor structures are simplified through VARIMAX rotation for ease of interpretation (Hair et al. 1998). As a rule of thumb, factor loadings greater than 0.55 is considered significant for a sample of 100 respondents (Hair et al. 1998). In this connection, variables with factor loadings less than 0.55 were discarded so as to achieve simpler structure with greater interpretability (Fava and Velicer 1992). As a result, four-factor and three-factor structures were obtained for trust-building mechanisms (Table 8.2) and trust expectations (Table 8.3) respectively. Table 8.4 shows the magnitudes of the measures and the statistical result of the principal component factor analysis.

Finally, the relationship model was examined by structural equation modeling (Fig. 8.1). The dashed boxes show the measurement models, the single-head arrows indicate the hypothesised relationships of the structural models and the double-head arrows represent covariance paths of exogenous latent variables and error terms in Fig. 8.1. The significant covariance paths of the exogenous latent

Table 8.2 Attributes of trust-building mechanisms

F	Coding	Indicators	RCM			
			1	2	3	4
1	ME_03	withholding/claiming money to secure profit (i.e. calculative)	**0.75**	0.110	0.099	0.208
1	ME_05	letting each other freely take own courses to fulfill the obligations	**0.747**	0.115	0.142	0.040
1	ME_07	building relational network (e.g. staff secondment)	**0.603**	0.513	−0.059	−0.029
1	ME_08	developing third party control, i.e. employing a third party to monitor (i.e. quadratic control)	**0.602**	0.516	0.358	−0.025
1	ME_10	establishing a data bank of the counterpart and updating the data bank from time to time (i.e. "knowledge of the trustee")	**0.697**	0.059	0.338	0.141
1	ME_13	investing time and rooms for mutual understanding (e.g. retreat, sharing seminar or caucus)	**0.638**	0.387	0.208	0.192
1	ME_14	developing a contiguous management styles (i.e. contiguity)	**0.707**	0.215	0.409	−0.057
1	ME_15	creating common membership (e.g. formation of coterie)	**0.548**	0.108	0.065	0.534
1	ME_16	negotiating common value	**0.634**	0.588	0.060	−0.093
1	ME_20	establishing pain-share (e.g. dividend or bonus) (i.e. products)	**0.770**	0.232	0.099	0.013
1	ME_21	developing shared meaning (e.g. setting up procedural manuals and/or contract terms specifically)	**0.658**	0.421	0.345	−0.055
1	ME_22	providing opportunity for staff of both parties to socialise together	**0.652**	0.475	0.046	−0.224
2	ME_11	developing communication systems (i.e. information-sharing systems)	0.403	**0.673**	0.04	0.152
2	ME_12	having channels for discourse (e.g. senior management of both parties has to conduct regular site meetings and inspections together)	−0.060	**0.595**	0.209	0.434
2	ME_17	establishing common goals (e.g. setting action plan)	0.182	**0.679**	0.389	0.152
2	ME_18	internalising and integrating one's other values (i.e. internalisation)	0.385	**0.730**	0.064	0.152
3	ME_04	preparing a list of alternatives and letting the counterpart know the organisation has the power to replace one another (i.e. selection)	0.139	0.225	**0.718**	0.108
3	ME_06	developing authority and control (i.e. fate control)	0.55	−0.070	**0.579**	0.488
3	ME_09	collecting pervious performance data of the counterpart (e.g. from external sources)	0.331	0.082	**0.700**	0.238

(continued)

Table 8.2 (continued)

F	Coding	Indicators	RCM			
			1	2	3	4
3	ME_19	aligning strategic directions (i.e. strategic alignment)	0.526	0.355	**0.537**	−0.37
4	ME_01	establishing rules of punishment and compensation for violation of agreement (i.e. deterrence)	0.013	0.010	0.228	**0.785**
4	ME_02	strictly administering out rules of punishment and compensation of the agreement (i.e. enforcement)	0.06	0.280	0.042	**0.847**

All scales are anchored with a 7-point Likert-scale (*1* = Strongly Disagree, *7* = Strongly Agree); *ME*: Trust-building Mechanism; *1*: networking; *2*: procedural measure; *3*: credit rating; *4*: calculativeness; *F*: Principal Component Factor; *RCM*: Rotated Component Matrix

Reference: Barney and Hansen (1994); Butler (1999); Cheung (2007); Costa and Bijlsma-Frankema (2007); Das and Teng (1998); Hardy et al. (1998); Jones and George (1998); Khalfan et al. (2007); Krishnan et al. (2006); Lewicki and Stevenson (1997); Mayer et al. (1995); McAllister (1995); Mishra (1996); Morrow et al. (2004); Nooteboom et al. (1997); Ross and LaCroix (1996); Rousseau et al. (1998); Sheppard and Sherman (1998); Verbeke and Greidanus (2009); Wright (2010)

Table 8.3 Attributes of trust expectations

F	Coding	Indicators	RCM		
			1	2	3
1	EP_01	My counterpart is reliable that h/she will fulfill the obligations without frequent reminds	**0.815**	0.201	0.262
1	EP_02	My counterpart is competent that h/she will use his/her skills and knowledge to complete the task	**0.743**	0.002	0.433
1	EP_03	My counterpart is honest that h/she will tell me what is gone wrong in the project and proactively rectify his/her mistakes or may even make reasonable compensation to me of his/her own faults	**0.678**	0.251	0.460
1	EP_07	My counterpart is prudent that h/she will rarely relay on my monitoring and crosschecking	**0.763**	0.414	0.265
1	EP_08	My counterpart is discreet that h/she will follow through the contract specification	**0.746**	0.444	−0.023
2	EP_05	My counterpart is conventional that h/she will place reliance on the existing institutional arrangements	0.273	**0.756**	0.375
2	EP_06	My counterpart is predictable whenever h/she will act opportunistically	0.212	**0.749**	0.178
2	EP_09	My counterpart is considerate that h/she will care about my interests in his/her decisions	0.555	**0.647**	0.155
2	EP_10	My counterpart is altruistic that h/she will sacrifice his/her benefit for my sake	0.573	**0.620**	0.247
2	EP_12	My counterpart is farsighted that h/she will propose alternative suggestions to cater for contingency	0.065	**0.804**	0.365
2	EP_14	My counterpart is affectionate that h/she will share his/her feelings and experiences with me	0.399	**0.700**	0.349
3	EP_04	My counterpart is consistent with his/her words that h/she will deal with me fairly	0.348	0.219	**0.751**
3	EP_11	My counterpart is intuitive that h/she will adopt my perspective in his/her decision	0.203	0.395	**0.764**
3	EP_13	My counterpart is empathetic that h/she will act effectively for me	0.210	0.358	**0.758**

All scales are anchored with a 7-point Likert-scale (1 = Strongly Disagree; 7 = Strongly Agree); *EP*: Trust expectation, *1*: self-awareness, *2*: responsiveness, *3*: value congruence; *F*: Principal Component Factor; *RCM*: Rotated Component Matrix

Reference: Alvarez et al. (2003); Cheung (2007); Colquitt et al. (2007); Costa and Bijlsma-Frankema (2007); Das and Teng (1998); Geyskens et al. (1996); Hardy et al. (1998); Jones and George (1998); Khalfan et al. (2007); Kjærnes (2006); Lewicki et al. (1998); Lewis and Weigert (1985); Mayer et al. (1995); McAllister (1995); Mishra (1996); Morrow et al. (2004); Nooteboom et al. (1997); Ross and LaCroix (1996); Rousseau et al. (1998); Shaw (1997); Sheppard and Sherman (1998); Sitkin and Roth (1993); Verbeke and Greidanus (2009); Wright (2010); Zucker (1986)

Table 8.4 Principal component factor analysis result of trust-building mechanism and trust expectation

	KMO	Bartlett's test of sphericity				F			
		χ^2	DF	Sig.		1	2	3	4
Trust-building mechanism	0.876	1439.77	231	0.000	Eigenvalue	9.672	2.519	1.432	1.101
					% of Variance	43.965	11.451	6.509	5.005
Trust expectation	0.922	1079.14	91	0.000	Eigenvalue	8.256	1.298	1.031	
					% of Variance	58.969	9.274	7.364	

KMO: Kaiser–Meyer–Olkin Measure of Sampling Adequacy; χ^2: Chi square; *DF*: degree of freedom; *Sig.*: significance; *F*: Principal Component Factor; *ME*: Trust-building Mechanism; *EP*: Trust expectation; *F1_ME*: networking; *F2_ME*: procedural measure; *F3_ME*: credit rating; *F4_ME*: calculativeness; *F1_EP*: self-awareness; *F2_EP*: responsiveness; *F3_EP*: value congruence

variables suggested that the variables are correlated, for example, $r = 0.88$ between F1_ME and F3_ME. With the covariance paths, the variances of the latent factors are maximised and the differentiation of the causal effects is made easier. The set of model fit parameter values of the final structural equation model is presented in Table 8.5. The standardised regression weights of the model are presented in Table 8.6. The computed Multivariate Kurtosis Value is 175.606, which is far greater than the threshold value of 1.96 (Kline 1998). Furthermore, the regression weights of the variables of the default model all fall within the upper and lower bounds of the regression weights generated from the 1000 bootstrapped samples at $p \leq 0.05$. These results collectively indicate that the parameter estimates obtained from the structural equation modeling analysis of this study are statistically significant. To summarise, all the relationship paths as specified in the structural equation model are positive and significant at $p \leq 0.05$.

8.5 Discussion

8.5.1 Taxonomies of Trust-Building Mechanism

Four factors (taxonomies) have been extracted for trust-building mechanism. Eleven attributes were included for factor 1 (F1_ME) which is called "networking" and refers to the trustor's influence on the trustee through third-party network. According to Sheppard and Sherman (1998), a trustor unavoidably establishes psychological contract and builds social network with the prospective trustees and the organisations linked to the trustees. Networking is developed through the trustee's effort in facilitating trustor to be connected in his/her network, thereby increasing the value and quality of the network. Factor 2 (F2_ME) consists of four attributes and is labeled as "procedural measure", which generally describes the process control of information. Considering that written and spoken words are the

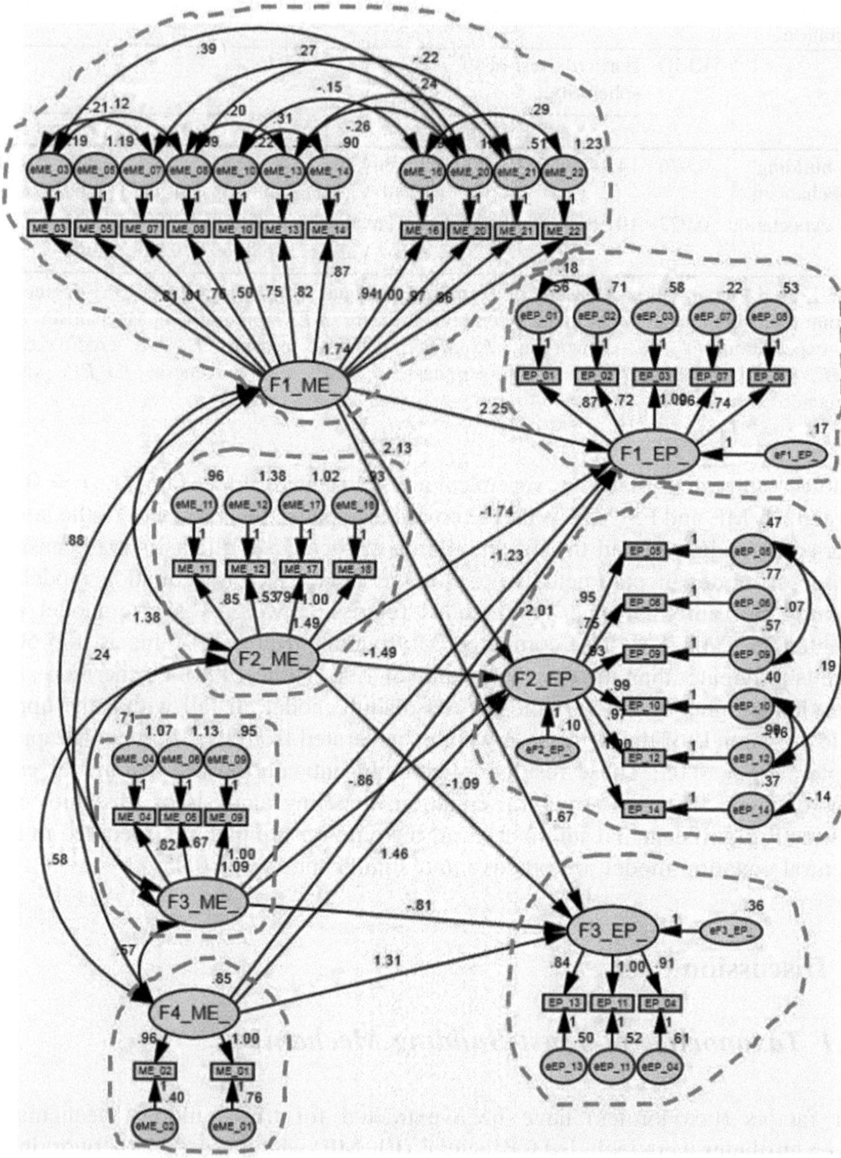

Note Standardized Regression Weight (β); ME: Trust-building Mechanism; EP: Trust expectation; F1_ME: networking, F2_ME: procedural measure, F3_ME: credit rating, F4_ME: calculativeness; F1_EP: self-awareness, F2_EP: responsiveness, F3_EP: value congruence; detail of the attributes of ME and EP can be found in **Table 2** and **Table 3** respectively; dashed boxes: measurement models; single-head arrows: causal relationship; double-head arrow: covariance path; e: error term; all the relationship paths are significant at $p \leq 0.05$

Fig. 8.1 Structural equation model of the relationship framework

Table 8.5 Goodness of fit indices and their threshold values

GOF	GOF range	Threshold	Final model
χ^2/DF	0 or above	1.00–3.00	1.884
GFI	0 (no fit) to 1 (perfect fit)	0.7 or above	0.703
TLI	0 (no fit) to 1 (perfect fit)	0.7 or above	0.840
CFI	0 (no fit) to 1 (perfect fit)	0.7 or above	0.859
MKV	nonzero	1.96 or above	175.606
RMSEA	0 (perfect fit) to 1 (no fit)	0.1 or below	0.083

GOF: Goodness Of Fit indices; χ^2/DF: Chi square/Degree of Freedom; *GFI*: Goodness of Fit Index; *TLI*: Tucker-Lewis Index; *CFI*: Comparative Fit Index; *MKV*: Multivariate Kurtosis Value; *RMSEA*: Root Mean Square Error of Approximation
References: Arrindell et al. (1999); Brennan and Brannan (2005); Bollen and Long (1992); Browne and Cudeck (1993); Chou and Bentler (1990); Hair et al. (1998); Jashapara (2003)

most common vehicles of communication among project team members, broader social conditions other than the contractual setting allow a trustor and a trustee to develop common language (Butler 2008) (e.g. jargon; cant; shoptalk; buzzwords). In the long run, behaviour related to discourse is built (e.g. custom; mores; convention). This language represents a set of social relations in which the trustor and the trustee are nested and determines the authority or legitimacy conferred with the communication. Caucus is one of the most effective ways to solicit confidential information. "Credit rating" is the name of factor 3 (F3_ME) and has three attributes. Credit rating mechanism embraces identity management that supports and integrates behavioural authenticity of the trustee. The establishment of credit rating allows the trustor to evaluate the credentials of the trustee. Recent record of the trustee is the most persuasive. Credit rating is a sensible and rational approach to incite reliable behaviour in which the trustor can judgmentally decide the trustworthiness of the trustee on the strength of his/her goodwill and reputation (Morrow et al. 2004). Two attributes are extracted for factor 4 (F4_ME) which is described as "calculativeness". Calculativeness focuses on how a trustor secures the anticipated outcome and imposes negotiated rules on a trustee. The prerequisite for the success of the mechanism is that the trustor perceives the intended action of the trustee as beneficial and the trustee is likely to maintain the need of the relationship. Under such circumstances, the trustor conducts cost-and-benefit analysis of the proposed negotiated agreement. If the agreement is profitable and cost effective, the trustor would use command and control strategies to preclude alternatives available to the trustee and impose significant anticipated termination or switching costs associated with leaving (Geyskens et al. 1996). These strategies allow the trustor to recover any loss arising thereof if the trustee does not abide with the agreement (Sheppard and Sherman 1998). The inclusion of liquidated damages in contract is a typical calculated measure to compensate the trustor for the respective foregone investments (Costa and Bijlsma-Frankema 2007).

Table 8.6 Regression weights and bootstrapping estimate

			Estimate	S.E.	C.R.	Lower	Upper	P
F1_EP_	←	F1_ME_	2.246	0.749	2.996	0.978	9.943	0.003
F2_EP_	←	F1_ME_	2.129	0.611	3.485	1.025	8.557	***
F3_EP_	←	F1_ME_	1.666	0.543	3.067	0.795	6.334	0.002
F1_EP_	←	F2_ME_	**−1.745**	0.745	−2.342	−9.007	−0.074	0.019
F2_EP_	←	F2_ME_	**−1.493**	0.604	−2.471	−9.028	−0.05	0.013
F3_EP_	←	F2_ME_	**−1.095**	0.538	−2.037	−6.171	−0.089	0.042
F1_EP_	←	F3_ME_	**−1.228**	0.564	−2.177	−7.555	0.795	0.03
F2_EP_	←	F3_ME_	**−0.861**	0.436	−1.975	−6.739	0.973	0.048
F3_EP_	←	F3_ME_	**−0.813**	0.405	−2.01	−4.734	0.444	0.044
F1_EP_	←	F4_ME_	2.009	0.657	3.058	0.901	8.104	0.002
F2_EP_	←	F4_ME_	1.459	0.511	2.856	0.563	7.654	0.004
F3_EP_	←	F4_ME_	1.315	0.468	2.813	0.57	5.06	0.005
EP_01	←	F1_EP_	0.871	0.094	9.246	0.664	1.105	***
EP_02	←	F1_EP_	0.716	0.094	7.581	0.535	0.896	***
EP_03	←	F1_EP_	1			1	1	
EP_07	←	F1_EP_	0.963	0.084	11.503	0.797	1.141	***
EP_08	←	F1_EP_	0.738	0.086	8.538	0.557	0.934	***
EP_05	←	F2_EP_	0.952	0.083	11.518	0.793	1.125	***
EP_06	←	F2_EP_	0.755	0.089	8.45	0.569	0.936	***
EP_09	←	F2_EP_	0.928	0.086	10.764	0.758	1.098	***
EP_10	←	F2_EP_	0.988	0.094	10.558	0.818	1.2	***
EP_12	←	F2_EP_	0.966	0.103	9.393	0.759	1.194	***
EP_14	←	F2_EP_	1			1	1	
EP_04	←	F3_EP_	0.914	0.106	8.603	0.722	1.172	***
EP_11	←	F3_EP_	1			1	1	
EP_13	←	F3_EP_	0.841	0.097	8.656	0.635	1.068	***

(continued)

Table 8.6 (continued)

			Estimate	S.E.	C.R.	Lower	Upper	P
ME_03	←	F1_ME_	0.814	0.097	8.353	0.632	1.058	***
ME_05	←	F1_ME_	0.81	0.115	7.021	0.59	1.073	***
ME_07	←	F1_ME_	0.756	0.099	7.606	0.577	0.975	***
ME_08	←	F1_ME_	0.498	0.092	5.395	0.328	0.712	***
ME_10	←	F1_ME_	0.752	0.113	6.674	0.549	1.054	***
ME_13	←	F1_ME_	0.817	0.115	7.08	0.617	1.114	***
ME_14	←	F1_ME_	0.867	0.112	7.769	0.644	1.136	***
ME_16	←	F1_ME_	0.945	0.118	8.035	0.754	1.21	***
ME_20	←	F1_ME_	1			1	1	
ME_21	←	F1_ME_	0.968	0.109	8.857	0.775	1.253	***
ME_22	←	F1_ME_	0.863	0.12	7.197	0.647	1.154	***
ME_11	←	F2_ME_	0.854	0.115	7.44	0.639	1.113	***
ME_12	←	F2_ME_	0.527	0.113	4.667	0.254	0.805	***
ME_17	←	F2_ME_	0.787	0.113	6.966	0.561	1.061	***
ME_18	←	F2_ME_	1			1	1	
ME_04	←	F3_ME_	0.821	0.155	5.279	0.516	1.253	***
ME_06	←	F3_ME_	0.675	0.144	4.672	0.358	1.123	***
ME_09	←	F3_ME_	1			1	1	
ME_01	←	F4_ME_	1			1	1	
ME_02	←	F4_ME_	0.957	0.147	6.503	0.662	1.365	***

Note: *EP*: Trust expectation; *F1_ME*: networking; *F2_ME*: procedural measure; *F3_ME*: credit rating; *F4_ME*: calculativeness; *F1_EP*: self-awareness; *F2_EP*: responsiveness; *F3_EP*: value congruence detail of trust-building mechanisms and trust expectations attributes refers to Tables 8.2 and 8.3
***: sig. < 0.001

8.5.2 Taxonomies of Trust Expectation

There are three taxonomies for trust expectation. Factor 1 (F1_EP) is described as "self-awareness". It addresses a trustee's attitude to faithful adherence to work rules and procedures as generalised compliance (Smith et al. 1983). In construction, a developer (trustor) relies on a main-contractor (trustee) to build a facility. This reliance is distinctly different from other indicators where incompetence goes beyond disappointment and anger to a feeling of betrayal (Butler 1991; Khalfan et al. 2007; Mishra 1996). Nevertheless, working competently does not necessarily create a sense of gratification than the fulfillment of other indicators of trust expectation (Wright 2010). Thus, competence builds on a foundation of expectation about a trustee's ability to complete the task assignments reliably (Sitkin and Roth 1993). The trustor assesses the competence of the trustee by assessing the trustee's qualities, skills and knowledge for accomplishing the task. Apart from the "can do" characteristic of competence, self-awareness also captures integrity that is the "will-do" part. Integrity is a complex concept allied with truth telling, honesty and fairness (Barney and Hansen 1994; Mayer et al. 1995; Ross and LaCroix 1996). Integrity can be expressed as an expectation to perform morally that a trustee will act consistently with his/her words (Lau and Rowlinson 2009; Shaw 1997). For example, a trustee is willing to tell a trustor what has gone wrong in a project and to rectify his/her mistakes proactively. In extreme cases, a trustee may even offer reasonable compensation to a trustor for his/her own faults (Shaw 1997). Factor 2 (F2_EP) is called "responsiveness". It is said to be a trustee's readiness to respond to any of the trustors' inquiry. He/She starts with trust and builds up the network using his/her expertise. The network does more than complementing the trustor's trusting initiatives, but also raises standards and showcase best practices (Mayer et al. 1995). The trustor is particularly interested in the extent to which a trustee can discharge his/her discretionary powers effectively and the extent to which a trustee can refrain from exploiting any advantage (Kilpatrick and Lapsley 1996). Thus, responsiveness refers to the expectation of a high degree of prudence exercised by the trustee in carrying out the tasks (Sheppard and Sherman 1998). The case that the trustor requires high levels of the discretion rests on the need to monitor the trustee. Factor 3 (F3_EP) is named as "value congruence" which refers to the ability a trustee to do good to a trustor. This ability is not only "doing thing right" but also "doing right thing". The trustor's perspective has been embedded in that of the trustee if value congruence is present. Thus, the trustee can act effectively and efficiently for the trustor (Krishnan et al. 2006). The trustor gains knowledge of the trustee through repeated interactions and/or simply relies on the existing institutional arrangements (Ratnasingam 2005). The belief in the trustee's consistent behaviours underpins predictability. The trustor assumes that the trustee will live up to shared norms and expectations associated with the tasks and in this regard to what should be done and take responsibility for (Kjærnes 2006). Foresightedness can be developed in extensive interactions, if the trustee can forecast possible difficulties or problems that the trustor may face and propose to jointly resolve them if these eventuate. High degree of foresightedness to deal with contingencies reinforces positive expectation to trust (Sengün and Wasti 2007). In the long run, intuition is built based upon the identification with a trustor's desires and intentions (Lewicki and Stevenson 1997). The trustee is said to be intuitive if he/she can foresee or guess what the trustor's actions would take in certain situations to prevent unwanted outcomes (Sheppard and Sherman 1998). It is commonly manifested as "perspective taking" (Evans and Krueger 2011). In other words, value congruence leads to the formation of "emotional bonds" between parties, which may eventually provide the trust base (Lewis and Weigert 1985; McAllister 1995; Morrow et al. 2004).

8.5.3 The Relationship Framework

The relationship framework between trust-building mechanisms and trust expectation is examined through a structural equation modeling analysis. The results of the analysis provide empirical support to the contingent nature of trust-building mechanisms. Based on the taxonomies developed in PCFA, trust building mechanism is represented by networking (F1_ME), procedural measure (F2_ME), credit rating (F3_ME) and calculativeness (F4_ME). Networking is found to be the most powerful type of trust-building mechanism to bring positive trust expectation (Fig. 8.1 and Table 8.6). Networking engenders social contracts that treasure the value of sustaining a given relational form (Kimber and Raghunath 2002; Sheppard and Sherman 1998). Tying in a social contract implies that (1) the trustee has a duty to conform by reciprocating the trustor's expectation; (2) the trustor is obliged to respond when the trustee fails to meet the obligation; and (3) third parties with social links to the trustor and the trustee have a duty to modulate the social relationships if the latter is in not complying (Sheppard and Sherman 1998). Thus, the trustor believes that the trustee in the network respects the agreement and other people with the social links to the trustee will monitor his/her performance. On the other hand, as calculativeness is based on the perceived structural constraints that bind the trustee to the trustor, it reflects a rather negative motivation for continuing the relationship. The constraints imply sanctions, which promote limited cooperation based upon deterrence. Thus, this mechanism has been labeled as "low trust" governance in sociology and rational choice in economics (Lewicki et al. 2006). In extreme cases, strict calculative controls actually appear to signal the absence of trust and therefore may be closer to low level of distrust (Rousseau et al. 1998).

An interesting finding is that procedural measure and credit rating are found to have negative influence on all trust expectations, while procedural measure has a higher negative impact than credit rating in general (Fig. 8.1; Table 8.6). Procedural measure helps contracting organisations to systematically present their ideas about the project and/or the nature of their relationship. Nevertheless, the easiest way for a trustor to know about a trustee is through direct communication.

Table 8.7 Means, standard deviations, internal consistency reliabilities, correlations and t-value of the principal component factors of trust-building mechanisms (n = 100)

Variable		μ	SD	1	2	3	4
1	F1_ME	4.75	1.03	0.927	0.731**	0.452**	0.157
2	F2_ME	4.50	1.11	−4.514**	0.785	0.458**	0.335**
3	F3_ME	4.58	1.10	−4.238**	−1.079	0.700	0.431**
4	F4_ME	4.06	1.15	−6.348**	−4.375**	−3.713**	0.761

Note: *ME*: Trust-building Mechanism; *F1_ME*: networking; *F2_ME*: procedural measure; *F3_ME*: credit rating; *F4_ME*: calculativeness; *Upper-diagonal*: Correlation coefficient *r*; *Diagonal*: Cronbach's alpha; *Lower-diagonal*: t-value
significant level *p*; * $p < 0.05$; ** $p < 0.01$

Table 8.8 Means, standard deviations, internal consistency reliabilities, correlations and t-value of the principal component factors of trust expectation (n = 100)

Variable		μ	SD	1	2	3
1	F1_EP	4.43	1.09	0.900	0.725**	0.656**
2	F2_EP	4.55	1.07	3.205*	0.918	0.710**
3	F3_EP	4.98	1.07	1.936*	−1.011	0.797

Note: *EP*: Trust expectation; *F1_EP*: self-awareness; *F2_EP*: responsiveness; *F3_EP*: value congruence; *Upper-diagonal*: Correlation coefficient *r*; *Diagonal*: Cronbach's alpha; *Lower-diagonal*: t-value
Significant level *p*, * $p < 0.05$; ** $p < 0.01$

If the organisations overemphasise on the procedures, ineffective communication would result and cause misunderstanding, bad feelings and even distrust. According to Hardy et al. (1998: 69) and Zucker (1986: 93), trust is founded on common knowledge base because this underpins the predictability of the behaviours of the trustee. To facilitate the trustee in presenting relevant information promptly and frequently, the trustor also has an obligation to transact for a variety of reasons and exchange different contents. This knowledge base is described as multiplexity of network relations (Burt 1982). Only the trustee who is able to be explicit and precise on the dispute matters and simultaneously communicates these matters with the trustor can kick-start the trusting cycle. Credit rating is having subjective and perceptive elements. It can be as direct as first impression (McAllister 1995). Nevertheless, in most cases, subjectivity creeps in because the evaluation of the trustee against objective criteria is not easy for the lack of track records (Geyskens et al. 1996). The trustee can be more proactive in this regard notwithstanding that the trustor seldom relies singularly on the information provided by the trustee.

The findings suggest that levels of trust may grow or diminish with respect to certain mechanisms. While some mechanisms provide incentives to collaborate and trust, there are some of the mechanisms that stimulate distrust in business relationship (Lewicki et al. 1998). Networking is rated the highest among the other three trust-building mechanisms (Table 8.7). The results suggest that construction practitioners have readily used networking strategies in their contracting relationship. More importantly, self-awareness is the least commonly found trust expectation among the other three (Table 8.8). In construction contracting, a trustor, for example, incorporates the desirable level of control through contract specification within the terms of a negotiated agreement (Kilpatrick and Lapsley 1996). It is suggested that the more elaborated the rules under which the project is to be operated are specified, the easier it is for the trustee to circumvent these rules and frustrate his/her self-awareness.

The research finding is consistent with the notion of conditional trust in which "*sufficient positive affect and a relative lack of negative affect*" (Jones and George 1998) acts to reinforce trustworthy attitudes and leads to conditional trust (Morrow et al. 2004). The respondents' attitude toward trust building runs counter to the current trend of building closer contracting network (Lau and Rowlinson 2009). They distance themselves from their business partners and maintain their business efficacy relying solely on the contractual link (Khalfan et al. 2007). It prompts to a further research question whether particular trust-building mechanisms can lead to distrust contingently (Luo 2007). A trustful trustee should be able to embrace the knowledge from and the ability to understand a trustor's feeling immediately without too much thinking about it, learning it or discovering it by using reasoning skills. Bringing these conditions and the findings together, the challenges of the modern construction market-place center on the simultaneous management of trust and distrust in a hostile environment in which certain mechanism may be just as inclined to distrust even they are to trust.

8.6 Chapter Summary

Trust is important in construction contracting as a means to suppress the practice of opportunism arising from the inherent of risk and information asymmetries in delivering construction projects (Wong and Cheung 2004). Trust facilitates cooperation in project teams in which members are having their own interests (Cheung 2007; Mayer et al. 1995). A trusting contracting environment is a pre-requisite condition for dispute avoidance. Although a great deal of interest in trust has been expressed, developing trust remains a kind of lip service in the construction industry where confrontational and litigious culture prevails (Egan 1998; Latham 1994). The study extends the study of trust in construction in examining the inter-relationships between trust building mechanisms and trust expectations. In this connection, the attributes of trust-building mechanisms and trust expectations are identified. These attributes are categorised into taxonomies through PCFA. Four and three taxonomies are developed for trust-building mechanisms and trust expectations respectively. With these taxonomies, a relationship framework between trust-building mechanisms and trust expectations is developed and arranged under a structural equation model format. With data collected from construction professionals working in Hong Kong, the structural equation model is examined. Only two types of mechanism, networking and calculativeness, are found to be positively related to the trust expectations. Furthermore, a networking mechanism builds closer business relationship and is found to be the most effective mechanism in deriving reciprocating trusting behaviours. Calculativeness represents the strategies that seek to contain undesired conduct and capitalise on the opportunities made possible because of the desired conduct (Lewicki et al. 1998). Thus, this mechanism reinforces the positive expectation regarding the trustee's conduct. On the other hand, conventional methods like procedural measure and credit rating, which are used to install orders, are found to be negatively related to trust development. Arguably this would drive trust. However, when organisational structure and culture are taken into consideration, these may pose a distrusting outlook as the trustee is put under surveillance. It may not be seen as a sincere gesture (Kramer 1999). In sum, it is recommended that managers, in an effort to cultivate a trusting and cooperative business partnership, should utilise approaches that promote and maintain trust by enhancing network and promoting initiatives (i.e. networking and calculativeness). The findings prompt to further research on the versatility of or the conditions conducive for certain trust-building mechanisms in terms of the trusting behaviours that can be reciprocated. The study also finds that excessive use of procedural measure and credit rating would lead to trust deterioration.

Acknowledgments The content of this chapter has been published in Volume 30(8) of International Journal of Project Management and is used with the permission from Elsevier.

References

Akgün, A. E., Byme, J. C., Lynn, G. S., & Keskin, H. (2007). Organizational unlearning as changes in beliefs and routines in organizations. *Journal of Organizational Change Management, 20*(6), 794–812.

Alvarez, S., Barney, J., & Bosse, D. (2003). Trust and its alternatives. *Human Resource Management, 42*(4), 393–404.

Arrindell, W. A., Sanavio, E., Aguilar, G., Sica, C., Hatzichristou, C., Eisemann, M., et al. (1999). The development of a short form of the EMBU: Its appraisal with students in Greece, Guatemala, Hungary and Italy. *Personality and Individual Differences, 27*(4), 613–628.

Aulakh, P. S., Kotabe, M., & Sahay, A. (1996). Trust and performance in cross-border marketing partnerships: A behavioral approach. *Journal of International Business Studies, 27*(5), 1005–1032.

Ba, S., & Pavlou, P. A. (2002). Evidence of the effect of trust building technology in electronic markets: Price premiums and buyer behavior. *MIS Quarterly: Management Information Systems, 26*(3), 243–268.

Barney, J. B., & Hansen, M. H. (1994). Trustworthiness as a source of competitive advantage. *Strategic Management Journal, 15*, 175–190.

Bigley, G. A., & Pearce, J. L. (1998). Straining for shared meaning in organization science problems of trust and distrust. *Academy of Management Review, 23*(3), 405–421.

Bollen, K. A., & Long, J. S. (1992). Tests for structural equation models: Introduction. *Sociological Methods & Research, 21*(2), 123–131.

Brennan, E. M., & Brannan, A. M. (2005). Participation in the paid labor force by caregivers of children with emotional and behavioral disorders. *Journal of Emotional and Behavioral Disorders, 13*(4), 237–246.

Browne, M. W., & Cudeck, R. (1993). Alternative ways of assessing model fit. In K. A. Bollen & J. S. Long (Eds.), *Testing structural equation models* (pp. 445–455). Newbury Park, CA: Sage Publications.

Burt, R. S. (1982). *Toward a structural theory of action: Network models of social structure, perception, and action*. New York: Academic Press.

Butler, J. (2008). An account of oneself. In B. Davies (Ed.), *Judith Butler in conversation: Analyzing the texts and talk of everyday life* (pp. 19–38). New York: Routledge.

Butler, J. K. (1991). Toward understanding and measuring conditions of trust: Evolution of a conditions of trust inventory. *Journal of Management, 17*(3), 643–663.

Butler, J. K. (1999). Trust expectation, information sharing, climate of trust, and negotiation effectiveness and efficiency. *Group and Organization Management, 24*(2), 217–236.

Cheung, S. O. (2007). *Trust in co-operative contracting in construction*. Hong Kong: City University of Hong Kong Press.

Cheung, S. O., Ng, T. S. T., Wong, S. P., & Suen, H. C. H. (2003). Behavioral aspects in construction partnering. *International Journal of Project Management, 21*(5), 333–343.

Chou, C. P., & Bentler, P. M. (1990). Model modification in covariance structure modeling: A comparison among likelihood ratio, Lagrange multiplier, and Wald tests. *Multivariate Behavioral Research, 25*(1), 115–136.

Colquitt, J. A., Scott, B. A., & LePine, J. A. (2007). Trust, trustworthiness, and trust propensity: A meta-analytic test of their unique relationships with risk taking and job performance. *Journal of Applied Psychology, 92*(4), 909–927.

Costa, A. C., & Bijlsma-Frankema, K. (2007). Trust and control interrelations: New perspectives on the trust-control nexus. *Group and Organization Management, 32*(4), 392–406.

Cummings, L. L., & Bromiley, P. (1996). The organizational trust inventory (OTI)—Development and validation. In R. Kramer & T. Tyler (Eds.), *Trust in organizations* (pp. 302–330). Thousand Oaks, CA: Sage Publications.

Currall, S. C., & Inkpen, A. C. (2006). On the complexity of organizational trust: A multi-level co-evolutionary perspective and guidelines for future research. In R. Backmann & A. Zaheer (Eds.), *Handbook of trust research* (pp. 235–246). Cheltenham: Edward Elgar.

Das, T. K., & Teng, B. S. (1998). Between trust and control: Developing confidence in partner cooperation in alliances. *Academy of Management Review, 23*(3), 491–512.

Doney, P. M., & Cannon, J. P. (1997). An examination of the nature of trust in buyer-seller relationships. *Journal of Marketing, 61*(2), 35–51.

Doz, Y. (1996). The evolution of cooperation in strategic alliances: Initial conditions or learning processes? *Strategic Management Journal, 17*(1), 55–84.

Egan, J. (1998). *Rethinking construction*. London: Department of the Environment, Transport and the Regions, HMSO.

Eriksson, P. E. (2008). Procurement effects on competition in client-contractor relationships. *Journal of Construction Engineering and Management, 134*(2), 103–111.

Evans, A. M., & Krueger, J. I. (2011). Elements of trust: Risk and perspective-taking. *Journal of Experimental Social Psychology, 47*(1), 174–177.

Far East Trade (Ed.). (2003). *Hong Kong builder directory*. Hong Kong: Far East Trade Press (35).

Fava, J. L., & Velicer, W. F. (1992). The effects of overextraction on factor and component analysis. *Multivariate Behavioral Research, 27*(3), 387–415.

Gefen, D. (2003). Trust and tam in online shopping: An integrated model. *MIS Quarterly: Management Information Systems, 27*(1), 51–90.

Geyskens, I., Steenkamp, J.-B. E. M., Scheer, L. K., & Kumar, N. (1996). The effects of trust and interdependence on relationship commitment: A trans-Atlantic study. *International Journal of Research in Marketing, 13*(4), 303–317.

Gummesson, E. (Ed.). (2002). *Total relationship marketing* (2nd ed.). Oxford: Butterworth-Heinemann.

Hair, J. F., Anderson, R. E., Tatham, R. L. & Black, W. C. (1998). *Multivariate data analysis*, (5th ed., pp. 88–92). Upper Saddle River, NJ: Prentice Hall.

Hardy, C., Philips, N., & Lawrence, T. (1998). Distinguishing trust and power in interorganizational relations: Forms and facades of trust. In L. Christel & B. Richard (Eds.), *Trust within and between organizations: Conceptual issues and empirical applications* (pp. 64–87). Oxford: Oxford University Press.

Houde, L. D., Sherman, D. M., White, T. B., & Sheppard, B. H. (2004). The four faces of trust: An empirical study of the nature of trust in relational forms. In N. Haslam (Ed.), *Relational models theory: A contemporary overview* (pp. 287–308). Mahwah, NJ: Lawrence Erlbaum.

Jashapara, A. (2003). Cognition, culture and competition: An empirical test of the learning organization. *Learning Organization, 10*(1), 31–50.

Jeffries, F. L., & Reed, R. (2000). Trust and adaptation in relational contracting. *Academy of Management Review, 25*(4), 873–882.

Johnson, M. D., & Selnes, F. (2004). Customer portfolio management: Toward a dynamic theory of exchange relationship. *Journal of Marketing, 68*(2), 1–17.

Jones, G. R., & George, J. M. (1998). The experience and evolution of trust: Implications for cooperation and teamwork. *Academy of Management Review, 23*(3), 531–546.

Khalfan, M. M. K., McDermott, P., & Swan, W. (2007). Building trust in construction projects. *Supply Chain Management: An International Journal, 12*(6), 385–391.

Kilpatrick, K., & Lapsley, I. (1996). Trust us'—Regulators and the governance of privatized utilities. *Public Money and Management, 16*(2), 39–46.

Kimber, D., & Raghunath, S. (2002). Discovering the needle of trust in the haystack of distrust: International and corporate citizenship alliances—Indian/Australian experiences. In P. V. Rao (Ed.), *Contributed Papers Presented at an International Seminar Held at Hyderabad on February 20–21* (Vol. IV(14), pp. 279–302).

Kjærnes, U. (2006). Trust and distrust: Cognitive decisions or social relations? *Journal of Risk Research, 9*(8), 911–932.

Kline, R. B. (Ed.). (1998). *Structural equation modeling*. New York: Guilford Press.

Kramer, R. M. (1999). Trust and distrust in organizations: Emerging perspectives, enduring questions. *Annual Review of Psychology, 50*(1), 569–597.

Krishnan, R., Martin, X., & Noorderhaven, N. (2006). When does trust matter to alliance performance? *Academy of Management Journal, 49*(5), 894–917.

Kuriyan, R., Kitner, K., & Watkins, J. (2010). ICTs, development and trust: An overview. *Information Technology & People, 23*(3), 216–221.

Latham, M. (1994). *Constructing the team. joint review of procurement and arrangements in the United Kingdom construction industry: Final report*. London: HMSO.

Lau, E., & Rowlinson, S. (2009). Interpersonal trust and inter-firm trust in construction projects. *Construction Management and Economics, 27*(6), 539–554.

Lewicki, R. J., & Stevenson, M. A. (1997). Trust development in negotiation: Proposed actions and a research agenda. *Business & Professional Ethics Journal, 16*(1), 99–132.

Lewicki, R. J., McAllister, D. J., & Bies, R. J. (1998). Trust and distrust: New relationships and realities. *Academy of Management Review, 23*(3), 438–458.

Lewicki, R. J., Tomlinson, E. C., & Gillespie, N. (2006). Models of interpersonal trust development: Theoretical approaches, empirical evidence, and future directions. *Journal of Management, 32*(6), 991–1022.

Lewis, J. D., & Weigert, A. (1985). Trust as a social reality. *Social Forces, 63*(4), 967–985.

Luo, Y. (2007). The independent and interactive roles of procedural, distributive, and interactional justice in strategic alliances. *Academy of Management Journal, 50*(3), 644–664.

Mayer, R. C., Davis, J. H., & Schoorman, F. D. (1995). An integrative model of organizational trust. *Academy of Management Review, 20*(3), 709–734.

McAllister, D. J. (1995). Affect- and cognition-based trust as foundations for interpersonal cooperation in organizations. *Academy of Management Journal, 38*(6), 24–59.

McKnight, H., & Chervancy, N. L. (2000). *An extended trust building model: Comparing experiential and non-experiential factors*. Hershey, PA: Idea Group Publishing.

Mishra, A. K. (1996). Organizational responses to crisis: The centrality of trust. In R. M. Kramer & T. R. Tyler (Eds.), *Trust in organizations: Frontiers of theory and research* (pp. 261–287). Thousand Oaks, CA: Sage Publications.

Molenaar, K., Washington, S., & Diekmann, J. (2000). Structural equation model of construction contract dispute potential. *Journal of Construction Engineering and Management, 126*(4), 268–277.

Moorman, C., Deshpande, R., & Zaltman, G. (1993). Factors affecting trust in market research relationships. *Journal of Marketing, 57*(1), 81–101.

Morrow, J. L., Hansen, M. H., & Pearson, A. W. (2004). The cognitive and affective antecedents of general trust within cooperative organizations. *Journal of Managerial Issues, 14*(1), 48–64.

Nooteboom, B. H., Berger, H., & Noorderhaven, N. G. (1997). Effects of trust and governance on relational risk. *Academy of Management Journal, 40*(3), 308–338.

Paiva, E. L., Roth, A., & Fensterseifer, J. E. (2008). Organizational knowledge and the manufacturing strategy process: A resource-based view analysis. *Journal of Operations Management, 26*(1), 115–132.

Parkhe, A. (1998). Understanding trust in international alliances. *Journal of World Business, 33*(3), 219–240.

Pavlou, P. A., & Gefen, D. (2004). Building effective online marketplaces with institution-based trust. *Information Systems Research, 15*(1), 37–59, 108.

Peters, T. J., & Waterman, R. H. (2004). *In search of excellence: Lessons from America's best-run companies*. London: Profile Books.

Pretty, J., & Ward, H. (2001). Social capital and the environment. *World Development, 29*(2), 209–227.

Ratnasingam, P. (2005). Trust in inter-organizational exchanges: A case study in business to business electronic commerce. *Decision Support Systems, 39*(3), 525–544.

Ratnasingham, P. (1998). The importance of trust in electronic commerce. *Internet Research, 8*(4), 313–321.

Ross, W., & LaCroix, J. (1996). Multiple meanings of trust in negotiation theory and research—A literature review and integrative model. *International Journal of Conflict Management, 7*(4), 314–360.

Rousseau, D. M., Sitkin, S. B., Burt, R. S., & Camerer, C. (1998). Not so different after all: A cross-discipline view of trust. *Academy of Management Review, 23*(3), 393–404.

Rousseau, D. M. (Ed.). (1995). *Psychological contracts in organization*. Thousand Oaks, CA: Sage Publications.

Schoorman, F. D., Mayer, R. C., & Davis, J. H. (2007). An integrative model of organizational trust: past, present, and future. *Academy of Management Review, 32*(2), 344–354.

Sengün, A. E., & Wasti, S. N. (2007). Trust, control and risk—A test of Das and Teng's conceptual framework for pharmaceutical buyer-supplier relationships. *Group and Organization Management, 32*(4), 430–464.

Sharma, S. (Ed.). (1996). *Applied multivariate techniques* (pp. 116–123). New York: Wiley.

Shaw, R. B. (Ed.). (1997). *Trust in the balance: Building successful organizations on results, integrity, and concern*. San Francisco, CA: Jossey-Bass Publishers.

Sheppard, B. H., & Sherman, D. M. (1998). The grammars of trust: A model and general implications. *Academy of Management Review, 23*(3), 422–437.

Sitkin, S. B., & Roth, N. L. (1993). Explaining the limited effectiveness of legalistic "remedies" for trust/distrust. *Organization Science, 4*(3), 367–392.

Smith, C. A., Organ, D. W., & Near, J. P. (1983). Organizational citizenship behavior: Its nature and antecedents. *Journal of Applied Psychology, 68*(4), 653–663.

Smith, R. L., & Smith, J. K. (2004). *Entrepreneurial finance*. New York: Wiley.

Velicer, W. F., & Jackson, D. N. (1990). Component analysis versus common factor-analysis—Some further observations. *Multivariate Behavioral Research, 25*(1), 97–114.

Verbeke, A., & Greidanus, N. S. (2009). The end of the opportunism vs trust debate: Bounded reliability as a new envelope concept in research on MNE governance. *Journal of International Business Studies, 40*(9), 1471–1495.

Walker, D. H. T. (2003). Implications of human capital issues. In D. Walker & K. Hampson (Eds.), *Procurement strategies a relationship-based approach*, (pp. 258–296). Malden, MA: Blackwell Science.

Williamson, O. E. (1993). Calculativeness, trust, and economic organization. *Journal of Law and Economics, 36*(1), 453–486.

Wong, P. S. P., & Cheung, S. O. (2004). Trust in construction partnering: views from parties of partnering dance. *International Journal of Project Management, 22*(6), 437–446.

Wong, P. S. P., & Cheung, S. O. (2005). Structural equation model on trust and partnering success. *Journal of Management in Engineering, 21*(2), 70–80.

Wong, P. S. P., Cheung, S. O., Yiu, T. W., & Pang, H. Y. (2008). A framework for trust in construction contracting. *International Journal of Project Management, 26*(8), 821–829.

Wright, S. (2010). Trust and trustworthiness. *Philosophia, 38*(3), 615–627.

Zaheer, A., McEvily, B., & Perrone, V. (1998). Does trust matter? Exploring the effects of interorganizational and interpersonal trust on performance. *Organization Science, 9*(2), 141–159.

Zucker, L. G. (1986). Production of trust: Institutional sources of economic structure. *Research in Organizational Behavior, 8*(1), 53–111.

Chapter 9
Developing a Trust Inventory for Construction Contracting

Sai On Cheung, Wei Kei Wong, Tak Wing Yiu and Hoi Yan Pang

Abstract Trust is central to every transaction that demands contributions from the parties involved. A trusting contracting environment facilitates problem solving, thus reduces happening of dispute. In construction, trust has been identified to be the key driver in fostering cooperation. Moreover, how to measure trust is inherently difficult. By operationalising a trust framework that includes system-based, cognition-based and affect-based trust, a trust inventory is proposed. The reliability and stability of the inventory were then validated through the test-retest methodology. The proposed trust inventory can be used to assess trust pattern akin to the assessment of conflict handling style and the measurement of depression through the use of the Rahim's Organisational Conflict Inventory and the Inventory of Measuring Depression respectively. Supporting view on the appropriateness of the trust framework and the potential uses of the trust inventory were confirmed with two senior construction professionals.

S. O. Cheung (✉) · W. K. Wong · H. Y. Pang
Construction Dispute Resolution Research Unit, Department of Civil and Architectural Engineering, City University of Hong Kong, Hong Kong, People's Republic of China
e-mail: saion.cheung@cityu.edu.hk

W. K. Wong
e-mail: wong_vickie@hotmail.com

H. Y. Pang
e-mail: karen-pang@hotmail.com

T. W. Yiu
Department of Civil and Environmental Engineering, Faculty of Engineering, The University of Auckland, Auckland, New Zealand
e-mail: k.yiu@auckland.ac.nz

S. O. Cheung (ed.), *Construction Dispute Research*,
DOI: 10.1007/978-3-319-04429-3_9,

9.1 Introduction

Trust is central to every transaction that demands contributions from the parties involved (Lewicki and Bunker 1995, 1996; Williamson 1975, 1981, 1993). The 2008 financial turmoil resulting from the credit crunch that has troubled the global financial markets is an illustrative example of how the absence of trust paralysed the banking credit system. The British Prime Minister, Gordon Brown, wrote in The Times on 10th October 2008 … *"Until only a few weeks ago, few, if any, appreciated the real significance of the money markets within the wider global financial crisis and the importance of trust in these markets. But the freezing of the market for medium-term funding reflects a total loss of trust between banks. …… This paralysis of lending from loss of confidence jeopardises the flow of money to every family and every business in the country."* Analogously, if parties in the construction supply chain do not trust each other, their skepticism may not completely paralyse the operation but will certainly create unnecessary enquiries and checking procedures, resulting in serious bottlenecks and inefficiency (Latham 1994). These deficiencies may cause disputes.

Changes are common during the construction phase of a project. In a distrusting environment, developers always assess the submission of the contractor with respect to change orders with an opportunistic lens. Likewise, contractors often inflate their submission in anticipation of hostile and skeptical evaluation. This scenario is extremely common in construction with a dispute being the ultimate outcome.

Trust is a fundamental ingredient or lubricant of social interaction (Gambetta 1998) and its positive impact on communication (Giffin 1967), leadership (Atwater 1988), management (Scott 1980), negotiation (Bazerman 1994), game theory (Milgrom and Roberts 1992), performance (Cummings 1983), labor relations (Taylor 1989), self-managed work teams (Lawler 1992), construction project management (Kadefors 2004; Wong and Cheung 2004, 2005, 2006; Wong et al. 2003, 2005; Wood and McDermott 1999) and owner/contractor relationships (Pinto et al. 2009), has been well documented. Notwithstanding, having a trusting contracting environment is an exception rather than the norm. In the connection, creating a trusting contracting environment has been identified as a major reform that can revitalise the construction industry. The notable progress in this regard is the use of partnering. By promoting trust, partnering has been identified as an effective tool to alleviate adversarial relationships (Bayliss et al. 2004; Cheung et al. 2003; Kwan and Ofori 2001; Wong and Cheung 2004, 2005; Wong et al. 2005). In fact, trust has been named as one of the most important pillars supporting the success of partnering (Pinto et al. 2009; Wong and Cheung 2004; Wood and McDermott 1999). In construction contracting, trust has been generically operationalised as the willingness of project team members to share information. This definition highlights the mutual dependence of contracting parties and a trusting environment can foster appropriate information sharing so that both can honor their commitments. Furthermore, Hannah (1991) attributed trust as a contributing factor to participants' satisfaction in his study of U.S. construction projects. CII (1993) concluded that trust-related procedures could provide maximum cost savings in construction project. Wong et al. (2008) affirmed the correlation of trust to 'performance', 'acting with integrity' and 'demonstrating concern'. A trust model was thus developed to fit in the nature of risk allocation in the construction industry (Zaghloul and Hartman 2003). In this connection, trusting relationship can facilitate project cost reduction. Recent studies including some conceptual frameworks on trust and project relationships between client and contractor have been reported (Huemer 2004; Kadefors 2004; Smyth 2003). They suggested that the presence of trust is crucial to overcome the adversarial outlook of construction industry although there is little empirical study to support this view. Wong et al. (2008) proposed a 3-trust type framework for construction contracting. This study aims to develop this framework into a trust inventory (hereafter the proposed inventory). The three trust types included in the framework are system-based, cognition-based and affect-based. System-based trust found on performance and faith in the system. Cognition-based trust is built on knowledge and understanding. Affect-based trust addresses feelings and emotions, thus tends to be more personal. These three types of trust co-exist and are mutually dependent. A system is only as good as its weakest point; hence a trust building project manager must install credible system and care for the team members. Although trust has been advocated as the key factor in enhancing efficiency of the construction industry, there has yet reported attempt in developing an instrument to evaluate trust status. This study aims to fill this gap. Measurement instrument is often described as inventory. Notable examples include Organisational Conflict Inventory (Rahim 1983), Organisational Culture Inventory (Cooke and Szumal 1993) and Inventory of Interpersonal Problems (Horowitz 1988). Upon completion of the proposed trust inventory, the level of trust by types can be evaluated. At project level, regular periodic evaluations shall provide longitudinal data on inter-organisational trust status. This information shall inform management actions. Another use of the inventory is to provide trust status for studies where trust is to be assessed.

9.2 The Study

There are three stages of work to develop an inventory (Table 9.1). The same developmental process has been used to develop measurement scale for managerial trust (Butler 1991), organisational trust (Cummings and Bromiley 1996) and interpersonal trust (Johnson-George and Swap 1982). Stage I involves the development of a theoretical trust framework. Stage II operationalises the framework into an inventory to be tested in Stage III. In essence, the elements of trust in the framework are firstly operationalised into trust behaviour statements. The inventory is then validated by a test-retest (Stage III). The validation involves the checking of reliability and constructs validity.

Table 9.1 Summary of a trust inventory development framework

Stage	Research tasks	Objectives	Deliverables
Conceptual development			
I	Review the trust types	To retrieve the trust dimensions of traditional construction contracting	Contracting conceptual model that describes the trust dimensions and their sub-elements
II	Operational sing Trust in the Construction Industry	To develop measurement statements for the trust inventory by examining previous studies on trust	Statements that best describe the trust types and their sub-elements
Empirical development			
III	Validation of the Inventory	Statistical analyses for validating the trust inventory	Results of statistical tests on the reliability, validity and stability of the trust inventory

9.2.1 Stage I: A Trust Framework for Construction Contracting

Trust has been one of the key research areas in the fields of social science (Kramer 1999; Lewis and Weigert 1985; Luhmann 1979; Rousseau et al. 1998; Yamagishi 1988), economics (Glaser et al. 2000; Zucker 1986), and organisational behaviour (Farris et al. 1973; Hartman 2000; McAllister 1995; McKnight et al. 1998; Whitener et al. 1998). Key research foci include the bases upon which trust can grow and the developed scales. For the purpose of this study, a summary of the published trust scales is provided in Table 9.2.

McAllister (1995) developed a scale to assess the interpersonal trust. Ding and Ng (2007) investigated the reliability and validity of the Chinese version of McAllister's two-dimensional trust scale (1995) with a group of architectural design professions in Hong Kong. These scales are used at interpersonal setting. More recently, instead of adopting a single scale, the authors after reviewing a number of trust studies (e.g. Hartman 2000; Kramer 1999; Lewis and Weigert 1985; Luhmann 1979; McAllister 1995; Rousseau et al. 1998), proposed a framework for inter-organisational trust in construction contracting (Wong et al. 2008). The following sections detail the development work of the proposed trust inventory for construction contracting.

The authors have proposed a trust framework for construction contracting that includes three types of trust; system-based, cognition-based and affect-based (Wong et al. 2008). System-based trust focuses on formalised and procedural arrangements (Lewis and Weigert 1985). These arrangements can build trust and strengthen communication channel between contracting parties because of the certainties derived from the system. Cognition-based trust develops from the confidence built upon objective knowledge that reveals the trust-worthiness of

Table 9.2 A list of trust inventories/scales

Trust inventories/Scales	Descriptions	References
Interpersonal trust scale	This scale measures one's expectation on the behaviour and promises from statements made by the other	Rotter (1967)[a]
Scale of interpersonal trust at work	This scale measures the trust between mutually dependent work groups within an organisation	Cook and Wall (1980)[a]
Specific interpersonal trust scale	This scale measures trust in interpersonal relationships	Johnson-George and Swap (1982)[b]
Conditions of trust inventory (CTI)	This inventory measures the conditions of managerial trust and conditions of trust between people in other types of relationships	Butler (1991)[c]
Behavioural response and interpersonal trust measures	Instruments of assessing affect-based, cognition-based trust level and behavioural responses associated with trusting or distrusting peers were developed	McAllister (1995)[c]
Scale of measuring boundary role persons (BRP) trust	This scale assesses trust between BRPs	Currall and Judge (1995)[a]
The organisational trust inventory (OTI)	The OTI measures the affective and cognitive trust between units of commercial organisations	Cummings (1983)[a]
Trust inventory for social relations	This inventory measures individual's trust level in the three classes of social relations: romantic, partners, family and friends	Couch et al. (1996)[d]
Scale of measuring intra-organisational trust	This scale measures and operationalisation of intra-organisational trust that reflects the essential elements of trust inside workplace	Dietz and Hartog (2006)[a]
The propensity to trust survey (PTS) trust scale	This scale measures individual differences in trust and trustworthiness	Evans and Revelle (2008)[b]

[a] Organisational behaviour study
[b] Personality study
[c] Managerial study
[d] Social study

the contracting parties. The exchange of such objective knowledge can be attained by interaction or communication. Affect-based trust builds on a sentimental platform. It describes an emotional bond tying individuals who invest in personal attachment and being thoughtful to each other (Lewis and Weigert 1985). Furthermore, these trust types were further divided into seven elements and twenty-three characterising trust behaviours (Table 9.3).

The proposed trust inventory for construction contracting builds on this framework as this has been considered by construction professionals in Hong Kong for its suitability in use in construction contracting (Wong et al. 2008).

Table 9.3 A trust framework for construction contracting (Wong et al. 2008)

Trust types	Elements	Characterising trust behaviours
System-based	Organisational policy	*II8.* Sufficient organisational resources in response to contracting parties' needs increase the sense of belonging to the organisation
		II12. An organisation should clearly define the job tasks required of individuals
		II14. Good management of organisation system reinforces goal achievement such as continual improvement, profit making and business expanding
		II23. Organisation policy should be clearly specified for solving cost, time, risk and safety issues
	Communication system	*II4.* Using effective communication methods are essential at work
		II9. Formal communications with working partners should be documented in a systematic way
		II21. Keeping a good communication system would avoid ambiguous situations and discrepancies occurring at all times
	Contracts and agreements	*II2.* A clearly defined contract document brings confidence and comforts to all parties
		II15. Information in the contract document should be explainable to parties who may be affected
		II16. Clarification of contract terms and agreements is important before the commencement of work to minimise future arguments
Cognition-based	Communication/interaction	*II1.* Keeping a long-term relationship with the other party has the benefit of maintaining better communication between individuals
		II10. Good interaction allows me to obtain more information from the other party
		II13. Attending work-related interaction frequently facilitates better understanding between individuals
		II18. Open and honest communication enables more work-related information exchange between individuals
	Knowledge	*II3.* Track record is an essential tool to judge the other party's competence and consistency level
		II5. Financial stability is one of the factors in evaluating a company's reliability
		II7. The other party will have confidence to work with me if I have a good reputation of being honest
Affect-based	Being thoughtful	*II19.* Showing care and concern to my workmate at appropriate time impresses his/her feeling more comfortable to work with me
		II20. Being considerate is a tool to understand an individual's needs and feeling at work so as to achieve his/her maximum capacity
		II22. Taking each party's needs into account in decision-making process encourages a compromising and satisfactory outcome
	Emotional investments	*II6.* Having a good personal relationship with the other party may also improve working relationship with him/her
		II11. I am more likely to rely on a working partner whom I have good impression
		II17. Spending appropriate time, energy and effort to understand other party's personal details and work background eliminates frictions between each other at work

9.2.2 Stage II: Operationalising for a Trust Inventory

This stage aims at operationalising the trust framework by reducing the trust types into characterising trust behaviours. The process therefore involves transforming the trusting behaviours described in Table 9.3 into measurement statements. To avoid leading statements, simple statements without the leading words such as system-based, cognition-based and affect-based trust are used. Furthermore, to avoid multiple interpretations, approximately equal numbers of items are used for each of the trust dimension. In these regards, the trust behaviour: "*An organisation should clearly define the job tasks required of individuals*" has been transformed to "*We shall clearly define the job tasks required of individuals*" as a measurement statement. A total of 25 measurement statements were then developed for the proposed inventory (Table 9.4).

9.2.3 Stage III: Inventory Validation

The aim of this stage is to examine the reliability and construct validity of the proposed trust inventory. Table 9.5 summarises the steps involved in validating the proposed trust inventory. The statistical tools to be used are also listed for ease of reference.

In essence, an inventory is validated by the well-established test-retest approach that measures the stability and reliability of a survey instrument over time (Beck et al. 1988; Cooke and Szumal 1993; Horowitz 1988; Oshio 2009). In this case, the survey instrument is the proposed trust inventory. Thus, a good set of test-retest result suggests that the inventory is reasonably stable over time. Test-retest involves applying the same test to the same respondent at two points in time (known as test-retest interval). Previous researches have used a test-retest interval varying from two weeks to a few months (Garson 2006; Nunnally and Bernstein 1994). It should be noted that if the test-retest interval is too short, the retest is susceptible to memory effect. However, too long a time interval is also not preferred since the conditions being assessed may have changed significantly due to other natural courses of event. In this regard, this study applied a 3-month test-retest interval (Kline 1993). Accordingly, the data collection questionnaire is given to a respondent twice (i.e. the 1st and the 2nd questionnaire surveys) at the test-retest period respectively. With the data collected from the test and retest, the reliability and construct validity of the proposed inventory were analysed. The steps taken in the inventory validation process are detailed here follows:

Table 9.4 The measurement statements of the proposed trust inventory

Trust types	Elements	Measurement statements
System-based	Organisational policy	II8. In response to contracting parties' needs, we shall increase the sense of belonging to the organisation by providing sufficient organisational resources
		II12. We shall clearly define the job tasks required of individuals
		II14. We shall reinforce goal achievement such as continual improvement, profit making and business expanding
		II23. We shall clearly specify for solving cost, time, risk and safety issues
	Communication system	II4. We shall use effective communication at work
		II9. We shall document the formal communications with working partners in a systematic way
		II21. We shall keep a good communication system to avoid ambiguous situations and discrepancies
	Contracts and agreements	II2. We shall clearly define contract document so as to bring confidence and comforts to all parties
		II15. We shall explain the information in the contract document to the parties who may be affected
		II16. We shall clarify the contract terms and agreements before the commencement of work to minimise future arguments
Cognition-based	Communication/interaction	II1. We think that keeping a long-term relationship with the other party has the benefit of maintaining better communication between individuals
		II10. We think that a good interaction allows me to obtain more information from the other party
		II13. We think that attending work-related interaction frequently facilitates better understanding between individuals
		II18. We think that open and honest communication enables more work-related information exchange between individuals
	Knowledge	II3. We think that track record is an essential tool to judge the other party's competence and consistency level
		II5. We think that financial stability is one of the factors in evaluating a company's reliability
		II7. We think that the other party will have confidence to work with me if we have a good reputation of being honest
Affect-based	Being thoughtful	II19. We feel that showing care and concern to my workmate at appropriate time impresses his/her feeling more comfortable to work with me
		II20. We feel that being considerate is a tool to understand an individual's needs and feeling at work so as to achieve his/her maximum capacity
		II22. We feel that taking each party's needs into account in decision-making process encourages a compromising and satisfactory outcome
	Emotional investments	II6. We feel that having a good personal relationship with the other party may also improve working relationship with him/her
		II11. We feel that we are likely to rely on a working partner whom we have good impression
		II17. We feel that spending appropriate time, energy and effort to understand other party's personal details and work background eliminates frictions between each other at work

Table 9.5 Inventory validation

Development process	Methodology	
Reliability and validity analyses		
Reliability analysis-first test	Step 1:	1st questionnaire survey
	Step 2:	Reliability analyses—assessing the internal consistency
		Cronbach's alpha (Internal consistency)
		Inter-item correlation (Discriminatory ability)
		Item-total correlation (Homogeneity)
	Step 3:	Construct validity
		Structural equation modeling: goodness-of-fit tests
Reliability analysis-second test (Retest)	Step 4:	2nd questionnaire survey
	Step 5:	Repeat step 2
	Step 6:	Repeat step 3
	Step 7:	Stability analyses (consistency of the inventory measurement statements over time)
		Pearson's correlation

9.2.3.1 Steps 1 and 4: The First and the Second Questionnaire Surveys

A questionnaire consists of the twenty-three statements developed in Stage II of the study was used to collect construction professionals' degree of agreement of the statements in a Likert scale of 1 (strongly disagree) to 7 (strongly agree). Two sets of data were collected from the same group of respondents in a 3-month interval.

A total of 467 questionnaires were sent to project managers, architects, engineers, contract/legal advisers, quantity surveyors and project coordinators in Hong Kong. They represent a broad cross-section of roles and professions in the industry. The list was compiled by identifying key personnel from the government and professional directories and web sites of companies. There were 163 responses to both surveys with a response rate of 34.9 %. Over 70 % of the respondents were practicing professionals with over 10-year experience. Table 9.6 summarises the profiles of the respondents.

9.2.3.2 Steps 2 and 5: Reliability Analyses: Assessing the Internal Consistency of the Test and Retest Data

Reliability analysis is an assessment of the degree of consistency or repeatability of an item, scale or instrument (Hair et al. 1998). This can be achieved by the methods of (1) conducing test-retest and (2) assessing internal consistency. In Table 9.5, steps 2 and 5 are related to reliability analysis.

For each of step 2 and 5, the internal consistency of the dataset was assessed by *Cronbach's alpha*, *inter-item correlation* and *item-total correlation*. The rationale of these tests is that the statements of the inventory should all be measuring the same construct and thus be highly inter-correlated (Hair et al. 1998). As a rule of thumb suggested by Nunnally and Bernstein (1994), the Cronbach's alpha value of

Table 9.6 Profile of respondents (by organisational types and professions)

Organisational types	Percentage (%)
Client	29.45
Consultant	27.61
Contractor	42.94
	100.00
Professions	**Percentage (%)**
Project manager	17.79
Architect	7.97
Engineer	39.88
Contract/legal adviser	3.07
Quantity surveyor	26.38
Project coordinator	4.91
	100.00

0.70 is the threshold for acceptance. *Inter-item correlation* measures the relationships among all statements for assessing the consistency of the proposed inventory (Ferketich 1991). To assess the internal consistency reliability, the proportion of the statement with correlation coefficient with other statements within the range 0.20–0.70 is considered (Idvall et al. 2002). If the proportion is less than 50 %, that statement is considered lacking inter-correlation with other statements. In addition, *item-total correlation* measures the relationships between a statement and the total score from the collection of statements within the proposed inventory (Robinson et al. 1991). The item-total correlation should achieve a value >0.30 (the acceptance threshold) for inclusion in the proposed inventory (Nunnally and Bernstein 1994). These standards have been applied in assessing reliability of measurement scales as suggested by Robinson et al. (1991) and Knapp and Brown (1995).

Accordingly, data collected for the first test was analysed. The Cronbach's alpha value is 0.94, which is greater than the acceptance threshold of 0.70 (Hair et al. 1998). Figures 9.1 and 9.2 show the inter-item correlations for the Test and Retest respectively. In the Test (Fig. 9.1), the proportion of the statement correlating with the other statements is represented by P_T. For example, the P_T of the statement II2 of System-based Trust is 9/9 in the Test. That means the inter-item correlation of statement II2, with all other nine statements, II4, II8, II9, II12, II14, II15, II16, II21 and II23, are within the acceptance range of 0.20–0.70 ($p < 0.01$). For each statement in System-based Trust, the proportion of correlation with the other statements varied between 8/9 and 9/9. The P_T of the statements in Affect-based Trust varied between 4/5 and 5/5, while the P_T of the statements in the Cognition-based Trust is 6/6. This suggested that each of the statement has adequate inter-correlation (i.e. 50 % as the acceptable level) with other statements in the proposed inventory.

Furthermore, the reliability of the retest data was fulfilled as well. Cronbach's alpha value is 0.90. The proportion of the statement correlating with other statements between 0.20 and 0.70 in the Retest (P_R) varied between 5/9 and 9/9 for

System-based Trust

Items	II2	II4	II8	II9	II12	II14	II15	II16	II21	II23
II2	1.000									
II4	0.626*	1.000								
II8	0.294*	0.345*	1.000							
II9	0.568*	0.555*	0.421*	1.000						
II12	0.206*	0.258*	0.097	0.322*	1.000					
II14	0.430*	0.529*	0.420*	0.447*	0.429*	1.000				
II15	0.542*	0.466*	0.226*	0.457*	0.431*	0.526*	1.000			
II16	0.437*	0.451*	0.348*	0.457*	0.258*	0.478*	0.493*	1.000		
II21	0.462*	0.550*	0.363*	0.540*	0.305*	0.641*	0.549*	0.377*	1.000	
II23	0.257*	0.327*	0.313*	0.333*	0.299*	0.404*	0.394*	0.254*	0.489*	1.000
P_T	9/9	9/9	8/9	9/9	8/9	9/9	9/9	9/9	9/9	9/9

Cognition-based Trust

Items	II1	II3	II5	II7	II10	II13	II18
II1	1.000						
II3	0.485*	1.000					
II5	0.473*	0.309*	1.000				
II7	0.544*	0.413*	0.596*	1.000			
II10	0.442*	0.411*	0.449*	0.570*	1.000		
II13	0.408*	0.398*	0.448*	0.538*	0.551*	1.000	
II18	0.445*	0.299*	0.418*	0.514*	0.516*	0.497*	1.000
P_T	6/6	6/6	6/6	6/6	6/6	6/6	6/6

Affect-based Trust

Items	II6	II11	II17	II19	II20	II22
II6	1.000					
II11	0.349*	1.000				
II17	0.179†	0.325*	1.000			
II19	0.465*	0.508*	0.379*	1.000		
II20	0.481*	0.368*	0.392*	0.605*	1.000	
II22	0.347*	0.339*	0.358*	0.390*	0.592*	1.000
P_T	4/5	5/5	4/5	5/5	5/5	5/5

**Correlation is significant at the 0.01 level. ($p < 0.01$)*
†Correlation is significant at the 0.05 level. ($p < 0.05$)
P_T: The proportion of the statement correlates between 0.20 and 0.70 with other statements in the First Test.

Fig. 9.1 Inter-item correlation (Test)

system-based trust, 5/6 and 6/6 for cognition-based trust and 4/5 and 5/5 for affect-based trust (Fig. 9.2). As shown in Table 9.7, the item-total correlation ranged from 0.41 to 0.62. This also indicated that the twenty-three statements are consistent with each other. As a result, the reliability of the retest data was thus confirmed.

Finally, the item-total correlations of the Test fall in the range of 0.42 and 0.76 (Table 9.7), this shows that all the statements are consistent with each other and a

System-based Trust

Items	II2	II4	II8	II9	II12	II14	II15	II16	II21	II23
II2	1.000									
II4	0.359*	1.000								
II8	0.227*	0.404*	1.000							
II9	0.240*	0.269*	0.295*	1.000						
II12	0.273*	0.284*	0.126	0.140	1.000					
II14	0.257*	0.132	0.290*	0.207*	0.336*	1.000				
II15	0.339*	0.337*	0.115	0.205*	0.406*	0.253*	1.000			
II16	0.302*	0.339*	0.229*	0.273*	0.207*	0.191†	0.310*	1.000		
II21	0.297*	0.366*	0.244*	0.242*	0.099	0.279*	0.205*	0.287*	1.000	
II23	0.242*	0.188†	0.164†	0.174†	0.315*	0.202*	0.393*	0.179†	0.217*	1.000
P_R	9/9	7/9	6/9	7/9	6/9	7/9	8/9	7/9	8/9	5/9

Cognition-based Trust

Items	II1	II3	II5	II7	II10	II13	II18
II1	1.000						
II3	0.367*	1.000					
II5	0.378*	0.311*	1.000				
II7	0.366*	0.343*	0.424*	1.000			
II10	0.278*	0.186†	0.286*	0.368*	1.000		
II13	0.201*	0.301*	0.317*	0.335*	0.448*	1.000	
II18	0.314*	0.245*	0.312*	0.379*	0.472*	0.392*	1.000
P_R	6/6	5/6	6/6	6/6	5/6	6/6	6/6

Affect-based Trust

Items	II6	II11	II17	II19	II20	II22
II6	1.000					
II11	0.347*	1.000				
II17	0.239*	0.297*	1.000			
II19	0.356*	0.403*	0.326*	1.000		
II20	0.249*	0.185†	0.425*	0.488*	1.000	
II22	0.190†	0.259*	0.383*	0.371*	0.305*	1.000
P_R	4/5	4/5	5/5	5/5	4/5	4/5

**Correlation is significant at the 0.01 level. ($p < 0.01$)*
†Correlation is significant at the 0.05 level. ($p < 0.05$)
P_R: The proportion of the statement correlates between 0.20 and 0.70 with other statements in the Retest.

Fig. 9.2 Inter-item correlation (Retest)

satisfactory homogeneity of statements is achieved. Based on the results of the reliability analyses, the proposed inventory, which consists of twenty-three statements, achieved the reliability requirement.

Table 9.7 Item-total correlation coefficient (Test)

Statements	Item-total correlation		Statements	Item-total Correlation	
	Test	Retest		Test	Retest
II1	0.615	0.469	II13	0.699	0.559
II2	0.613	0.482	II14	0.722	0.454
II3	0.495	0.428	II15	0.673	0.550
II4	0.693	0.526	II16	0.541	0.428
II5	0.656	0.564	II17	0.421	0.460
II6	0.575	0.408	II18	0.635	0.603
II7	0.740	0.624	II19	0.649	0.615
II8	0.496	0.419	II20	0.692	0.518
II9	0.657	0.439	II21	0.756	0.543
II10	0.718	0.613	II22	0.609	0.501
II11	0.554	0.462	II23	0.495	0.424
II12	0.424	0.433			

9.2.3.3 Steps 3 and 6: Construct Validity Analyses of the Inventory

Having ensured that the proposed inventory meet the level of reliability, these two steps aim at analysing its construct validity, which is concerned with how well the concept is defined by the measure (Hair et al. 1998). To achieve this, confirmatory factor analyses were employed. Structural equation modeling (SEM) was performed (by the use of AMOS software) to assess the authenticity of the pre-specified relationships among the twenty-three statements. SEM is widely used to characterise relationships among observed and unobserved variables by way of path diagrams (Hair et al. 1998). In this study, SEM describes the structure of the proposed inventory that shows the relationships between the three trust types and their respective characterising measurement statements.

The construct validity was determined by appropriate goodness-of-fit indices of SEM, namely χ^2/df, Goodness of Fit Index (GFI), Comparative Fit Index (CFI), Tucker-Lewis Index (TLI) and Root Mean Square of Approximation (RMSEA). Satisfactory model fit is achieved by attaining $\chi^2/df < 2.00$ (Bollen 1989), GFI, CFI and TLI > 0.80 (Maskarinec et al. 2000) and RMSEA < 0.08 (Hair et al. 1998). Based on the Test data, the initial values obtained for the goodness-of-fit indices are: χ^2/df: 1.89, GFI: 0.81, CFI: 0.89, TLI: 0.88 and RMSEA: 0.07. To establish the more satisfactory goodness-of-fit indices, modifications were suggested by the model improvement function of the AMOS software. Without altering the substantive meaning of the structure, two modifications were made by adding relationships between the statements of II8/II12 and II19/II22. As such, results of the refined model show statistical improvement of goodness-of-fit indices: χ^2/df: 1.81, GFI: 0.82, CFI: 0.90, TLI: 0.89 and RMSEA: 0.07. The final structure of the proposed inventory is shown in Fig. 9.3.

Similarly, the same construct validity checked by SEM was performed with the Retest data. The final structure is shown in Fig. 9.4. With the goodness-of-fit indices of χ^2/df: 1.66, GFI: 0.84, CFI: 0.86, TLI: 0.84 and RMSEA: 0.06.

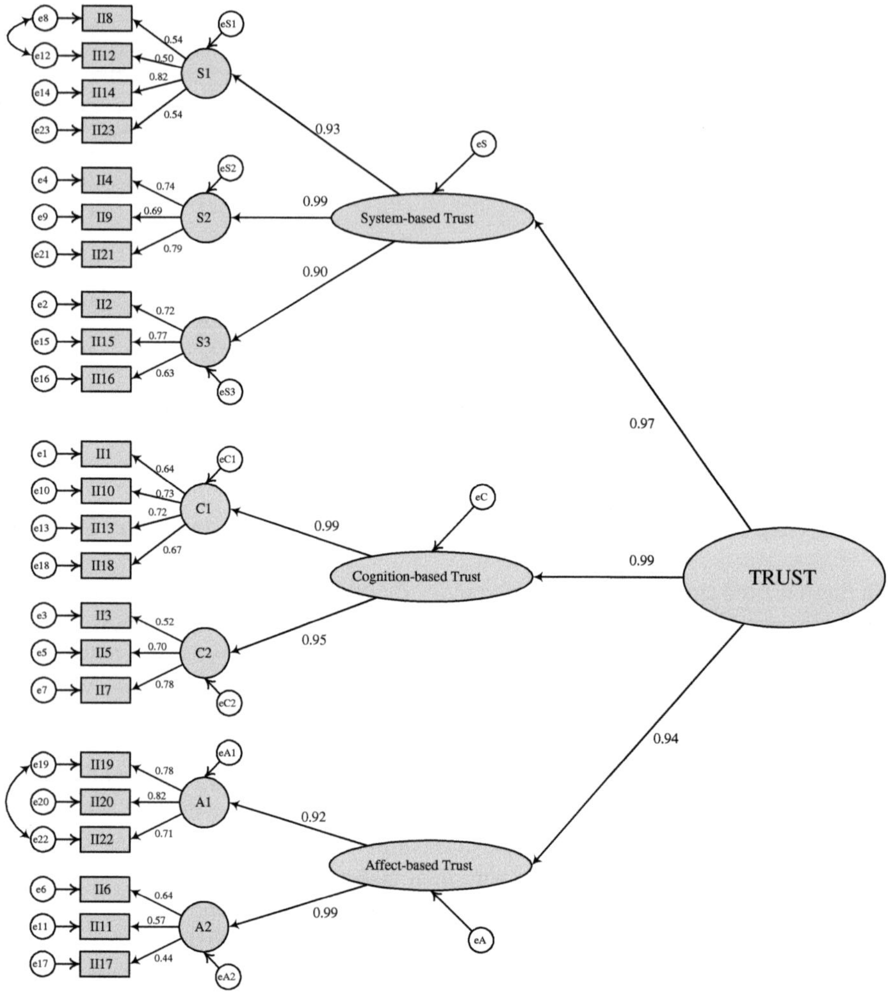

Fig. 9.3 The final structure of the proposed inventory (Test)

The construct validity was statistically supported in both Test and Retest datasets.

9.2.3.4 Step 7: Stability Analysis

The aim of stability analysis is to examine the consistency of the proposed inventory over time. Pearson's correlation coefficients of the Test and Retest

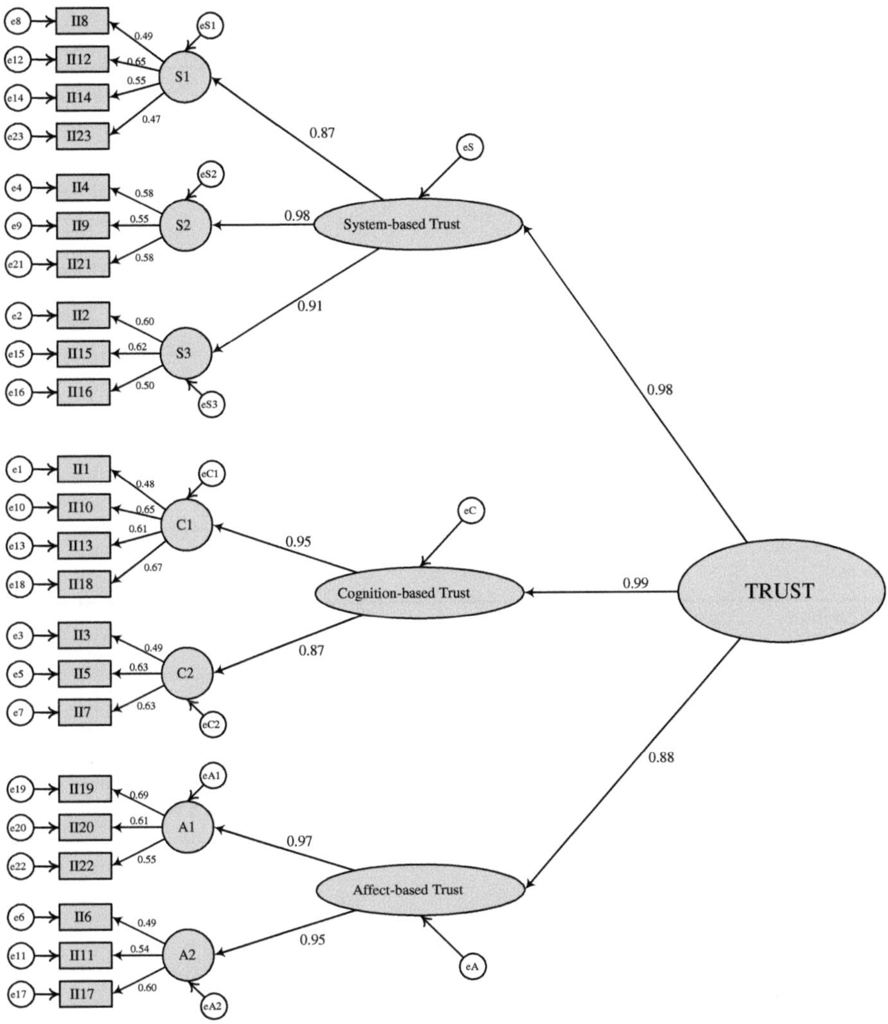

Fig. 9.4 The final structure of the proposed inventory (Retest)

datasets are used for this purpose. A Pearson's correlation coefficient of 0.50 or above is considered as reasonable and has been used in the studies by Bennett (1996), Torkzadeh and Doll (1991) and Parr et al. (2006). As shown in Table 9.8, the Pearson's correlation coefficients for the twenty-three statements of the proposed inventory range from 0.515 to 0.767 ($p < 0.01$) and thus meeting the statistical test on stability.

Table 9.8 Pearson's correlation coefficient

Statements	Pearson's coefficient	Statements	Pearson's coefficient
II1	0.672**	II13	0.608**
II2	0.658**	II14	0.663**
II3	0.772**	II15	0.690**
II4	0.731**	II16	0.648**
II5	0.767**	II17	0.659**
II6	0.680**	II18	0.677**
II7	0.595**	II19	0.735**
II8	0.683**	II20	0.515**
II9	0.657**	II21	0.677**
II10	0.677**	II22	0.559**
II11	0.684**	II23	0.678**
II12	0.597**		

** Correlation is significant at 0.01 ($p<0.01$).

Table 9.9 Summary of results of validation

Development process	Types of test	Statistical requirement	Results		Status
			Test	Retest	Passed
1. Reliability (reliability of the inventory measurement statements)	Cronbach's Alpha	>0.70	0.94	0.90	Passed
	Inter-item correlation	0.20–0.70	8/9 (8 out of 9) to 9/9 for system-based; 6/6 for cognition-based; 4/5 and 5/5 for affect-based	5/9 to 9/9 for system-based; 5/6 and 6/6 for cognition-based; 4/5 and 5/5 for affect-based	Passed
	Item-total correlation	>0.30	0.42–0.76	0.41–0.62	Passed
2. Construct validity (validity of the inventory structure)	*Structural equation modeling: goodness-of-fit tests*				
	χ^2/df	<2.00	1.81	1.66	Passed
	GFI	>0.80	0.82	0.84	Passed
	CFI	>0.80	0.90	0.86	Passed
	TLI	>0.80	0.89	0.84	Passed
	RMSEA	<0.08	0.07	0.06	Passed
3. Stability (stability of the inventory over time)	Pearson's correlation	>0.50	0.515–0.772		Passed

9.3 Discussion

The reliability, construct validity and stability of the proposed inventory are supported statistically by Test and Retest. For ease of reference, Table 9.9 summarised the results of the tests conducted.

The validated trust inventory can be used in two major ways. Firstly, it can be used to conduct industry–wide study on organisational trust in construction organisations. The results thereof can be used as benchmarks of trust indices. Sub-indices in terms of System, Cognition and Affect types of trust also allow identification of good practices and/or areas for improvement. Secondly, at project level, longitudinal data of trust status can be collated by regular periodic evaluation using the inventory. In this respect, the inventory can be adopted to suit the characteristics of the organisations by the incorporation of the organisational specificities while retaining the theoretical constructs.

Further evaluations of the inventory include its ease of use, the appropriateness of the proposed trust types in construction contracting and the ways to enhance the three types of trust. Two in-depth interviews were conducted, one with a chief quality surveyor and another one with a senior project manager. Both of them have over 20 years experience managing various complex projects. The Chief Quantity Surveyor is in charge of the commercial management arm of an international consulting firm that has offices around the world. The Senior Project Manager also works for global engineering and project management organisation that is having the major market share in Hong Kong. Firstly, both confirmed that trust is the most effective tool to improve efficiency in construction contracting. Nonetheless, both also identified that it is not easy to develop trust in the highly competitive construction market. Secondly, they both found the inventory easy to use and could be helpful as a means to understand the practice of trusting behaviours of project participants. In this connection, both concurred that the twenty-three statements in the proposed inventory are adequate to identify the basic trust types in construction contracting. They opined that construction practitioners should welcome the proposed inventory as its potential use in studies where trust is a variable. Notable examples include studies in procurement system, contract administration, project management, quality/safety management, risk management and claims/disputes management. To facilitate its implementation, the interviewees suggested investigating the possibility of incorporating the inventory as part of a project performance key performance indicators system. This shall formalise the assessment of trust level of contracting parties during project duration.

Both interviewees also offered their views on the applications of system-based trust, cognition-based trust and affect-based trust in construction contracting. One of the interviewees suggested that system-based trust and affect-based trust are both imperative to improve relationships among contracting parties. System-based trust can be developed through credible policies and respect of contracts. To develop system-based trust, having a well-developed system is essential, as it shall crystallise expectations against which performance can be gauged. As for affect-based trust, he witnessed how projects failed as a result of uncooperative acts of the key players. In other words, trust underpins cooperation that is critical to solve problems that demand contributions from the participants. He thought that cognition-based trust is comparatively less important because organisations without good track record, sound reputation or stable financial state would probably unable to survive the vigorous tender evaluation process. However, another interviewee had a slightly different view. He believed that system-based trust and cognition-based trust are more influential than affect-based trust in improving the relationships of contracting parties. The existence of system-based trust is fundamental if the contracting parties had no previously working relationship. The contract and project management system in place form the platform upon which understanding to be further enhanced. By the same token, cognition-based trust is probably the prerequisite before a contractual relationship can be established. Affect-based trust expresses personal relationship other than working relationship. He commented that the development of affect-based trust should be avoided in the workplace.

In sum, the proposed trust inventory builds on the theoretical constructs for trust in construction contracting advocated by Wong et al. (2008). The stability and reliability the proposed trust inventory has been tested rigorously with well-recognised inventory development methodology (Cooke and Szumal 1993; Horowitz 1988; Rahim 1983). The trust inventory shall be instrumental for use in studies where trust status is a variable. This adds another dimension to the wealth of studies on trust in construction that are informative on trust factors but little has been done on trust measurement.

9.4 Chapter Summary

Trust is central in every transaction and helps to avoid disputes. The financial turmoil in 2008 exemplified the negative consequences if there is no trust in the system and among business partners. Although trust has been a topical research area in construction, there has yet been reported any attempt in developing trust inventory for use in construction contracting. This chapter reports such a study. Three main stages of work are involved; (i) Developing a trust framework; (ii) operationalising the trust framework into an inventory and (iii) validating the inventory. The trust framework proposed by the authors (Wong et al. 2008) was used in the fulfillment of stage I of the research and operationalised into an inventory format. The test-retest methodology was applied to validate the inventory. The reliability, construct validity and stability of the proposed inventory were statistically tested. In addition, affirmative comments have been received from two very experienced senior construction professionals on the relevancy of the trust dimensions as well as the use of the inventory in construction contracting. To conclude, the proposed trust inventory can work well to assess the trust level among contracting parties. This inventory could be applied in a wide spectrum of studies in procurement system, contract administration, project management, quality/safety management, risk management and claims/disputes management where trust is an important and significant variable.

Acknowledgments The content of this chapter has been published in Volume 29(2) of the International Journal of Project Management and is used with the permission of Elsevier.

References

Atwater, L. E. (1998). The relative importance of situational and individual variables in predicting leader behavior. *Group and Organization Studies, 13*, 290–310.

Bayliss, R., Cheung, S. O., Suen, C. H., & Wong, S. P. (2004). Effective partnering tools in construction: A case study on MTRC TKE contract 604 in Hong Kong. *International Journal of Project Management, 22*(3), 253–263.

Bazerman, M. H. (1994). *Judgment in managerial decision making*. New York: Wiley.

Beck, A. T., Epstein, N., Brown, G., & Steer, R. A. (1988). An inventory for measuring clinical anxiety: Psychometric properties. *Journal of Consulting and Clinical Psychology, 56*(6), 893–897.

Bennett, M. (1996). *Pearson Correlation Coefficient. Pearson Correlation Coefficient*. Retrieved June 5, 2007, from http://web.indstate.edu/mary/N322/pearsonr.html

Bollen, K. A. (1989). *Structural equations with latent variables*. New York, NY: Wiley.

Butler, J. K. (1991). Toward understanding and measuring conditions of trust: Evolution of a condition of trust inventory. *Journal of Management, 17*, 643–663.

Cheung, S. O., Ng, S. T., Wong, S. P., & Suen, C. H. (2003). Behavioral aspects in construction partnering. *International Journal of Project Management, 21*(5), 333–343.

Construction Industry Institute (CII). (1993). *Cost trust relationship, construction industry institute, publication 24-1 bureau of engineering research*. Austin: The University of Texas.

Cook, J., & Wall, T. (1980). New work attitude measures of trust, organizational commitment and personal need non-fulfillment. *Journal of Occupational Psychology, 53*(1), 39–52.

Cooke, R. A., & Szumal, J. L. (1993). Measuring normative beliefs and shared behavioral expectations in organizations: The reliability and validity of the organizational culture inventory. *Psychological Reports, 72*, 1299–1330.

Couch, L. L., Adams, J. M., & Jones, W. H. (1996). The assessment of trust orientation. *Journal of Personality Assessment, 67*, 305–323.

Cummings, L. L. (1983). Performance-evaluation system in context of individual trust and commitment. In F. J. Landy, S. Zedrick, & J. Cleveland (Eds.), *Performance measurement and theory* (pp. 89–93). Hillsdale, NJ: Lawrence Earlbaum.

Cummings, L. L., & Bromiley, P. (1996). The organizational trust inventory (OTI)—Development and validation. In R. Kramer & T. Tyler (Eds.), *Trust in organizations* (pp. 302–330). Thousand Oaks, CA: Sage Publications.

Currall, S. C., & Judge, T. A. (1995). Measuring trust between organizational boundary role persons. *Organizational Behavior and Human Decision Processes, 64*, 151–170.

Dietz, G., & Hartog, D. N. D. (2006). Measuring trust inside organisations. *Personnel Review, 35*(5), 557–588.

Ding, Z. K., & Ng, F. F. (2007). Reliability and validity of the Chinese version of McAllister's trust scale. *Construction Management and Economics, 25*(11), 1107–1117.

Evans, A. M., & Revelle, W. (2008). Survey and behavioral measurements of interpersonal trust. *Journal of Research in Personality, 42*(6), 1585–1593.

Farris, G., Senner, E., & Butterfield, D. (1973). Trust, culture and organizational behavior. *Industrial Relations, 12*, 144–157.

Ferketich, S. (1991). Focus on psychometrics: Aspects on item analysis. *Research in Nursing & Health, 14*(2), 165–168.

Gambetta, D. G. (1988). Can we trust? In D. G. Gambetta (Ed.), *Trust* (pp. 213–237). New York: Basil Blackwell.

Garson, G. D. (2006). *Reliability analysis*. http://www2.chass.ncsu.edu/garson/pa765/

Giffin, K. (1967). The contribution of studies of source credibility to a theory of interpersonal trust in communication department. *Psychological Bulletin, 68*, 104–120.

Glaser, E. L., Laibson, D. I., Scheinkman, J. A., & Soutter, C. L. (2000). Measuring trust. *Quarterly Journal of Economics, 115*(3), 811–846.

Hair, J. F., Anderson, R. E., Tatham, R. L., & Black, W. C. (1998). *Multivariate data analysis*. Englewood Cliffs, N.J.: Prentice Hall.

Hannah, L. (1991). *Conditions of trust in the construction industry and their relevance to project success construction industry co-operative alliance: Research implementation report 91-01*. Clemson University.

Hartman, F. (2000). The role of trust in project management. In *Proceeding of the PMI Research Conference*. Pennsylvania, USA: Project Management Institute.

Horowitz, L. M. (1988). Inventory of interpersonal problems: Psychometric properties and clinical applications. *Journal of Consulting and Clinical Psychology, 56*(6), 885–892.

Huemer, L. (2004). Activating trust: The redefinition of roles and relationships in an international construction project. *International Marketing Review, 21*(2), 187–202.

Idvall, E., Hamrin, E., & Unosson, M. (2002). Development of an instrument to measure strategic and clinical quality indicators in postoperative pain management. *Journal of Advanced Nursing, 37*(6), 532–540.

Johnson-George, C., & Swap, W. C. (1982). Measurement of specific interpersonal trust: Construction and validation of a scale to assess trust in a specific other. *Journal of Personality and Social Psychology, 43*(6), 1306–1317.

Kadefors, A. (2004). Trust in project relationships—Inside the black box. *International Journal of Project Management, 22*(3), 175–182.

Kline, P. (1993). *Handbook of psychological testing*. London: Routledge.

Knapp, T. R., & Brown, J. K. (1995). Focus on psychometrics: Ten measurement commandments that often should be broken. *Research in Nursing & Health, 18*(6), 465–469.

Kramer, R. M. (1999). Trust and distrust in organizations: Emerging perspectives, enduring questions. *Annual Review of Psychology, 50*, 569–598.

Kwan, A. Y., & Ofori, G. (2001). Chinese culture and successful implementation of partnering in Singapore's construction industry. *Construction Management and Economics, 19*(6), 619–632.

Latham, M. (1994). *Constructing the team: Final report on joint review of procurement and contractual arrangements in the United Kingdom construction industry*. London: HMSO.

Lawler, E. (1992). *The ultimate advantage: Creating the high involvement organizations*. San Francisco: Jossey-Bass.

Lewicki, R. J., & Bunker, B. B. (1995). Trust in relationships: A model of development and decline. In B. B. Bunker & J. Z. Rubin (Eds.), *Conflict, cooperation, and justice* (pp. 133–174). San Francisco, CA: Jossey-Bass.

Lewicki, R. J., & Bunker, B. B. (1996). Developing and maintaining trust in work relationships. In R. M. Kramer & T. R. Tyler (Eds.), *Trust in organizations: Frontiers of theory and research* (pp. 114–139). Thousand Oaks, CA: Sage.

Lewis, J. D., & Weigert, A. (1985). Trust as a social reality. *Social Forces, 63*(4), 967–985.

Luhmann, N. (1979). *Trust and power*. New York: Wiley.

Maskarinec, G., Novotny, R., & Tasaki, K. (2000). Dietary patterns are associated with body mass index in multiethnic women. *Journal of Nutrition, 130*, 3068–3072.

McAllister, D. J. (1995). Affect- and cognition-based trust as foundations for interpersonal cooperation in organizations. *Academy of Management Journal, 38*(1), 24–59.

McKnight, D. H., Cummings, L. L., & Chervany, N. L. (1998). Initial trust formation in new organizational relationships. *The Academy of Management Review, 23*(3), 473–490.

Milgrom, P., & Roberts, J. (1992). *Economics, organization and management*. Englewood Cliffs, N.J.: Prentice Hall.

Nunnally, J. C., & Bernstein, I. H. (1994). *Psychometric theory*. New York: McGraw-Hill.

Oshio, A. (2009). Development and validation of the dichotomous thinking inventory. *Social Behavior and Personality, 37*(6), 729–742.

Parr, C. L., Veierød, M. B., Laake, P., Lund, E., & Hjartåker, A. (2006). Test-retest reproducibility of a food frequency questionnaire (FFQ) and estimated effects on disease risk in the Norwegian women and cancer study (NOWAC). *Nutrition Journal, 5*(4), 1–10.

Pinto, J. K., Slevin, D. P., & English, B. (2009). Trust in projects: An empirical assessment of owner/contractor relationships. *International Journal of Project Management, 27*(6), 638–648.

Rahim, M.A. (1983). Rahim organizational conflict inventories: Professional manual. Palo Alto, CA: Consulting Psychologists Press. Retrieved June 5, 2007, from, reliab.htm.

Robinson, J. P., Shaver, P. R., & Wrightsman, L. S. (1991). Criteria for scale selection and evaluation. In J. P. Robinson, P. R. Shaver & L. S. Wrightsman (Eds.), *Measures of personality and social psychological attitudes* (pp. 1–16). San Diego, CA: Academic Press.

Rotter, J. B. (1967). A new scale for the measurement of interpersonal trust. *Journal of Personality, 35*(4), 651–665.

Rousseau, D. M., Sitkin, S. B., Burt, R. S., & Camerer, C. (1998). Not so different after all: A cross-discipline view of trust. *Academy of Management Review, 23*(3), 393–404.

Scott, D. (1980). The casual relationship between trust and the assessed value of management by objectives. *Journal of Management, 6*, 157–175.

Smyth, H. J. (2003). *Developing customer-supplier trust: A conceptual framework for management in project working environments. CRMP working paper*, http://www.crmp.net/papers/index.htm

Taylor, R. G. (1989). The role of trust in labor-management relations. *Organization Development Journal, 7*(1), 85–89.

Torkzadeh, G., & Doll, W. J. (1991). Test-retest reliability of the end-user computing satisfaction instrument. *Decision Sciences, 22*(1), 26–37.

Whitener, E. M., Brodt, S. E., Korsgaard, M. A., & Werner, J. M. (1998). Managers as initiators of trust: An exchange relationship framework for understanding managerial trustworthy behaviour. *The Academy of Management Review, 23*(3), 513–530.

Williamson, O. E. (1975). *Markets and hierarchies*. New York: Free Press.

Williamson, O. E. (1981). The economics of organization: The transaction cost approach. *American Journal of Sociology, 87*, 548–577.

Williamson, O. E. (1993). Calculativeness, trust and economic organization. *Journal of Law and Economics, 30*, 131–145.

Wong, S. P., & Cheung, S. O. (2004). Trust in construction partnering: Views from parties of the partnering dance. *International Journal of Project Management, 22*(6), 437–446.

Wong, S. P., & Cheung, S. O. (2005). Structural equation model on trust and partnering success. *Journal of Management in Engineering, 21*(2), 70–80.

Wong, S. P., Cheung, S. O., & Ho, K. M. (2005). Contractor as trust initiator in construction partnering—A prisoner's dilemma perspective. *Journal of Construction Engineering and Management, ASCE, 131*(10), 1045–1053.

Wong, S. P., Cheung, S. O., Cheung, W. K., & Suen, C. H. (2003). Nature and function of trust in construction partnering in Hong Kong. In *Proceedings of the Joint International Symposium of International Council for Building (CIB) W55, W65 and W107, Cape Town, South Africa* (pp. 764–773).

Wong, W. K., & Cheung, S. O. (2006). Trust in organizational relationship and its delineation. In *Proceeding of International Conference on Construction and. Real Estate Management (ICCREM) 2006, 13th Rinker International Conference* (pp. 1–7) Florida: Oxford.

Wong, W. K., Cheung, S. O., Yiu, T. W., & Pang, H. Y. (2008). A framework for trust in construction contracting. *The International Journal of Project Management, 26*, 821–829.

Wood, G., & McDermott, P. (1999). Searching for trust in the UK construction industry: An interim view. In *Proceedings of the CIB W92 Procurement Systems Conference*. http://www.scpm.salford.ac.uk/trust/

Yamagishi, T. (1988). The provision of a sanctioning system in the United States and Japan. *Social Psychology Quarterly, 51*, 265–271.

Zaghloul, R., & Hartman, F. (2003). Construction contracts: The cost of mistrust. *International Journal of Project Management, 21*(6), 419–424.

Zucker, L. G. (1986). The production of trust: Institutional sources of economic structure. In B. M. Staw & L. L. Cummings (Eds.), *Research organizational behavior* (Vol. 8, pp. 55–111). Greenwich, CT: JAI Press.

Chapter 10
Interweaving Trust and Communication for Project Performance

Sai On Cheung and Tak Wing Yiu

Abstract Project performance is often regarded as the ultimate indicator of project success. Trust is considered the most important catalyst for performance. Thus, trust and project performance is intrinsically linked. Construction project development involves extensive information exchanges. Timely and effective transfer of relevant information is critical in view of the mutual dependent nature of construction activities. Effective communication reduces dispute arising from misunderstanding. Mediation analysis was performed to investigate the mediating role of communication in the trust-performance relationship. Five significant mediation models were identified. Effective information flow is found to be the versatile mediator of the trust-project performance relationship among all of the significant models. This implies that the improvement of information flow would likely achieve satisfactory performance.

10.1 Introduction

Construction project development involves extensive information exchanges among members of multi-disciplinary project teams. Such information be it financial, technical or administrative, are important for the proper completion of the project (Wong and Lam 2011). Furthermore, timely transfer of relevant

S. O. Cheung (✉)
Construction Dispute Resolution Research Unit,
Department of Civil and Architectural Engineering, City University of Hong Kong,
Hong Kong, People's Republic of China
e-mail: saion.cheung@cityu.edu.hk

T. W. Yiu
Department of Civil and Environmental Engineering, Faculty of Engineering,
The University of Auckland, Auckland, New Zealand
e-mail: k.yiu@auckland.ac.nz

S. O. Cheung (ed.), *Construction Dispute Research*,
DOI: 10.1007/978-3-319-04429-3_10,

information is critical in view of the mutual dependent nature of construction activities (Deng et al. 2001; Rojas and Songer 1999; Wong and Lam 2011). This points to the need for effective communication. For example, Scanlin (1998) found that ineffective communication is one of the root causes of many project problems including dispute (Cheung and Pang 2013). A trusting climate has also been advocated as a prerequisite for effective communication (Butler 1999). If project team members can trust each other, they will be more willing to share information, in particular those that are controversial. Hence it appears that there is a close relationship among trust, communication and project performance. Furthermore, the better the performance, the less chance of having disputes. There are few reported studies on the joint effect of trust and communication on project performance and this study aims to fill this gap. This study timely reminds the significance of developing a trusting climate to enhance timely and effective communication. The study therefore includes the following research activities: (1) measuring trust, (2) developing taxonomies of communication and project performance and (3) examining the relationships among trust, communication and project performance.

10.2 Project Performance

Performance is often regarded as the ultimate indicator of project success. However, defining and evaluating project performance vary with the perspective taken (Chan and Chan 2004; Shenhar et al. 1997). For instance, in a multi-disciplinary project team, the architect, engineer and surveyor are having different goals and thus would have different expectations on project performance (Bryde and Robinson 2005; Lim and Mohamed 1999). Project is ideally to be completed in a win–win fashion for all team members (Lim and Mohamed 1999). However, in reality, project performance is often benchmarked with the needs of the employer (Chan and Chan 2004; Shenhar et al. 1997). In this regard, project performance is commonly assessed by comparing employer's expectations and the actual attainments (Yu et al. 2005). This method is used to determine whether the project is 'successful or failing' and 'satisfactory or unsatisfactory' (Liu and Walker 1998). Success of a project can be identified as (1) achieving the cost within pre-determined budget, (2) satisfying project quality and (3) creating positive project value, while total failure of a project implies none of these are fulfilled (Yu et al. 2005). Alternatively, expected performance can be used as a reference to evaluate and measure project performance. The requirements of the employer are principally expressed in the brief. However, information on employer's brief is often inadequate (Leung and Liu 2003). If goals are not clearly defined, the conflict so induced would hamper project performance. As for the attributes of construction project performance, the fundamental elements are time, cost and quality. These elements, often regarded as 'golden triangle', are mutually dependent. Thus, managing the interrelationships among time, cost and quality is also crucial to address different goals (Westerveld 2003). In the context of construction dispute, Cheung et al. (2000) added dispute resolution satisfaction as a measure of project performance. Construction organisations therefore aim at achieving shortest project duration, lowest cost and highest quality in their projects so as to establish track record for future dealings. To achieve these, effective teamwork and strong leadership underpinned by seamless communication are the prerequisites.

10.3 The Role of Trust in Project Performance

The completion of capital projects often takes several years. Each contracting party has to establish the trustworthiness of his project team members. This is due to the fact that efficiently accomplishing project objectives is highly dependent on their cooperation, which is vulnerable without trust (Porta et al. 1996). Trust also lowers the risk taken by the contracting parties (Zaghloul and Hartman 2003) and facilitate management functions such as planning, organising, controlling and staffing. Efficient project execution is indeed cost saving (Zaghloul and Hartman 2003). In addition, trust can lubricate problem negotiation and reduce transaction costs (Diallo and Thuillier 2005). For example, in partnering projects, safety and quality conditions, innovative technology and chance of business are the benefits derivable from a trusting relationship (Cook and Hancher 1990). In this context, trust can be identified as the most important success factor in fostering cooperation (Cheung et al. 2003; Cook and Hancher 1990; Kwan and Ofori 2001; Wong and Cheung 2004), eliminating adversarial relationship by information sharing (Cook and Hancher 1990), improving productivity (Kwan and Ofori 2001) and establishing a trusting platform for dispute resolution (Cheung et al. 2003). Wong and Cheung (2004) found that competence, problem solving, communication, openness, alignment, information flow, unity, respect, compatibility, long-term relations, finance, reputation, use of ADR and satisfactory contract terms are trust attributes in partnering projects. Other studies on the classification of trust include those by Luhmann (1979), Lewis and Weight (1985), McAllister (1995), Hartman (1999) and Kramer (1999). Wong (2007) summarised these studies and proposed three forms of trust identifiable in the construction industry. These are System-based Trust, Cognition-based Trust and Affect-based Trust (Table 10.1 refers).

10.4 The Role of Communication in Project Performance

Communication is defined as a two-way process between sender(s) and receiver(s) (Cleland and Ireland 2002). A sender is defined as one who sends information to receiver(s) who can be a person, a group, or an organisation (Baguley 1994). It should not be a one-way process if an open and active communication platform is available (Thomas et al. 1998). In construction, complex organisational structures

Table 10.1 Forms of trust in construction (Wong 2007)

Forms of trust	Descriptions
1. System-based trust (SBT)	This type of trust is objective and is related to formalised system. It does not involve in personal/emotional matters. It has three key attributes: (1) contracts and agreements (CA), (2) communication system (CS) and (3) organisation policy (OP)
2. Cognition-based trust (CBT)	This type of trust relies on objective information and develops through communication between team members. It is concluded that if consistency, competence and integrity of knowledge are the three characteristics to build up trust relationship between team members. It has two key attributes: (1) knowledge (K) and (2) communication/interaction (CI)
3. Affect-based trust (ABT)	This type of trust is highly related to personal feelings. It provides an emotional bond linking to all the participants in the project team. It has two key attributes: (1) emotional investments (EI) and (2) being thoughtful (BT)

Table 10.2 Major communication problems in the construction industry (Higgin and Jessop 2001)

Type of communication	Problems
1. Communication with prospective clients	- Decision makers are not randomly scattered in the population - Many prospective users know little about the technical and professional services available to them
2. Communications between clients and advisers	- Advisers have not explored the needs and limitations of the client thoroughly - There is not sufficient discussion on all the possible means of meeting clients' needs - Matching the needs and possibilities of client are seldom fully achieved
3. Communications within the design team	- There is not sufficient time to establish understanding on the common objective of the design team - The design team is seldom aware of all necessary steps to realise an optimum overall outcome - The means for design coordination means are unclear
4. Communications related to contract	- There is a variety of wide and very complex ways to form a contractual relationship - There is concern over the degree of certainty on the expectations of the contracting parties
5. Communications within the construction team	- Decisions are incomplete and unduly rushed because of inadequate and insufficient information - There is a lack of understanding of the communication process

and fragmented supply chain often cause communication problems (Dainty et al. 2006). Higgin and Jessop (2001) summarised some of the communication problems and these are shown in Table 10.2. The consequences of communication break down include misleading and unclear information in the design drawings, report, contracts and work orders (Oberlender 2000). Extra efforts are necessitated to rectify the incorrect messages and project performance is thus inevitably negatively affected. If communication is improved at the initial stage of a project, the identified design criteria and solutions can ensure the fulfillment of the client's requirement and priorities (CIRC 2001).

When a project commences, most of the members of the contracting parties meet each other for the first time. It cannot be expected that they will trust each other unless they have previous dealings. Communication thus plays a pivotal role to bring them together and begin to share information. At the initial stage, getting to have a consensus view on to the project objectives, client's requirements, specifications, priorities, constraints, etc. is crucial to make the first step out to develop mutual understanding and trust (Muller and Turner 2005).

Furthermore, communication is not just information exchange but can also enable project team members to fully express their concerns (Jarvenpaa and Leidner 1999). When a member is able to have better knowledge of the others, their relationship can be fostered (Pietroforte 1997). In this regard, effective communication can be seen as sign of enthusiasm and optimism within a project team. Project team members are beginning to feel like friends rather than just teammates (Jarvenpaa and Leidner 1999). Within the project duration (Harrison 1985), the earlier a trusting relationship is established, the better the collaboration among the project team members. Both formal and informal communications are in fact important to the proper functioning of an organisation (Dainty et al. 2006). In construction, different procurement approach would need to be dovetailed with an appropriate organisation structure. The structure sets the formal framework directing the relationship as far as formal communication is concerned. Formal communication is systematic but often thin in contents (Emmitt and Gorse 2003). On the other hand, informal communication does not follow rigid rules and guidelines. It may, however, be useful for problem-solving, decision-making and enhancing information exchange (Pietroforte 1997; Pinto and Pinto 1991). The communication media, either formal or informal, are the key channels through which messages can be successfully transmitted. Different modes of communication media, frequency and effectiveness may trigger different responses (Santoro and Saparito 2003). In sum, three main aspects of communication are proposed. These are communication effectiveness, communication method and communication frequency.

10.4.1 Communication Effectiveness

Effective communication would lead to a better trust level among team members (Das and Teng 1998). Church (1996) opined that good communication is open, honest, participative and direct. It is essential for construction professionals to have a platform to exchange information during the course of construction (Emmitt and Gorse 2003). A common platform guarantees that messages can be understood and processed correctly (Gorse et al. 1999). It is also a catalyst to understand the other parties' needs and difficulties (Thomas et al. 1998). However, effective communication is not easy to be achieved. The diverging goals of the project team members are also the major obstacle against effective communication (Pietroforte 1997). Team members who fail to communicate are incapable to share their views with the other team members (Cleland and Ireland 2002). Drawings, bills of quantities and specification are the key commonly used media to establish common understanding of project requirements (Higgin and Jessop 2001).

10.4.2 Communication Method

A message may get distorted during transmission (Dainty et al. 2006). The longer the transmission chain is, the greater the distortion could be. Message distortion in communication between project team members would jeopardise their relationship for the misunderstanding, extra-workload or even conflict that may arise. In fact, informal supportive forms of communication underpin project collaboration (Gorse et al. 1999). To this ends, face-to-face meeting, fax, email and telephone can be used. Employing appropriate communication methods respective to the nature of the message would reduce the chance of miscommunication (Gorse et al. 1999). Different communication media are having their own characteristics and abilities, thus their functionalities in terms of speed, richness and volume (Mcdonough et al. 1999). It is suggested that using e-mail is the fastest way to send simple messages. When the message is complex, face-to-face communication is more appropriate despite more time is needed (McDonough et al. 1999).

10.4.3 Communication Frequency

Communication frequency is about how often project team members communicate with each other. Santoro and Saparito (2003) advocated that frequent communication is essential to build inter-organisational relationship because of the acute updating of the parties' intentions and capabilities of providing more information. The frequency of using different communication media including face-to-face communication, e-mail, faxes and telephone to transmit information is related to communication effectiveness (McDonough and Kahn 1996). The higher the communication frequency, the better the team overall performance. High frequency of communication might bring new ideas and solutions. It offers opportunities for the project team to deploy resources synergistically and make informed decisions through analysing information from different sources (Daft and Lengel 1986). In this manner, trust can be developed, in particular collective effort beings successful experience. Furthermore, active and open communication signifies commitments (Jarvenpaa and Leidner 1999). Proper management of the communication process does have positive effect on the quality of the information exchange (Dainty et al. 2006).

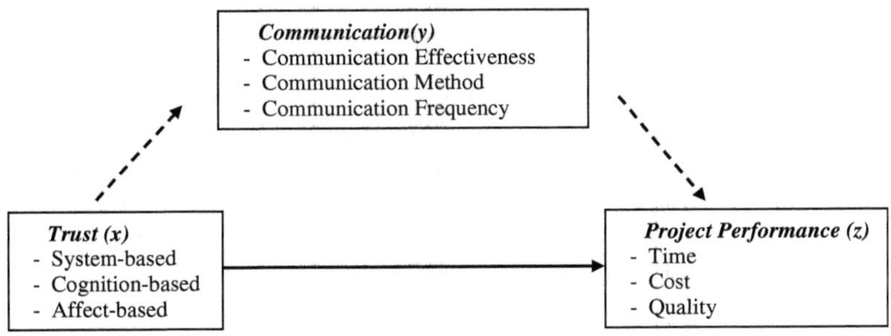

Fig. 10.1 A conceptual model

10.5 Research Model and Hypothesis

Once communication problems surface, effective information exchanges become difficult. Trusting relationship would be hampered because of skepticism (Wong and Cheung 2006). If uncontrolled, an adversarial relationship among contracting parties will be developed. It is therefore suggested that trust, communication and project performance are linked. This study looks into this relationship. Explicitly, building a trusting relationship among project team members is the key elements leading to project success. The effect of a trusting relationship on project performance may be direct and/or indirect. Indirect effect may be mediated by another variable. This study explores *communication*, as one of the possible mediators between trust and project performance. The use of mediation analysis seeks to understand the effect of *trust* and *project performance* via a mediator, *communication*. In this connection, trust is hypothesised to help predict and explain variability in the mediator (i.e. communication), which in turn is anticipated to help predict and explain variability in project performance (Iacobucci 2008). A hypothesised framework is shown in Fig. 10.1. It has the form $x \rightarrow y \rightarrow z$, where x is the antecedent (i.e. Trust); y is the mediator (i.e. communication); and z is the consequence (i.e. the project performance) (James and Brett 1984).

10.5.1 Measures

10.5.1.1 Trust

Trust is measured using the construction based trust inventory developed by Cheung et al. (2011). This inventory consists of 23 statements that serve as trust indicators in construction. These statements assess the trust indicators for system-based trust (10 statements), cognition-based trust (7 statements) and affect-based trust (6 statements). Each respondent was requested to rate the degree of agreement against each of the questions with reference to a recently completed project.

10.5.1.2 Communication

Communication effectiveness, frequency and method were measured using 12 statements developed from a literature search (Higgin and Jessop 2001; McDonough et al. 1999; Santoro and Saparito 2003; Thomas et al. 1998). The respondents were required to indicate the degree of their agreement with the listed statements on a Likert scale from 1 (strongly disagree) to 7 (strongly agree), also with reference to the same recently completed project.

10.5.1.3 Project Performance

Similarly, the performance of the project was measured using 12 statements developed from a literature search (Chan and Kumaraswamy 1996; McDonough et al. 1999; Thomas et al. 1998; Xiao and Proverbs 2002a, b). The respondents were required to indicate the degree of their achievement with the listed statements on a Likert scale from 1 (not achieved) to 7 (highly achieved).

10.6 Data Collection

A questionnaire survey was conducted to collect data from project practitioners in Hong Kong. The target respondents include engineers, architects and surveyors in public and private sectors. They were randomly identified from the web pages of telephone directory from the government departments and the local professional institutes. Simple random sampling was then used to draw a sample from each of the target populations. They were contacted, and if they agreed to participate in this survey, questionnaire was sent to them by post, fax or email, according to their preference.

10.7 Results and Discussions

A total of 273 questionnaires were sent, and 103 valid responses were obtained giving a response rate of 38 %. Among the respondents, more than 45 % of them had more than 5 years' experience in construction. With the collected data, reliability analyses were firstly performed to indicate the degree of internal consistency (Ferketich 1990; Hair et al. 1998). If the result is in a regular pattern, the analysed data is considered reliable. In this study, three well-known approaches, Cronbach's alpha, inter-item correlation and item-total correlation, are used to assess the internal consistency of the dataset. The following steps were then preformed:

Table 10.3 The reliability analyses—thresholds for acceptance (Ferketich 1990; Nunnally and Bernstein 1994; Knapp and Brown 1995)

Reliability analyses	Thresholds
Cronbach's alpha	Shall be greater than 0.70
Inter-item correlation	Shall fall in the range of 0.20–0.70
Item-total correlation	Shall be greater than 0.30

(a) With the use of Structural Equation Modeling (SEM), confirm the structure of trust indicators developed by Cheung et al. (2011);
(b) Develop the taxonomies for communication and project performance, and
(c) Based on the results obtained from (a) and (b) above, explore the effect of communication on Trust-Project Performance relationship by mediation analyses.

10.7.1 Reliability Analyses

As a rule of thumb, the Cronbach's alpha value of 0.70 is the threshold for acceptance. Inter-item correlation measures the relationships among all items for assessing the consistency (Ferketich 1990). To assess the internal consistency reliability, the proportion of the item with correlation coefficient with other items within the range 0.20–0.70 is considered (Idvall et al. 2002). If the proportion is <50 %, that statement is considered lacking inter-correlation with other items. In addition, item-total correlation measures the relationships between an item and the total score from the collection of items (Robinson et al. 1991). The item-total correlation should achieve a value >0.30 (the acceptance threshold) for inclusion in the analysis (Nunnally and Bernstein 1994). A summary of the thresholds for acceptance is shown in Table 10.3. Due to the fact that not all the items in the questionnaire can satisfy the requirements as shown in Table 10.3, items for measuring trust (T8, T9, T15), communication (C2, C5, C6, C7, C8, C9) and project performance (PP7) were discarded (Table 10.3).

10.7.2 The Trust Indicators: Confirmatory Factor Analysis

Confirmatory factor analysis was performed to confirm the factor structure of these indicators. To achieve this, SEM was applied and the goodness-of-fit was assessed. The χ^2/df ratio of 1.496 is less than the threshold value of 2.000 (Cummings and Bromiley 1996; Havila et al. 2004; McAllister 1995). However, the GFI, CFI and TLI values are slightly below the recommended value of 0.900 (Cummings and Bromiley 1996; Garson 2006; Havila et al. 2004; Maskarinec et al. 2000). The RMSEA value is 0.070, which is within the threshold value of 0.080 (Cummings and Bromiley 1996; Garson 2006; Havila et al. 2004). The goodness-of-fit tests are satisfactory.

Table 10.4 Taxonomies for communication

Factors		Factor loadings	
		1	2
Factor 1: Effective information flow (CF1)			
C4	The project provided adequate access to people with necessary information	**0.883**	0.073
C3	I could get sufficient information at the appropriate time	**0.872**	0.028
C1	I understood the expectation of the other party	**0.548**	0.135
Factor 2: Communication methods (CF2)			
C11	Email is the best communication method to facilitate information transfer	−0.051	**0.813**
C10	Telephone contact is the best communication method to facilitate information transfer	0.076	**0.723**
C12	Written/fax is the best communication method to facilitate information transfer	0.315	**0.691**
	% of variance	37.223	23.351

10.7.3 Communications and Project Performance: The Taxonomies

Exploratory factor analyses were employed to develop the taxonomies of communication and project performance. This can be achieved by performing Principal Component Factor Analyses (PCFA). The suitability of the data was tested by the Kaiser–Meyer–Olkin (KMO) Measure of Sampling Adequacy with the threshold value of 0.500 (Cheung et al. 2000; Holt 1997; Wong and Cheung 2004). The KMO measure for the taxonomies of communication and project performance are 0.646 and 0.850 respectively, which are greater than the threshold value. These taxonomies are shown in Tables 10.4 and 10.5.

Two factors were generated for communication. 60 % of the total variance was explained by these two factors. Factor 1 consists of the items C4, C3 and C1 and represents effective information flow (CF1). This factor explains 37.22 % of the total variance. As previously mentioned, effective information flow is critical to project delivery. Contractor may suffer from not getting sufficient information at the appropriate time (Higgin and Jessop 2001). Such an effective information flow can clarify client's requirement throughout the project so that the gaps between his expectations and the actual project performance can be minimised. In addition, a substantive and timely response of the messages can also maintain a trusting relationship in a project team (Jarvenpaa and Leidner 1999), and lead to inter-organisational success (Santoro and Saparito 2003). Factor 2 consists of the items C11, C10 and C12 representing communication methods. This factor explains 23.35 % of the total variance. In the course of project delivery, it is important to select the most suitable method to communicate with different contracting parties (Christensen and Bailey 1997). The distance between the parties is also a key

Table 10.5 Taxonomies for project performance

Factors		Factor Loadings		
		1	2	3
Factor 1: Minimised project time, cost and maximised quality control (PPF1)				
PP1	Generally, the project was completed on schedule.	**0.778**	−0.005	0.248
PP10	I believe the construction project progress was under a good performance	**0.725**	0.261	0.342
PP2	The project was completed within budget	**0.715**	0.040	0.315
PP6	Cost control during construction stage was efficient	**0.695**	0.381	0.042
PP8	Defects were kept minimised in the project	**0.648**	0.189	0.027
PP9	I believe that the construction cost of the project was reasonable	**0.608**	0.365	0.165
Factor 2: Satisfactory and worthwhile quality (PPF2)				
PP5	The quality of the building was satisfactory	0.275	**0.735**	−0.126
PP12	I believe that the building satisfied the client's special requirements	0.040	**0.726**	0.280
PP11	I believe that money spent on the project was worthy	0.267	**0.580**	0.437
Factor 3: Effective time control (PPF3)				
PP3	The proper use of effective scheduling smoothened the construction stage	0.122	0.133	**0.865**
PP4	The claim of extension of time was reasonable	0.386	0.087	**0.624**
	% of variance	42.46	9.813	9.274

factor while selecting the most suitable method (Cheng et al. 2001). As the nature of each communication method is unique, the best use of the nature can transmit the information effectively and efficiently (Cheng et al. 2001).

The taxonomies for project performance are shown in Table 10.5. It consists of three factors. Factor 1, entitled 'maximised project time, cost and quality control', explains 42.46 % of variance. It consists of six items that item PP1 representing time, items PP10 and PP8 representing quality and items PP2, PP6 and PP9 representing cost. Thus this factor represents the maximisation of project time, cost and quality control. Factor 2, namely 'satisfactory and worthwhile quality', explains nearly 10 % of the variance. There are three items with a factor loading above 0.500. Items PP5 and PP12 represent quality and item PP11 represents cost. Factor 3 is entitled as 'effective time control'. It consists of two items, PP3 and PP4, which represents the effective management of project schedule. This factor explains 9.27 % of the total variance.

Based on the factors obtained from the PCFA, factor scales were devised for use in the next section. These scales are the composite measures created for each observation on each factor extracted in the PCFA (Hair et al. 1998). Therefore, two and three factor scales were developed for the taxonomies of communication and project performance respectively.

10.7.4 The Effect of Communication on Trust-Project Performance Relationship

The most commonly used approach to test mediating effect is that suggested by Baron and Kenny (1986). With this procedure, the following three regression models shall be considered:-

$$y = \beta_1 + ax + \varepsilon_1 \quad (10.1)$$

$$z = \beta_2 + cx + \varepsilon_2 \quad (10.2)$$

$$z = \beta_3 + c'x + by + \varepsilon_3 \quad (10.3)$$

where x is the independent variable (i.e. trust), y is the mediator (i.e. communication), z is the dependent variable (i.e. project performance), the betas are the intercepts, the epsilons are the model fit errors, and the a, b, c, and c' terms are the regression coefficients capturing the relationships between three focal variables. Evidence for mediation is said to be likely if:

(1) "the term a in Eq. 10.1 is significant. This is evidence of a linear relationship between the independent variable (x) and the mediator (y)";
(2) "the regression coefficient c in Eq. 10.2 is significant. This gives a linear relationship between the independent variable (x) and the dependent variable (z)"; and
(3) "the term b in Eq. 10.3 is significant. This indicates that the mediator (y) helps predict the dependent variable (z), and also c', the effect of the independent variable (x) directly on the dependent variable (z), becomes significantly smaller in size relative to c in Eq. 10.2".

The third criterion, i.e. the comparison of size between c in Eq. 10.2 and c' in Eq. 10.3 can be conducted by z test (Sobel 1982):-

$$z = \frac{a \times b}{\sqrt{b^2 s_a^2 + a^2 s_b^2}}$$

where a and s_a^2 are obtained from Eq. 10.1, and b and s_b^2 from Eq. 10.3. The possible outcomes for testing mediation are summarised in Table 10.6.

Applying the 3-step approach developed by Baron and Kenny (1986), the mediating impacts of communication on the trust-project performance relationship were investigated. This approach first regresses the mediator (i.e. the communication) on the independent variable (i.e. the trust). If the regression is significant, the dependent variables (i.e. the project performance) are regressed on the independent variable in a second regression model. Finally if the second regression is found to be statistically significant, the dependent variable is regressed on the mediator and the independent variable using the same regression. The reduction in the effect size of the independent variable in this final regression will support mediation; an insignificant independent variable in this final regression will

Table 10.6 Possible outcomes for testing of mediating effect (Baron and Kenny 1986; Lacobucci 2008)

Possible outcomes	Implications
A. If neither *a* nor *b* is significant,	Then there is no evidence on mediation, all the variance in *z* attribute to *x* is direct, not mediated through the mediator (*y*)
B. If *a* is not significant, or *b* is not significant,	
C. If (1), (2) and (3) hold,	Then there is, at least, partial mediation, the variance in *z* attributable to *x* is partly a direct effect, and partly an indirect effect mediated through *y*
D. If (1), (2) and (3) hold, and *c'* is not significantly different from zero,	Then the effect is said to be perfect or complete mediation. All the variance in *z* explained by *x* is indirect, mediated through *y*

indicate perfect or complete mediation (Gregory and Albritton 2010). In this study, a total of 42 models (devised from the combination of seven trust variables, two factor scales for communication and three factor scales for project performance) were identified for mediation analysis. Five models are finally detected partial mediation (Table 10.7 refers). The description of Model 1 is as follows:-

"Increased Cognition-based Trust (CBT)—Communication/Interaction (CI) significantly associated with increased Satisfactory and Worthwhile Quality (PPF2) ($\beta = 0.505$, $p < 0.001$) *and increased effective information flow (CF1)* ($\beta = 0.423$, $p < 0.01$). *Effective information flow (CF1) significantly associated with Satisfactory and Worthwhile Quality (PPF2)* ($\beta = 0.191$, $p < 0.01$). *The three-step approach of Baron and Kenny* (1986) *was met, and the standardised beta coefficient was reduced by* 0.08 *from* 0.505 *(without mediation) to* 0.425 *(with mediation controlled), suggesting a partial mediation effect of Effective Information Flow (CF1) between Cognition-based Trust (CBT)—Communication/Interaction (CI) and Satisfactory and Worthwhile Quality (PPF2). Results of Sobel test indicated Effective Information Flow (CF1) as a mediator of this relationship* ($z = 2.034$, $p < 0.05$), *with* 15.8 % *of the effect of Cognition-based Trust (CBT)—Communication/Interaction (CI) on Satisfactory and Worthwhile Quality (PPF2) going through the mediator*". The other four models can be described in a similar manner. Generally, 'effective information flow' (CF1) is a versatile mediator for these five models. 35.2 % of the effect of CBT-K on PPF1 (Model 2), 18.1 % of the effect of CBT-K on PPF2 (Model 3), 30.5 % of the effect of ABT-BT on PPF1 (Model 4) and 20.5 % of the effect of ABT-BT on PPF2 (Model 5) are significantly going through this mediator.

The current study examines the nature of the relationship between trust, communication and project performance in construction industry. The findings indicate that there are relationships between trust that affects communication and thus influences project performance. In practice, project performance is the key concern of construction organisations. The time, cost and quality performance are the ingredients of project performance and are influential to project success. Due to the accelerating population in Hong Kong, the demand on infrastructures and buildings increase rapidly. The time factor becomes significant for the project

Table 10.7 Tests of mediating effect of communication on the trust-project performance relationship

Variables	$x \to y$	$x \to z$	$y \to z$;	$x \to z$; mediator (y) controlled	Sobel test statistic
Model 1: x: Cognition-based trust (CBT)—communication/interaction (CI); y: Effective information flow (CF1); z: Satisfactory and worthwhile quality (PPF2)	0.423**	0.505***	0.191**	0.425***	2.034*
Model 2: x: Cognition-based trust (cbt)—knowledge (K); y: Effective information flow (CF1); z: Minimised project time, cost and maximised quality control (PPF1)	0.326*	0.332**	0.361***	0.215*	2.258*
Model 3: x: Cognition-based trust (CBT)—knowledge (K); y: Effective information flow (CF1); z: Satisfactory and worthwhile quality (PPF2)	0.326*	0.382***	0.213**	0.313***	1.980*
Model 4: x: Affect-based trust (ABT)—being thoughtful (BT); y: Effective information flow (CF1); z: Minimised project time, cost and maximised quality control (PPF1)	0.276*	0.321***	0.353***	0.223*	2.218*
Model 5: x: Affect-based trust (ABT)—being thoughtful (BT); y: Effective information flow (CF1); z: Satisfactory and worthwhile quality (PPF2)	0.276*	0.297***	0.221**	0.236**	1.975*

*** $p < 0.001$
** $p < 0.01$
* $p < 0.05$

completion in order to meet the huge demand of the society. Taking into account the interests of the clients and contractors, they often expect spending the least money but making the greatest profit. In order to achieve this win–win strategy the cost factor is also important. However, nowadays the end-users look for not only the value but also the quality. They prefer good quality and therefore the quality factor should not be neglected while evaluating project success. Consequently, construction project teams shall avoid diminishing the quality of the end product, increased costs and delayed completion or failure to facilitate project success. To achieve this, competent team members and harmonious relationship are preferred. Trust and communication are believed to be the important elements that contribute to project success. From Table 10.7, it can be revealed that 'effective information flow' (CF1) is the versatile mediator that mediates all these significant models. In this aspect, the improvement of information flow can likely maximise project time, cost and quality. Similar observations were reported by Albino et al. (2002), Fok et al. (2001), Ndekugri and McCaffer (1988) and Lari (2002). These studies suggested that information flow in an organisation can improve production and quality for the organisations.

Furthermore, results of Sobel tests revealed that considerable variances (for example, Model 2: 35.2 % and Model 4: 30.5 %) go through the mediator (i.e. 'effective information flow'). This implies that these considerable variances in project performance could be due to the effect of information flow. These findings also imply that, to minimise project time, cost and maximise quality control (PPF1) or to achieve satisfactory and worthwhile quality (PPF2), project practitioners can target at promoting effective information flow. Perhaps this may be a pragmatic way to achieve these project performance as effective management of information flow can minimise project risk and mitigate project delays/ uneconomical decisions. In the presence of effective information flow, potential dispute could be identified and solved earlier. Collaborative work environment can also be promoted if project practitioners are keen to share information. Thus it is important for construction organisations to promote effective information flow in managing projects.

10.8 Chapter Summary

Effective management of information flow can minimise project risk and uneconomical decisions. Problems can be solved before being protracted and become disputes. Collaborative working environment can also be promoted when project practitioners are keen to share information. Communication study is under-researched in construction engineering and management. Two main streams of communication study can be summarised from a review of construction management literature: (i) document communication pattern (Shohet and Frydman 2003; Thomas et al. 1998) and (ii) computerised communication system (Naresh and Jahren 1997; Luiten and Tolman 1997). These studies focus on communication patterns and means to shorten the communication time. The main goal of project manager is to achieve desired project performance. It is therefore of both academic and practical value to advance our understanding of how desired project performance can be accomplished. This study proposed that trust and communication are factors contributing to the efficient achievement of project objectives, which means good project performance. In addition, their relationship was examined by the use of mediation analysis. A total of five significant mediation models were identified. These models reveal that trust affects communication and then influences project performance. In this study, 'effective information flow' is found to be the versatile mediator to the trust-project performance relationships among all of these significant models. This implies that the improvement of information flow would likely improve the achievement of project performance (i.e. 'achieve satisfactory and worthwhile quality' and 'maximise project time, cost and quality'). It is advised that project practitioners can target at promoting effective information flow. This may be a pragmatic way to achieve these project performance. This study extends the study of communication in examining its roles in a trust-performance relation and has the following contributions; (i) articulating communication as a mediator in the trust-performance relation and (ii) analytical examination of the mediating function of communication. This study shall be a valued methodological addition to the application of analytical tool for in-depth analysis of the specific role of intervening variable.

Acknowledgments Special thanks to Mr. Man Chung Lam for collecting data for the study. The content of this chapter has been published in Volume 139(8) of the Journal of Construction Engineering and Management and is used with the permission from ASCE.

References

Albino, V., Pontrandlfo, P., & Scozzi, B. (2002). Analysis of information flows to enhance the coordination of production processes. *International Journal of Production Economics, 75*(1/2), 7–19.

Baguley, P. (1994). *Effective communication for modern business*. London: McGraw-Hill.

Baron, R. M., & Kenny, D. A. (1986). The moderator-mediator variable distinction in social psychological research: Conceptual, strategic, and statistical considerations. *Journal of Personality and Social Psychology, 51*(6), 1173–1182.

Bryde, D. J., & Robinson, L. (2005). Client versus contractor perspectives on project success criteria. *International Journal of Project Management, 23*, 622–629.

Butler, J. K. (1999). Trust expectations, information sharing, climate of trust and negotiation effectiveness and efficiency. *Group and Organization Management, 24*(2), 217–238.

Chan, A. P. C., & Chan, A. P. L. (2004). Key performance indicators for measuring construction success. *Benchmarking: An International Journal, 11*(2), 203–221.

Chan, D. W. M., & Kumaraswamy, M. M. (1996). An evaluation of construction time performance in the building industry. *Building and Environment, 31*(6), 569–578.

Cheng, E. W. L., Li, H., Love, P. E. D., & Irani, Z. (2001). Network communication in the construction industry. *Corporate Communications, 6*(2), 61–70.

Cheung, S. O., & Pang, H. Y. (2013). Anatomy of construction disputes. *The Journal of Construction Engineering and Management (ASCE), 139*(1), 15–23.
Cheung, S. O., Ng, S. T., Wong, S. P., & Suen, C. H. (2003). Behavioral aspects in construction partnering. *International Journal of Project Management, 21*(5), 333–343.
Cheung, S. O., Tam, C. M., Ndekugri, I., & Harris, F. C. (2000). Factors affecting clients' project dispute resolution satisfaction in Hong Kong. *Construction Management and Economics, 18*, 281–294.
Cheung, S. O., Wong, W. K., Yiu, T. W., & Pang, H. Y. (2011). Developing a trust inventory for construction contracting. *The International Journal of Project Management, 29*(2), 184–196.
Christensen, E. W., & Bailey, J. R. (1997). A source accessibility effect on media selection. *Management Communication Quarterly, 10*(3), 373–387.
Church, A. H. (1996). Giving your organizational communication C-P-R. *Leadership and Organization Development Journal, 17*(7), 4–11.
CIRC (2001). *Construct for excellence, report of the construction industry review committee*. Retrieved from http://www.etwb.gov.hk/FileManager/EN/archives/rpt_construction_ind_review_committee/reporte.pdf.
Cleland, D. I., & Ireland, L. R. (2002). *Project management: Strategic design & implementation*. New York: Mc-Graw Hill.
Cook, E. L., & Hancher, D. E. (1990). Partnering: Contracting for the future. *Journal of Management in Engineering, 6*(4), 431–446.
Cummings, L. L., & Bromiley, P. (1996). The Organizational trust inventory (OTI): Development and validation. In R. M.Kramer & T. R. Tyler (Eds.), *Trust in organizations: Frontiers of theory and research* (pp. 302–330). New York: SAGE Publications.
Daft, R. L., & Lengel, R. H. (1986). Organizational information requirements, media richness and structural design. *Management Science, 32*(5), 554–571.
Dainty, A., Moore, D., & Murray, M. (2006). *Communication in construction: Theory and practice*. New York: Taylor & Francis.
Das, T. K., & Teng, B. S. (1998). Between trust and control: Developing confidence in partner cooperation in alliances. *The Academy of Management Review, 23*(3), 491–512.
Deng, Z. M., Li, B., Tam, C. M., Shen, Q. P., & Love, P. E. D. (2001). An application of the internet-based project management system. *Automation in Construction, 10*(2), 239–246.
Diallo, A., & Thuillier, D. (2005). The success of international development projects, trust and communication: An African perspective. *International Journal of Project Management, 23*, 237–252.
Emmitt, S., & Gorse, C. A. (2003). *Construction communication*. Oxford: Blackwell Polishing.
Ferketich, S. (1990). Focus on psychometrics: Internal consistency estimates of reliability. *Research in Nursing & Health, 13*, 437–440.
Fok, L. Y., Fok, W. M., & Hartman, S. J. (2001). Exploring the relationship between quality management and information systems development. *Information & Management, 38*(6), 335–371.
Garson, G. D. (2006). *Structural equation modeling*. Retrieved from http://www2.chass.ncsu.edu/garson/pa765/structur.htm.
Gorse, C. A., Emmitt, S., & Lowis, M. (1999). Problem solving and appropriate communication medium. *Association of Researchers in Construction Management, 2*, 511–518.
Gregory, B. T., & Albritton, M. D. (2010). The mediating role of psychological empowerment on the relationships between P-O fit, job satisfaction and in-role performance. *Journal of Business and Psychology, 25*, 639–647.
Hair, J. F., Anderson, R. E., Tatham, R. L., & Black, W. C. (1998). *Multivariate data analysis* (5th ed.). New Jersey: Prentice Hall.
Harrison, F. L. (1985). *Advanced project management*. Aldershot: John Wiley.
Hartman, F. (1999). *The role of trust in project management: Proceeding of the PMI Research Conference*, September 1999.
Havila, V., Johanson, J., & Thilenius, P. (2004). International business-relationship triads. *International Marketing Review, 21*(2), 172–186.

Higgin, G., & Jessop, N. (2001). *Communication in the building industry: The report of a pilot study*. London: Tavistock Publications.

Holt, G. (1997). Construction research questionnaire and attitude measurement: Relative index or mean. *Journal of Construction Procurement, 3*(2), 88–94.

Iacobucci, D. (2008). *Mediation Analysis*. Los Angeles: Sage.

Idvall, E., Hamrin, E., & Unosson, M. (2002). Development of an instrument to measure strategic and clinical quality indicators in postoperative pain management. *Journal of Advanced Nursing, 37*(6), 532–540.

James, L. R., & Brett, J. M. (1984). Mediators, moderators, and test for mediation. *Journal of Applied Psychology, 69*, 307–321.

Jarvenpaa, S. L., & Leidner, D. E. (1999). Communication and trust in global virtual teams. *Organization Science, 10*(6), 791–815.

Knapp, T. R., & Brown, J. K. (1995). Ten measurement commandments that often should be broken. *Research in Nursing & Health, 18*, 465–469.

Kramer, R. M. (1999). Trust and distrust in organizations: Emerging perspectives, enduring questions. *Annual Review of Psychology, 50*, 569–598.

Kwan, A. Y., & Ofori, G. (2001). Chinese culture and successful implementation of partnering in Singapore's construction industry. *Construction Management and Economics, 19*(6), 619–632.

Lari, A. (2002). An integrated information system for quality management. *Business Process Management Journal, 8*(2), 169–182.

Leung, M. Y., & Liu, A. M. M. (2003). Analysis of value and project goal specificity in value management. *Construction Management and Economics, 21*, 11–19.

Lewis, J. D., & Weigert, A. (1985). Trust as a social reality. *Social Forces, 63*(4), 967–985.

Lim, C. S., & Mohamed, M. Z. (1999). Criteria of project success: An exploratory re-examination. *International Journal of Project Management, 17*(4), 243–248.

Liu, A. M. M., & Walker, A. (1998). Evaluation of project outcomes. *Construction Management and Economics, 16*, 209–219.

Luhmann, N. (1979). *Trust and power*. New York: Wiley.

Luiten, G. T., & Tolman, F. P. (1997). Automating communication in civil engineering. *Journal of Construction Engineering and Management, ASCE, 123*(2), 113–120.

Maskarinec, M. P., Jenkins, R. A., Counts, R. W., & Dindal, A. B. (2000). Determination of exposure to environmental tobacco smoke in restaurant and tavern workers in one US city. *Journal of Exposure Science and Environmental Epidemiology, 10*, 36–49.

McAllister, D. J. (1995). Affect- and cognition-based trust as foundations for interpersonal cooperation in organizations. *Academy of Management Journal, 38*(1), 24–59.

McDonough, E. F., & Kahn, K. B. (1996). Using hard and soft technologies for global new product development. *Research and Development Management, 26*(2), 241–253.

McDonough, F. E., Kahn, K. B., & Gribbin, A. (1999). Managing communication in global product development teams. *IEEE Transactions on Engineering Management, 46*(4), 375–386.

Muller, R., & Turner, J. R. (2005). The impact of principal–agent relationship and contract type on communication between project owner and manager. *International Journal of Project Management, 23*, 398–403.

Naresh, A. L., & Jahren, C. T. (1997). Communication and tracking for construction vehicles. *Journal of Construction Engineering and Management (ASCE), 123*(3), 261–268.

Ndekugri, I., & McCaffer, R. (1988). Management information flow in construction companies. *Construction Management and Economics, 6*(4), 273–294.

Nunnally, J. C., & Bernstein, I. H. (1994). *Psychometric Theory*. New York: Mc-Graw Hill.

Oberlender, G. D. (2000). *Project management for engineering and construction*. Boston: Mc-Graw Hill.

Pietroforte, R. (1997). Communication and governance in the building process. *Construction Management and Economics, 15*, 71–82.

Pinto, M. B., & Pinto, J. K. (1991). Determinants of cross-functional cooperation in the project implementation process. *Project Management Journal, 22*, 13–20.

Porta, R. L., Lopez-de-Silane, F., Shleifer, A. & Vishny, R. W. (1996). *Trust in large organizations, NBER working paper no. 5864*. The National Bureau of Economic Research.

Robinson, P. B., Stimpson, D. V., Huefner, J. C., & Hunt, H. K. (1991). An attitude approach to the prediction of entrepreneurship. *Entrepreneurship Theory Practice, 15*, 13–30.

Rojas, E. M., & Songer, A. D. (1999). Web-centric systems: A new paradigm for collaborative engineering. *Journal of Management in Engineering, 15*(2), 39–45.

Santoro, M. D., & Saparito, P. A. (2003). The firm's trust in its university partner as a key mediator in advancing knowledge and new technologies. *IEEE Transactions on Engineering Management, 50*(3), 362–373.

Scanlin, J. (1998). The internet as an enabler of the Bell Atlantic project office. *Project Management Journal, 9*(2), 6–7.

Shenhar, A. J., Levy, O., & Dvir, D. (1997). Mapping the dimensions of project success. *Project Management Journal, 28*(2), 5–13.

Shohet, I. M., & Frydman, S. (2003). Communication patterns in construction at construction manager level. *Journal of Construction Engineering and Management (ASCE), 129*(5), 570–577.

Sobel, M. E. (1982). Asymptotic confidence intervals for indirect effects in structural equation models, sociological methodology. In S. Leinhardt (Ed.), American Sociological Association. Information Age Publishing, Washington, DC.

Thomas, S. R., Tucker, R. L., & Kelly, W. R. (1998). Critical communication variables. *Journal of Construction Engineering and Management, 124*(1), 58–66.

Westerveld, E. (2003). The project excellence model: linking success criteria and critical success factors. *International Journal of Project Management, 21*, 411–418.

Wong, F. W. H., & Lam, P. T. I. (2011). Difficulties and hindrances facing end users of electronic information exchange systems in design and construction. *Journal of Management in Engineering, 27*(1), 28–39.

Wong, S. P., & Cheung, S. O. (2004). Trust in construction partnering: Views from parties of the partnering dance. *International Journal of Project Management, 22*(6), 437–446.

Wong, W. K. (2007). A trust inventory for use in the construction industry. M.Phil. Thesis, Department of Building and Construction, City University of Hong Kong, Hong Kong, China.

Wong, W. K. & Cheung, S. O. (2006). *Trust in organizational relationship and its delineation: Proceeding of ICCREM 2006, 13th Rinker International Conference*. Oxford, Florida.

Xiao, H., & Proverbs, D. (2002a). The performance of contractors in Japan, the UK and the USA: A comparative evaluation of construction cost. *Construction Management and Economics, 20*, 425–435.

Xiao, H., & Proverbs, D. (2002b). The performance of contractors in Japan, the UK and the USA: An evaluation of construction quality. *International Journal of Quality & Reliability Management, 19*(6), 672–687.

Yu, A. G., Flett, P. D., & Bowers, J. A. (2005). Developing a value-centred proposal for assessing project success. *International Journal of Project Management, 23*, 428–436.

Zaghloul, R., & Hartman, F. (2003). Construction contracts: The cost of mistrust. *International Journal of Project Management, 21*(6), 419–424.

Part III
Negotiation: The Gateway to Dispute Resolution

Chapter 11
The Behavioural Dimensions of Construction Dispute Negotiation

Sai On Cheung, Tak Wing Yiu and Pui Ting Chow

Abstract Negotiating dispute involves finding common grounds over disagreements, thus sometimes is viewed as an art. Construction disputes are always negotiated first before other resolution methods come to service. Reaching a settlement through negotiation helps to maintain a harmonious relationship between the disputants. In these regards, negotiation is the most cost efficient method to resolve construction dispute. Negotiation skill therefore is essential to all construction professionals. This chapter first gives a brief introduction of the traditions of negotiation studies. Then, the causes of negotiation failure are discussed with emphasis on the behavioural factors. A study on the relation between negotiating style and negotiation outcome is presented to illustrate the impact of behavioural factor. It is found that the use of obliging, dominating and avoiding styles appear to be less influential in achieving functional negotiation outcomes than using integrating style. The use of compromising style is also found to be a practical approach in resolving dispute.

S. O. Cheung (✉) · P. T. Chow
Construcion Dispute Resolution Research Unit, Department of Civil and Architectural Engineering, City University of Hong Kong, Hong Kong, People's Republic of China
e-mail: saion.cheung@cityu.edu.hk

P. T. Chow
e-mail: ronnie.chow.pt@hotmail.com

T. W. Yiu
Department of Civil and Environmental Engineering, Faculty of Engineering, The University of Auckland, Auckland, New Zealand
e-mail: k.yiu@auckland.ac.nz

S. O. Cheung (ed.), *Construction Dispute Research*,
DOI: 10.1007/978-3-319-04429-3_11,

11.1 Introduction

Construction contracting environment is dispute-laden (Rhys Jones 1994). This can be attributed to the fact that construction project management requires the coordinated effort of a temporarily assembled project team comprising professionals of various disciplines. Project team members have to pursue their own goals and maximise their own benefits and sometimes sacrifice those of the others (Newcombe 1996; Walker 2002). In addition, managing design changes during the course of construction laid futile ground for the germination and manifestation of construction disputes. Dispute once crystallised requires a proactive resolution that prevents aggravation of the negative impact on project performance (Brown and Marriott 1999; Fenn et al. 1997). Disputes are always negotiated first before other resolution methods come to service (Brown and Marriott 1999; Cheung et al. 2002a; Cheung and Yeung 2002; Goldberg et al. 1992; Hibberd and Newman 1999). Reaching a settlement through negotiation helps to maintain a harmonious relationship among disputants. Thus, negotiation is the most cost efficient method to resolve construction dispute (Ren et al. 2003). Negotiation skill therefore is essential to all construction professionals, in particular those at managerial position. In fact, individuals have to negotiate with one another to establish common grounds, arrange their affairs in commerce and everyday life, and reconcile areas of disagreement (Brown and Marriott 1999). As such, negotiation has been a topical issue in the management research. This chapter first gives a brief introduction of these traditions of negotiation studies and then discusses the causes of negotiation failure from a behavioural point of view. As an illustration, a study on the relation between negotiators' style and negotiation outcome is presented.

11.2 Traditions of Negotiation Studies

Carnevale and Pruitt (1992) identified three main traditions of negotiation studies. The first consists of guidelines to international and industrial negotiators (Lewicki and Litterer 1985; Murnighan 1991). These guidelines are valuable references but must be used with contextual and cultural caveats. The second tradition involves the use of mathematical models based on rational approaches advocated by economists and game theorists (Nash 1950; Kagel and Roth 1991; Zeuthen 1930). These models are both descriptive, in the sense of specifying the parameters within which negotiators operate or some of the dynamics of negotiation behaviour, and prescriptive in recommending rational policies. However, these models tend to have specific focus with selective indicators and restrictive assumptions. Rapid advancement in information technology in the last two decades has paved the way for the development of computer-based decision support systems to assist negotiators (e.g. Carmel et al. 1993). The accessibility of the World Wide Web has further released the geographical obstacles that prohibit face to face negotiation (Cheung et al. 2004; Druckman et al. 2002; Griffith et al. 2002; Holsapple et al. 1998).

The third tradition emphasises on empirical studies in both laboratory and field settings (Douglas 1962; Steven 1963; Walton and McKersie 1965). This tradition is useful in explaining the hurdles and difficulties faced by negotiators. This tradition helps in highlighting causes of negotiation failure.

11.3 Causes of Negotiation Failure

Four types of negotiation failure are identified, including contract zone conundrum, negotiator's selection, political pressure and withdrawal (Downie 1991; Mnookin 1993; Sebenius 1992; Underdal 1983). Examples of each type of negotiation failure in construction dispute negotiation are given in Table 11.1.

11.3.1 Contract Zone Conundrum

Sebenius (1992) explains that people negotiate in order to satisfy their collective interests better through joint decision-making. In this connection, the normative approach of decision-making in negotiation is that negotiators evaluate their alternatives with reference to both their interests and those of their counterparts. Moreover, it is important to distinguish parties' underlying interest from issues under negotiation. The efficiency of bargaining is calculated based on the size of the contract zone (Neale and Bazerman 1985). However, there is no readily available means to determine the contract zone. In real life situations, decisions are mix-motive with proposals being the aggregate utility covering several issues. Contradictory expectations on issues to be negotiated are more problematic as these may lead to a zero or small contract zone. The failure to ascertain the contract zone may eventually lead to deadlock and stalemate (Neale and Bazerman 1985). The availability of positive contract zone is a prerequisite for a settlement (Fig. 11.1). Misjudgment on the contract zone would hamper the prospect of a negotiated settlement.

11.3.2 Selection of Negotiator

Labour dispute often involves the union, which supposedly represent the members. Its presence makes dispute events more newsworthy. The dispute is then publicised whereby existing images, stereotypes and expectations are projected. The ruling of union may not meet with individual members' specific needs or aspirations. Different preferences on the disputing issues may end with internal diversification. The internal structure of a union may split into several groups of acute disagreement. In this connection, the inability caused by absence of unanimous decision may ultimately lead to negotiation cessation. The principal/agent situation is analogous to the union/member scenario. Problems can arise when the

Table 11.1 Possible causes of negotiation failure

Possible causes of negotiation failure
Contract zone conundrum
Dilemmas in concession-making and issue management
Unresolvable nature of issue
Strategic barrier
Reactive devaluation of compromises and concessions
Many solution concepts
Limitation in game's structure, rules, and possible moves are not common knowledge
Widely scattered negotiation outcomes in practice
Uncertainty
Inaccurate information
Selection of negotiator
Union preference, politics and internal structure
Principal/agent problem
Pressure
External pressure
Politically inadequate
Withdrawal
Cognitive barrier
Significant departures from full game-theoretic "rationality"
Insensitivity to behavioural expectations

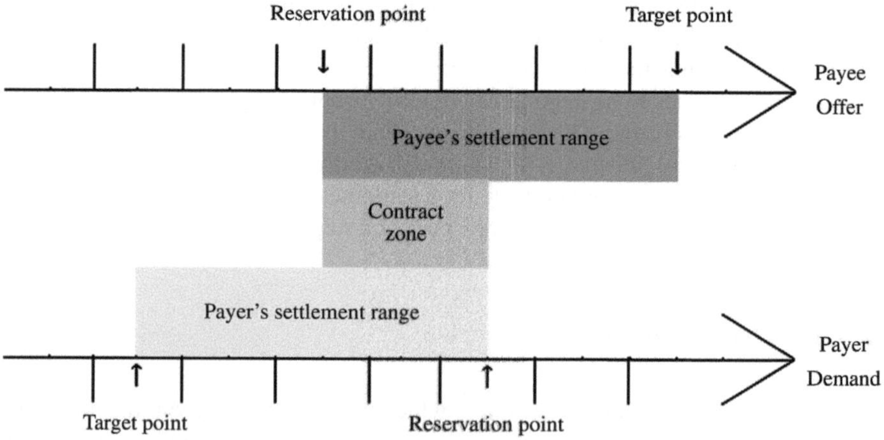

Fig. 11.1 Contract zone

interest of the agent conflicts with that of his/her principal (Lewicki and Litterer 1985). Whilst the use of professional advisors like claim consultants is quite common, prudent caveat against the negotiation being hijacked should be exercised (Cheung et al. 2000). The importance of having the 'right' of negotiator therefore cannot be undervalued.

11.3.3 Pressure

Pressure affects negotiating behaviour in two ways. The first is external. The continuity of an organisation depends on not only its own capability but also the external market conditions. Organisation who fails to deal with external pressure is unlikely to develop and sustain its competitive edge. The time and resources invested in the negotiation can be used in other more productive way. The second factor is political. Unduly strong public expectations on a conflict may hinder its resolution. A politically inadequate solution is a misfit between what a theoretically desirable solution and what is expected. Having an outcome that meets technical, economic and political expectations may well be just an ideal.

11.3.4 Withdrawal

Cognitive barrier can be explained in the context of information processing. Human decision-making employs inferences and judgments. Loss aversion and framing effects can have critical influence on negotiator's behaviour. Departure from the canons of rationality is possible. Insensitive to behavioural expectations is another possible cause of negotiation failure. Behavioural expectations are those related to the image and reputation of the negotiator. In other words, "*negotiation is not simply a decision-making process; it is also to some extent a matter of fame and reputation*" (Iklé 1964). Iklé (1964) described three possible options for a negotiator when confronted with critical decisions; (1) accepting currently available terms in the proposals from counterparts; (2) continual negotiating in the hope of securing better terms; and (3) breaking off talks with no intention of resuming them in short-run. Option (3) denotes withdrawal and is considered as a negotiation failure.

In this regard, the relationship between negotiation failure and negotiating behaviour has been a major concern under the behavioural tradition (Cohen 2001; Stevenson 1991; Tor and Bazerman 2003; Underdal 1983). These studies advocate that negotiators are critical in the success or otherwise of a negotiation. Adding to this collection, the relationships between negotiating behaviours and negotiation outcomes is further examined.

11.4 The Behavioural Tradition of Negotiation Studies

Majority of negotiation studies of the behavioural tradition investigate the effects of personality on negotiating behaviours (Allred 2000; Lytle et al. 1999; Mintu-Wimsatt and Calantone 1996; Shell 2001). These studies are useful in suggesting guides of good negotiation practices. Inefficient negotiation discourages early

Fig. 11.2 The conceptual framework for the study

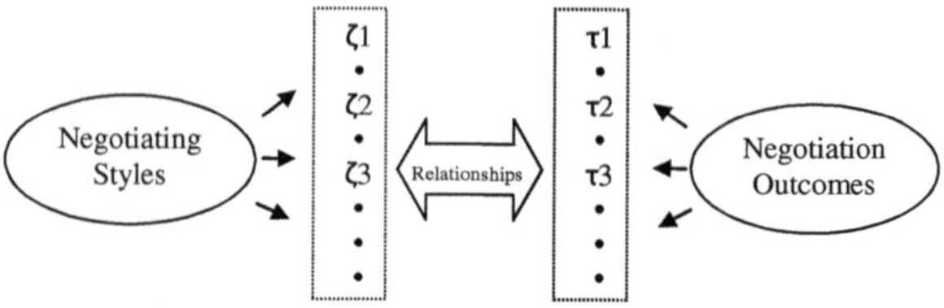

ζ_i = *Negotiating Styles* (i=1,2,3...)
τ_i = *Negotiation Outcomes* (i=1,2,3...)

settlement, and creates an adversarial contracting environment and thus renders the use of expensive arbitration or litigation (Ren et al. 2003; Zack 1994). One of the reasons for such inefficiency is due to the misinterpreting the negotiating style of the negotiators. The following five steps are applied to investigate the relationships between the negotiating styles and negotiation outcomes.

(i) Identify a style classification framework from literature and select a measurement tool;
(ii) Collect data from experienced construction professionals on negotiating styles and negotiation outcomes;
(iii) Test the authenticity of the tool with the data collected;
(iv) Develop taxonomies for the negotiation outcomes; and
(v) Investigate the relationships between the negotiating styles and negotiation outcomes.

The conceptual framework of this study is given in Fig. 11.2.

11.5 Step 1: Negotiating Styles and Its Measurement

People negotiate every day. To study the habitual negotiating characteristics of construction professionals, reference is made to previously completed models (Perdue et al. 1986). Negotiating style is often framed by conflict management style. Follett (1940) suggested five ways to handle conflict: domination, compromise, integration, avoidance and suppression. Blake and Mouton (1964, 1970) presented a conceptual framework for the classification of interpersonal conflict handling styles. These are forcing, withdrawing, smoothing, compromising, and problem solving. In fact, this framework has widely been used in negotiation and conflict management studies (Chakrabarty et al. 2002; Gross et al. 2000; Hammock et al. 1990; Oetzel 1998). In addition, these five styles can also be gauged by the degrees of concern for production and people. Thomas (1976) modified this scheme and took into account the intentions of the party. Similar to the approach of Blake and Mouton (1964), Rahim and Bonoma (1979) and Rahim (1983) differentiated the styles of handling interpersonal conflict by two basic dimensions:

concern for self and concern for others. Concern for self represents the degree to which a person attempts to satisfy his or her own concerns. Concern for others represents the degree to which a person wants to satisfy the concerns of others. These two dimensions describe the motivational orientations of an individual when exposed to a conflict. Supported by the studies of Ruble and Thomas (1976) and Van de Vliert and Kabanoff (1990), the two-dimensional model was refined. The integrated model has five conflict handling styles: integrating, obliging, compromising, dominating and avoiding. This refined model is called Dual Concern Model (Rahim 1992).

An effective negotiator adopts a negotiating style that fits the circumstances (Rahim 2002). A style is considered appropriate if its use can result in effective solution formulation to a problem. In this respect, the predominant view is that integrating or problem-solving style is most appropriate for achieving "win–win" solution (Blake and Mouton 1964; Likert and Likert 1976). However, Rahim (2001), Rahim and Bonoma (1979) and Thomas (1976) suggested that one style might be more appropriate than another depending upon the situation. In general, integrating and to some extent compromising styles are appropriate for dealing with strategic issues. The other styles can be used to deal with tactical or day-to-day problems (Rahim 2002).

The next task in this Step 1 is to select an appropriate style measurement instrument for use in this study. The selected instrument should be commonly adopted in similar studies. In this regard, the conceptual underpinnings suggested by Blake and Mouton (1964, 1970) and the dual concern model of Rahim and Bonoma (1979) and Rahim (1983) are widely used in the study of conflict-handling styles (Friedman et al. 2000; Hammock et al. 1990; Lee et al. 2003; Oetzel 1998; Rahim 1983; Rahim et al. 2000; Rahim 2002; Van De Vliert and Kabanoff 1990). Hence, an instrument that has been developed basing on the above observations is considered appropriate for this study. Accordingly, the Rahim Organisational Conflict Inventory-II (ROCI-II) was used to measure negotiating styles. This instrument has been designed to measure the five conflict management styles suggested by Blake and Mouton (1964, 1970): integrating, obliging, dominating, avoiding and compromising.

11.6 Step 2: Data Collection

A questionnaire survey was used to collect data. Two types of data were collected from each response; one on negotiating behaviours of the respondent and the second on the negotiation outcome with reference to a recent negotiation completed by the respondent. For data on negotiating style, the ROCI-II, which consists of 28 statements on negotiating behaviours, was modified to suit the construction context. A 7-point Likert scale was used to measure the degree of agreement on the practice of the behaviour during the negotiation. A high score represents stronger agreement. As for the second type of data, it is based on a

Table 11.2 Composition of respondents

Grouping	Organisation type	Number	Percentage
Clients	Government departments	11	15.7
	Private developers	6	8.6
	Consultants (Surveyors, Architects and Engineers)	26	37.1
Contractors	Main contractor (Building works)	25	35.7
	Sub-contractor	2	2.9
	Total	70	100

literature review on the possible negotiation outcomes under the influence of the five negotiating styles (Friedman et al. 2000; Gross and Guerrero 2000; Prein 1976; Rahim 1983; Rahim et al. 2000). The respondents were asked to assess the degree of achievement with respect to the itemised outcomes on a 7-point Likert scale. A total of 150 questionnaires were sent to construction professionals holding senior positions in Hong Kong. The list was compiled by identifying key personnel from the government and professional directories and web sites of companies. 70 of them responded and returned the questionnaire. The response rate was 47 % and 64 % of the respondents have more than 10 years experience in construction. As for employing organisations, 60 % of the respondents work for clients while the other 40 % are employees of contracting organisations. The composition of the respondents by organisation type is shown in Table 11.2.

11.7 Step 3: Testing the Authenticity of the ROCI-II Instrument

The identification of the five styles of Blake and Mouton (1964, 1970) by the ROCI-II Instrument is firstly analysed by the use of Principal Component Factor Analysis (PCFA). This technique examines the factor structure. Interpretation of variables can be accomplished by summarising the data according to the constructs (Hair et al. 1995).

Before performing a PCFA, the suitability of the data was first evaluated by examining the Kaiser–Meyer–Olkin (KMO) measure of sampling adequacy. The KMO value for the PCFA is 0.679, a figure above the threshold requirement of 0.5 (Holt 1997; Cheung and Yeung 1998; Cheung et al. 2000). In addition, the low significance of the Bartlett test of sphericity suggests the adequacy of the data set to perform PCFA. To shortlist factors, the eigenvalue-greater-than-1 principle, which is the commonly used criterion, was applied. Factors having an eigenvalue greater than 1 were considered significant, and those with eignvalue below 1 were discarded. In order to simplify the factor structures and obtain more meaningful factor solution, rotation of the factor matrices was performed to reduce the ambiguities that often accompany initial un-rotated factor solutions. Varimax rotation was employed in the present study. The final factor matrix for negotiating style after Varimax rotation is given in Table 11.3. The authenticity of the ROCI-II instrument was examined by comparing the items included for each factor with those indicated by the designer. But for items 3, 4 and 27, the items retrieved for the five factors were identical to the original design of the instrument (Chakrabarty et al. 2002; Hammock et al. 1990). The interpretations of the extracted five factors therefore generally fit well with the style classification of Blake and Mouton (1964, 1970). As such, it is reasonable to use the ROCI-II as the instrument to valid measure of negotiating styles of construction professionals.

By calculating the average scores respective to the five factors, the average scores for the five negotiating styles were obtained. Table 11.4 shows the summary. Although the differences between these scores are not significant, nonetheless, the style that displays a higher average score may be viewed as the one that is more often used as compared with the others. In this context, compromising, among others, is the style most often used by the respondents.

11.8 Step 4: Taxonomies of Negotiation Outcomes

Taxonomy is a system by which categories are related to one another by means of class inclusion (Rosch 1988). In this step, taxonomies of negotiation outcomes were developed for use in Step 5. This was also achieved by the use of Principal Component Factor Analysis (PCFA); the procedures for such are described in Step 3. Similarly, the Kaiser–Meyer–Olkin (KMO) measure of sampling adequacy was examined. The KMO values for this PCFA are 0.660, which is also above the threshold requirement of 0.5. Furthermore, the low significance of the Bartlett test of sphericity suggests the adequacy of the data set to perform PCFA. The factor matrix for negotiation outcomes after Varimax rotation is shown in Table 11.5. The seven factors extracted can be described as follows:

Factor 1 Problem Solving
Factor 2 Conflict Escalation
Factor 3 Relationship Deterioration
Factor 4 Inaction
Factor 5 Further Disagreement
Factor 6 Relationship Maintained
Factor 7 Conflict Reduction.

The following section describes each of the factors extracted. These factors are further identified as either functional or dysfunctional negotiation outcome. These identifications are needed to enable the working of Step 5.

Factor 1 is named as Problem Solving since the items are related to solution formulation, conflict reduction and achievement of compliance. Problem solving, a desirable negotiation outcome in conflict resolution, is no doubt the main goal of every negotiation. In construction industry, this negotiation outcome is ideal if a

Table 11.3 Rotated factor matrix for negotiating styles

Item no.		Factors[a]				
		IN	OB	DO	CO	AV
Factor 1: *Integrating*		I	II	III	IV	V
23	I collaborate with the other to come up with decision acceptable to us	**0.79**	−0.03	0.10	0.61	0.08
22	I try to bring all concerns out in the open so that the issues can be resolved in the best possible way	**0.74**	0.00	−0.08	0.01	0.18
12	I exchange accurate information with the other so that we can solve the problem together	**0.70**	0.27	0.15	0.22	−0.09
5	I try to work with the other to find solutions to a problem which satisfy our expectations	**0.65**	0.27	0.27	0.03	−0.07
1	I try to investigate an issue with the other to find a solution that will be acceptable to everyone involved	**0.63**	0.22	0.11	0.34	−0.14
28	I try to work with the other for a proper understanding of a problem	**0.46**	0.18	0.31	0.43	0.10
Factor 2: *Obliging*		I	II	III	IV	V
24	I try to satisfy the expectations of the other	0.19	**0.76**	−0.08	0.05	0.06
10	I usually try to accommodate the wishes of the other	0.23	**0.76**	0.09	0.19	0.24
2	I generally try to satisfy the needs of the other	0.07	**0.76**	−0.03	0.12	−0.20
11	I give into the wishes of the other	−0.17	**0.66**	0.12	0.28	0.25
19	I often go along with the suggestions of the other	0.09	**0.65**	0.00	0.10	0.30
13	I usually allow concession to the other	0.38	**0.55**	−0.16	−0.08	0.27
3	I attempt to avoid being "put on the spot" and try to keep my conflict with the other to myself	0.34	**0.42**	0.23	0.35	0.31
Factor 3: *Dominating*		I	II	III	IV	V
9	I use my authority to make a decision in my favor	−0.12	−0.05	**0.89**	0.12	0.05
8	I use my influence to get my ideas accepted	−0.05	0.13	**0.86**	0.19	0.01
25	I sometimes use my power to win a competitive situation	0.15	−0.07	**0.78**	−0.07	0.33
21	I am generally firm in pursuing my side of the issue	0.45	−0.08	**0.74**	0.01	0.12
18	I use my expertise to make a decision in my favour	0.36	0.01	**0.66**	0.12	0.08
Factor 4: *Compromising*[b]		I	II	III	IV	V
7	I try to find a middle course to resolve an impasse	0.16	0.05	0.00	**0.82**	0.16
14	I usually propose a middle ground to break deadlocks	0.01	0.14	0.17	**0.72**	0.13

(continued)

Table 11.3 (continued)

Item no.		Factors[a]				
		IN	OB	DO	CO	AV
15	I negotiate with the other so that compromise can be reached	0.49	0.15	−0.06	**0.62**	0.18
4	I try to integrate my ideas with the other to come up with a decision jointly	0.44	0.37	0.22	**0.49**	0.16
20	I use "give and take" so that a compromise can be reached	0.44	0.29	−0.02	**0.47**	0.18
27	I try to avoid unpleasant exchanges with the other	0.01	0.12	0.23	**0.47**	0.44
Factor 5: *Avoiding*		I	II	III	IV	V
16	I try to stay away from disagreement with the other	0.10	0.05	0.10	0.05	**0.85**
17	I avoid an encounter with the other	0.12	0.21	0.09	0.06	**0.80**
26	I try to keep my disagreements with the other to myself to avoid hard feelings	0.07	0.14	0.15	0.24	**0.63**
6	I usually avoid open discussion of my differences with the other	−0.14	0.11	0.05	0.29	**0.61**

[a] IN = Integrating; OB = Obliging; DO = Dominating; CO = Compromising; AV = Avoiding
[b] The most used style in construction negotiation

Table 11.4 Average scores of negotiating styles

Negotiating styles	Average scores
1. Integrating	4.16
2. Obliging	4.26
3. Dominating	4.17
4. Compromising	4.56
5. Avoiding	4.20

dispute is settled with a solution that satisfies the goals and needs of the parties. Thus, it can be identified as a functional negotiation outcome (Rahim 1992), i.e. an outcome generally favoured by the disputants. Factor 2 is described as Conflict Escalation because these outcomes are characterised by higher level of conflict. As such, Factor 2 is identified as a dysfunctional outcome (Rahim 1992), i.e. an outcome having a negative connotation as the dispute is unlikely to be resolved with escalating conflict. Factor 3 relates to the deterioration of relationship between the disputants. Relationship between the negotiators could be a critical factor in tackling the conflict. In Hong Kong where this study was conducted, relationship is a prime factor in business dealings. Except for Government projects, there is no requirement for open tender. As a matter of fact, private developers would not invite contractors with whom they have bad relationship to tender for their projects. Even in publicly funded projects, attitude to claim forms part of tender evaluation (EWTB 2002). Thus, with deteriorating relationship, chance of future cooperation becomes distant. This factor therefore is a

Table 11.5 Rotated factor matrix for negotiation outcomes

Item No.		Factors						
		I	II	III	IV	V	VI	VII
Factor 1: Problem solving								
11	Less conflict-laden environment was produced.	**0.71**	0.13	−0.31	0.30	−0.14	0.31	0.07
12	More behavioural compliance with both parties was achieved	**0.70**	−0.14	0.24	−0.06	0.26	−0.06	0.07
1	The solution found satisfied the goals and needs of both parties	**0.65**	−0.04	−0.41	0.04	−0.25	−0.13	0.07
2	Optimal and creative solution to problem was found	**0.65**	0.12	−0.25	0.11	0.16	−0.24	0.10
16	The levels of conflict were reduced	**0.58**	−0.21	−0.17	0.41	0.10	0.24	−0.06
Factor 2: Conflict escalation								
19	More task conflict was experienced	0.14	**0.78**	0.25	0.23	−0.06	0.00	0.06
18	A higher level of ongoing conflict was experienced	0.14	**0.76**	0.37	0.12	0.11	−0.02	−0.22
7	There was lack of basic information needed to construct solutions to the conflicts	−0.11	**0.71**	0.16	0.21	0.29	0.08	0.10
8	The dispute was difficult to resolve	−0.19	**0.66**	−0.00	−0.13	0.01	−0.02	−0.21
Factor 3: Relationship deterioration								
15	Task conflict was turned into relationship conflict	−0.10	0.26	**0.76**	−0.17	−0.11	0.03	−0.17
13	I ignored the needs and expectation of the other party	−0.13	0.17	**0.69**	0.10	0.10	−0.08	−0.01
21	The negotiation process was a one-side decision-making process	−0.09	0.07	**0.56**	0.24	0.00	−0.44	0.23
14	Solution development was likely to be sub-optimal, resulting in wasted resources	−0.26	0.42	**0.55**	0.28	0.11	0.05	−0.01
Factor 4: Inaction								
6	I withdrew from a threatening situation	0.24	0.14	0.09	**0.81**	−0.19	−0.13	−0.02
5	The issue was postponed until a better time	−0.08	0.19	0.07	**0.75**	0.30	0.08	0.05
Factor 5: Further disagreement								
3	There were further disagreements or escalations in conflict	−0.00	0.26	−0.02	−0.08	**0.77**	−0.13	0.14
4	Stalemate was aroused	0.13	−0.04	0.08	0.19	**0.70**	−0.16	−0.27

(continued)

Table 11.5 (continued)

Item No.		Factors						
		I	II	III	IV	V	VI	VII
Factor 6: *Relationship maintained*								
9	Some of each party's needs were satisfied, but not all of them	−0.11	0.12	−0.08	0.05	−0.22	**0.82**	0.02
10	Relationship between the parties was kept in tact for future interaction	0.60	−0.15	0.04	−0.03	−0.03	**0.62**	0.16
Factor 7: *Conflict reduction*								
20	Less future disputes were likely made	0.22	−0.02	−0.02	0.10	0.08	0.09	**0.84**
17	The agreement was difficult to reach	0.02	0.34	0.09	0.16	0.33	0.11	**−0.60**

dysfunctional outcome (Rahim 1992) due to its negative impact on conflict. Factor 4 is described as Inaction. This negotiation outcome is characterised by withdrawal from and postponement of the negotiation process. Inaction is often undesirable (Rahim 1992), as chances of getting the dispute resolved would become remote. Factor 5 includes the outcomes of further disagreement after the negotiation and the dispute reaches a stalemate. This outcome identified as Further Disagreement, is the least a negotiator wants and clearly a dysfunctional negotiation outcome (Rahim 1992). Factor 6 includes more positive negotiation outcomes such as some of the needs of the parties are satisfied and further interaction is kept. This functional outcome is described as Relationship Maintained. Finally, Factor 7 is interpreted as Conflict Reduction as this is a lesser chance for future dispute, a functional outcome favoured by the disputants.

The development of taxonomies is summarised in Fig. 11.3. This figure is in fact the enhanced version of the conceptual model for this study (Fig. 11.3 refers). The result of first step analysis indicated that the ROCI-II is a reasonable tool to be used to measure negotiating styles. The taxonomies of negotiation outcome developed in this part of the study reduce the number of variables into more manageable numbers for investigating the relationships between these two dimensions (presented as a narrow). This is to be reported in the following section.

11.9 Step 5: The Relationships Between the Negotiating Styles and Negotiation Outcomes

The relationships between negotiating styles and negotiation outcomes were explored by the use of multiple regression analyses (MRA), a statistical technique that can be used to analyse the relationship between a single dependent variable and several independent variables (Hair et al. 1995). In this study, for each of the

Fig. 11.3 The development of taxonomies

Negotiating styles

1. Integrating
2. Obliging
3. Dominating
4. Compromising
5. Avoiding

Taxonomy of Negotiation Outcomes

1. Problem Solving
2. Conflict Escalation
3. Relationship Deterioration
4. Inaction
5. Further Disagreement
6. Relationship Maintained
7. Conflict Reduction

regression model, the dependent variable is one of the seven outcome taxonomies and the independent variables are the negotiating styles. Therefore, a total of seven regression models were developed. Based on the factors identified by the PCFA described in Step 3 and Step 4, factor scales were calculated for the purpose of the multiple regressions. These scales are the composite measure created for each observation on each factor extracted in the PCFA (Hair et al. 1995). Therefore, new sets of variables for each of the negotiation outcome taxonomies were calculated for the multiple regression analysis. In sum, the equation of the multiple regressions is in the following form:

$$O = a_0 + a_1 S_1 + a_2 S_2 + a_3 S_3 + a_4 S_4 + a_5 S_5 \tag{11.1}$$

where O = Dependent variable (Taxonomy of Negotiation outcome); S_i = Independent variables (Negotiating styles).

As described in Step 4, the identification of the factors in the taxonomies is either functional or dysfunctional. For ease of discussion, the statistical results of these two types of negotiation outcomes are presented in Tables 11.6 and 11.7. In the tables, the R^2 values represent the combined effect of the entire variate in prediction and range from 0.144 to 0.504. Comparable results were reported by a number of similar studies in the study of self-reported conflict style (Oetzel 1998), organisational conflict styles (Gross and Guerrero 2000) and styles on buyer–supplier negotiations (Sharland 2001). The relative contribution of the negotiating styles towards the negotiation outcome can be compared by normalising the coefficients of the regression equation. The higher the normalised coefficient, the greater its contribution towards the prediction of the outcomes.

It appears that the use of obliging, dominating and avoiding styles are less influential in achieving functional negotiation outcome. Therefore, relying on the power position to control others, self-sacrifice and withdrawal from conflict does not mean that the conflict can be resolved. Using these types of negotiating style may even result in conflict escalation and relationship deterioration. More tedious and costly conflict resolution method may then become necessary (Cheung 1998; Hills 1992).

Generally, project practitioners are suggested to use integrating style in the conflict resolution process, as this style was found having positive contribution towards functional outcomes and negatively correlated with dysfunctional outcomes.

Table 11.6 Overall results of multiple regression analysis (for functional negotiation outcomes)

Dependent variables (Outcomes)	Independent variables (Negotiating styles)	R^2	Standardised regression coefficients	Normalised regression coefficients
Problem solving		0.306		
	Integrating		0.536	0.602
	Obliging		−0.083	0.093
	Dominating		0.143	0.160
	Compromising		0.012	0.013
	Avoiding		−0.117	0.132
Relationship maintained		0.504		
	Integrating		0.549	0.366
	Obliging		−0.192	0.128
	Dominating		−0.172	0.115
	Compromising		0.407	0.271
	Avoiding		−0.181	0.120
Conflict reduction		0.144		
	Integrating		0.318	0.250
	Obliging		0.168	0.133
	Dominating		−0.078	0.062
	Compromising		−0.424	0.335
	Avoiding		0.278	0.220

These can be shown from the results obtained by MRA as shown in Tables 11.6 and 11.7. In Table 11.6, the normalised regression coefficients of integrating style are 0.602, 0.366, and 0.250 for the three functional outcomes: problem solving, relationship maintained and conflict reduction respectively, which are the highest in magnitude among the other styles. These results suggested that use of integrating style contributes to problem solving and conflict reduction with improved relationship, an observation in line with findings in previous studies (Friedman et al. 2000, Gross and Guerrero 2000, Rahim et al. 2000). The integrating style, which locates at high concern for self as well as the other party, has also been described as problem solving, collaboration, cooperation, solution-orientation, win–win, or positive-sum style. The use of integrating style is also regarded as highly effective, as it provides the chance to address the other side's concerns and goals such that the disputing parties can strive for a win–win solution (Tutzauer and Roloff 1988). Furthermore, integrating style also carries two distinctive elements: consultation and problem solving (Prein 1976). Consultation involves open and direct communication to address a problem. Problem solving includes the ability to devise creative solutions. Therefore, this style emphasises the concerns of both parties by finding mutually acceptable solutions unique to the problem. This also involves active collaboration between the parties such as open exchange of information and examination of differences (Rahim et al. 2000). In sum, less conflict-laden environment can be formed using an integrating style and the exploration of mutual interests can result in creative and efficient outcomes (Fisher and Ury 1991). In experimental researches, this style has proved to be able to achieve the highest

Table 11.7 Overall results of Multiple Regression Analysis (for dysfunctional negotiation outcomes)

Dependent variables (Outcomes)	Independent variables (Negotiating styles)	R^2	Standardised regression coefficients	Normalised regression coefficients
Conflict escalation		0.234		
	Integrating		−0.180	0.195
	Obliging		−0.018	0.019
	Dominating		0.062	0.067
	Compromising		−0.172	0.186
	Avoiding		0.493	0.533
Relationship deterioration		0.303		
	Integrating		−0.624	0.459
	Obliging		0.045	0.033
	Dominating		0.267	0.197
	Compromising		0.255	0.188
	Avoiding		0.167	0.123
Inaction		0.229		
	Integrating		0.102	0.120
	Obliging		−0.077	0.090
	Dominating		−0.133	0.156
	Compromising		−0.018	0.021
	Avoiding		0.523	0.613
Further disagreement		0.462		
	Integrating		0.021	0.012
	Obliging		0.437	0.243
	Dominating		0.451	0.252
	Compromising		−0.643	0.358
	Avoiding		0.243	0.135

levels of joint gain for the negotiating parties (Ben Yoav and Pruitt 1984a, b; Pruitt et al. 1983). Reported field studies also showed that supervisors using an integrating style achieved more behavioural compliance with their requests (Rahim and Bonoma 1979) and a sign of low conflict level.

Incidentally, the concept of partnering is also based on the use of integrating style in project management. Moreover, the use of compromising style appears both correlated with functional and dysfunction outcomes. The results suggest that compromising style does not lead to further disagreement but the level of conflict is unlikely to be reduced. This was supported by Rahim and Bonoma (1979), who suggested that a moderate amount of conflict, handled in a constructive manner, is instrumental in attaining and maintaining an optimum level of organisational effectiveness. Hence, the use of compromising style to retain a little amount of conflict can actually be beneficial to the projects, provided that the dispute does not worsen. In this regard, compromising may therefore be regarded as a practical approach in resolving dispute (Rahim 1992).

The findings are supportive to the notion of contingent use of negotiating styles. It is interesting to note that the use of power, withdrawal and accommodation would not bring positive negotiation outcome. These results are similar to the suggestion of Follett (1940) who advocated the need for an integrative method (problem-solving) for managing organisational conflict and believed that other methods of handling conflict were ineffective in dealing with conflict. In terms of career and professional achievement, this may be the most important finding from the study.

The study employs the ROCI-II instrument to measure negotiating style. The instrument has been tested and uses of it have been widely reported (Friedman, et al. 2000; Hammock et al. 1990; Lee et al. 2003; Oetzel 1998; Rahim 1983; Rahim et al. 2000; Rahim 2002; Van De Vliert and Kabanoff 1990). Notwithstanding, styles obtained are based on self-evaluation by the negotiators and bias is possible. An alternative is to have data derive from observation. As dispute negotiations are mostly conducted privately, style identification through observation may not always be possible. This could well be used in another study where permissions of the disputants are obtained for observation. In such cases, dispute specific contextual factors such as type, magnitude or complexity of the dispute can also be taken into account.

11.10 Chapter Summary

The construction industry is perceived to be dispute laden. This can be attributed to many factors such as the lack of common goals, competing needs of the project team members, inequitable risk allocation, changes in construction plan and specification and contradictory and erroneous information. All these factors contribute to the germination and manifestation of construction disputes. Dispute is always negotiated first before other resolution methods are considered. During negotiation, characteristics specific to the disputants such as personality plays the key role in framing how the negotiation was conducted, hence the negotiation outcome. The interest in understanding the negotiating behaviours and negotiation outcomes is therefore immense. In this context, this chapter first gives a brief introduction of the traditions of negotiation studies, then discusses the causes of negotiation failure with emphasis on the behavioural dimensions and finally seeks to investigate the relationships between the negotiating styles and negotiation outcomes in a five-step process. Step 1 identifies a style classification framework from literature and selects a measurement tool. The Rahim Organisational Conflict Inventory-II (ROCI-II) that measures the five negotiating styles: integrating, obliging, avoiding, dominating and compromising portrayed by Blake and Mouton (1964, 1970), was selected for use in the study. Step 2 involved the use of a questionnaire to collect data. In Step 3, the authenticity of

ROCI-II was tested with the technique of Principal Component Factor Analyses (PCFA). ROCI-II appears to be a reliable and valid measure of negotiating styles. In Step 4, taxonomies of negotiation outcomes, three functional negotiation outcomes (problem solving, relationship maintained and conflict reduction) and four dysfunctional negotiation outcomes (conflict escalation, relationship deterioration, inaction and further disagreement) were identified. Based on these results, the final step of the study was to investigate the relationships between the negotiating styles and negotiation outcomes by conducting Multiple Regression Analyses (MRA). It is found that the use of obliging, dominating and avoiding styles appear to be less influential in achieving functional negotiation outcomes than using integrating style. The use of compromising style is also found to be a practical approach in resolving dispute. These findings are supportive to the notion of contingent use of negotiating styles.

Acknowledgments Special thanks to Miss Sau Fung Yeung for collecting data for the study. Part of the content has been published in Volume 132(8) of the Journal of Construction Engineering and Management and is used with the permission from ASCE.

References

Allred, K. (2000). Distinguishing best and strategic practices: A framework for managing dilemma between creating and claiming value. *Negotiation Journal, 16*(4), 387–397.

Ben-Yoav, O., & Pruitt, D. G. (1984a). Accountability to constituents: A two-edged sword. *Organizational Behaviour and Human Performance, 34*, 283–295.

Ben-Yoav, O., & Pruitt, D. G. (1984b). Resistance to yielding and the expectation of cooperative future interaction in negotiation. *Journal of Experimental Social Psychology, 20*(4), 323–335.

Blake, R. R., & Mouton, J. S. (1970). The fifth achievement. *Journal of Applied Behavioural Science, 6*, 413–436.

Blake, R. R., & Mouton, J. S. (1964). *The Managerial Grid*. Houston: Gulf.

Brown, H., & Marriott, A. (1999). *ADR Principles and Practice*. London: Sweet and Maxwell.

Carmel, E., Herniter, B. C., & Nunamaker, J. F. (1993). Labour-management contract negotiation in an electronic meeting room: A case study. *Group Decision and Negotiation, 2*(1), 27–60.

Carnevale, P., & Pruitt, D. (1992). Negotiation and mediation. *Annual Review of Psychology, 43*, 531–582.

Chakrabarty, S., Brown, G., & Gilbert, A. H. (2002). Evaluation of Rahim's organizational conflict inventory-II as a measure of conflict-handling styles in a sample of Indian salespersons. *Psychological Reports, 90*, 549–567.

Cheung, S. O. (1998). Values of alternative dispute resolution in construction. *Construction Law Journal, 14*(2), 101–110.

Cheung, S. O., & Yeung, Y. W. (1998). The effectiveness of the dispute resolution advisor system: A critical appraisal. *The International Journal of Project Management, 16*(6), 367–374.

Cheung, S. O., & Yeung, Y. W. (2002). The effectiveness of the dispute resolution advisor system: A critical appraisal. *International Journal of Project Management, 16*(6), 367–374.

Cheung, S. O., Suen, C. H., & Lam, T. L. (2002). Fundamentals of alternative dispute resolution processes in construction. *Journal of Construction Engineering and Management, 128*(5), 409–417.

Cheung, S. O., Yiu, T. W., & Suen, C. H. (2004). CoNegO: Construction negotiation online. *Journal of Construction Engineering and Management, 130*(6), 844–852.
Cheung, S. O., Tam, C. M., Ndekugri, I., & Harris, F. C. (2000). Factors affecting clients' project dispute resolution satisfaction in Hong Kong. *Construction Management and Economics, 18*(3), 281–294.
Cohen, R. (2001). Breaking the deadlock: Guarantees in international mediation. *Cambridge Review of International Affairs, 14*(2), 39–52.
Douglas, A. (1962). *Industrial peacemaking*. New York: Columbia University Press.
Downie, B. M. (1991). When negotiation fail: Causes of breaksoen and tactics for breaking the stalemate. *Negotiation Journal, 7*(2), 175–186.
Druckman, D., Harris, R., & Ramberg, B. (2002). Computer-assisted international negotiation: A tool for research and practice. *Group Decision and Negotiation, 11*, 231–256.
EWTB. (2002). Works Bureau Technical Circular No. 22/2002 – Marking scheme in tender evaluation for works contracts. Development Bureau, the Government of Hong Kong Special Administrative Region.
Fenn, P., Lowe, D., & Speck, C. (1997). Conflict and dispute in construction. *Construction Management and Economics, 15*(6), 513–518.
Fisher, R., & Ury, W. (1991). *Getting to yes: Negotiating agreement without giving in*. New York: Penguin Books.
Follett, M. P. (1940). Constructive conflict. In H. C. Metcalf & L. Urwick (Eds.), *Dynamic administration: The collected papers of Mary Parker Follett*. New York: Harper. (Original work published 1926).
Friedman, R. A., Currall, S. C., & Tsai, J. C. (2000). What goes around comes around: The impact of personal conflict style on work conflict and stress. *The International Journal of Conflict Management, 11*(1), 32–55.
Goldberg, S. B., Sander, F. E. A., & Roger, N. H. (1992). *Dispute resolution: Negotiation, mediation, and other processes*. Brown: Little.
Griffith, T., Tansil, D., & Benson, L. (2002). Negotiating technology implementation: An empirical investigation of a website introduction. *Group Decision and Negotiation, 11*(1), 1–22.
Gross, M. A., & Guerrero, L. K. (2000). Managing conflict appropriately and effectively: An application of the competence model to Rahim's organizational conflict styles. *The International Journal of Conflict Management, 11*(3), 200–226.
Hair, A., Tatham, R. L., & Black, W. C. (1995). *Multivariate data analysis* (5th ed.). Englewood Cliffs NJ: Prentice Hall.
Hammock, G. S., Richardson, D. R., Pilkington, C. J., & Utley, M. (1990). Measurement of conflict in social relationships. *Personality and Individual Differences, 11*(6), 577–583.
Hibberd, P., & Newman, P. (1999). *ADR and adjudication in construction disputes*. London: Blackwell Science.
Hills, M. J. (1992). A case for an alternative approach to the resolution of disputes under the jct 80 standard form of building contract.
Holsapple, C., Lai, H., & Whinston, A. (1998). A formal basis for negotiation support system. *Group Decision and Negotiation, 7*(3), 203–227.
Holt, G. (1997). Construction research questionnaire and attitude measurement: Relative index or mean. *Journal of Construction Procurement, 3*(2), 88–94.
Iklé, F. C. (1964). *How nations negotiate*. Millwood, NY: Kraus Reprint Co.
Kagel, J., & Roth, A. E. (1991). *Handbook of experimental economics*. Princeton NJ: Princeton University Press.
Lee, Y., Zoonky, L., & Larsen, K. R. T. (2003). Coping with internet channel conflict. *Communications of the ACM, 46*(7), 137–142.
Lewicki, R. J., & Litterer, J. (1985). *Negotiations*. Homewood, IL: Richard D Irwin.
Likert, R., & Likert, J. G. (1976). *New ways of managing conflict*. New York: McGraw-Hill.
Lytle, A., Brett, J., & Shapiro, D. (1999). The strategic use of interests, rights, and power to resolve disputes. *Negotiation Journal, 15*(1), 31–51.

Mintu-Wimsatt, A., & Calantone, R. (1996). Exploring factors that affect negotiator's problem-solving orientation. *Journal of Business and Industrial Marketing, 11*(6), 61–73.
Mnookin, R. (1993). Why negotiations fail: An exploration of barriers to the resolution of conflict. *Ohio State Journal of Dispute Resolution, 8*(2), 235–249.
Murninghan, J. K. (1991). *The dynamics of bargaining games*. Homewood, IL: Prentice Hall.
Nash, J. F. (1950). Equilibrium points in n-person games. *Proceedings of national academy science, 36*, 48–49.
Neale, M. A., & Bazerman, M. H. (1985). The effects of framing and negotiation overconfidence on bargaining behaviours and outcomes. *Academy of Management Journal, 28*(1), 34–49.
Newcombe, R. (1996). Empowering the construction project team. *International Journal of Project Management, 14*(2), 75–80.
Oetzel, J. G. (1998). The effects of self-control and ethnicity on self-reported conflict styles. *Communication Reports, 11*(2), 133–143.
Perdue, B. C., Day, R. L., & Michaels, R. E. (1986). Negotiating styles of industrial buyers. *Industrial Marketing Management, 15*(3), 171–176.
Prein, H. C. M. (1976). Stijlen van conflicthanteing (Styles of handling conflict). *Nederlands Tijdschrift voor de Psychologie, 31*, 321–346.
Pruitt, D. G., Carnevale, P. J., Ben-Yoav, O., Nochajski, T. H., & Van Slyck, M. (1983). Incentives for cooperation in integrative bargaining. In R. Tietz (Ed.), *Aspiration levels in bargaining and economic decision making* (pp. 22–34). Berlin: Springer.
Rahim, M. A. (1983). *Rahim organization conflict inventories: Professional manual*. Palo Alto: Consulting Psychologists Press.
Rahim, M. A. (1992). *Managing conflict in organizations* (2nd ed.). London: Praeger.
Rahim, M. A. (2001). *Managing conflict in organizations* (3rd ed.). Westport, CT: Quorum Books.
Rahim, M. A. (2002). Toward a theory of managing organizational conflict. *The International Journal of Conflict Management, 13*(3), 206–235.
Rahim, M. A., & Bonoma, T. V. (1979). Managing organizational conflict: A model for diagnosis and intervention. *Psychological Reports, 44*, 1323–1344.
Rahim, M. A., Manger, N. R., & Shapiro, D. L. (2000). Do justice perceptions influence styles of handling conflict with supervisors? What justice perceptions, precisely? *The International Journal of Conflict Management, 11*(1), 9–31.
Ren, Z., Anumba, C. J., & Ugwu, O. O. (2003). The development of multi-agent system for construction claims negotiation. *Advances in Engineering Software, 34*(11–12), 683–696.
Rhys Jones, S. (1994). How constructive is construction law? *Construction Law Journal, 10*(1), 28–38.
Rosch, E. (1988). Principles of categorization. In A. M. Collins & E. E. Smith (Eds.), *Readings in cognitive science*. San Mateo: Morgan Kaufmann Publishers Inc.
Ruble, T. L., & Thomas, K. W. (1976). Support for a two-dimensional model of conflict behaviour. *Organizational Behaviour and Human Performance, 16*, 143–155.
Sebenius, J. K. (1992). Negotiation analysis: A characterisation and review. *Management Science, 38*(1), 18–38.
Sharland, A. (2001). The negotiation process as a predictor of relationship outcomes in international buyer-supplier arrangements. *Industrial Marketing Management, 30*(7), 551–559.
Shell, G. R. (2001). Bargaining styles and negotiation: The Thomas-Kilmann conflict mode instrument in negotiation training. *Negotiation Journal, 17*(2), 155–174.
Steven, C. M. (1963). *Strategy and collective bargaining negotiation*. New York: McGraw Hill.
Stevenson, D. (1991). The failure of peace by negotiation in 1917. *Historical Journal, 34*(1), 65–86.
Thomas, K. W. (1976). Conflict and conflict management. In M. D. Dunnette (Ed.), *Handling of industry and organization psychology* (pp. 889–935). Chicago: Rand McNally.

Tor, A., & Bazerman, M. H. (2003). Focusing failure in competitive environments: Explaining decision errors in the Monty Hall game, the acquiring a company problem, and multiparty ultimatums. *Journal of Behavioural Decision Making, 16*(5), 353–374.

Tutzauer, F., & Roloff, M. E. (1988). Communication processes leading to integrative agreements: Three paths to joint benefits. *Communication Research, 15*, 360–380.

Underdal, A. (1983). Causes of negotiation failure. *European Journal of Political Research, 11*(2), 183–195.

Van de Vliert, E., & Kabanoff, B. (1990). Toward theory-based measures of conflict management. *Academy of Management Journal, 33*(1), 199–209.

Walker, A. (2002). *Project management in construction*. London: Blackwell Science.

Walton, R., & McKersie, R. (1965). *A behavioural theory of labour negotiation: An analysis of a social interaction system*. New York: McGraw Hill.

Zack, J. G. (1994). The negotiation of settlements - a team sport. *Cost Engineering, 36*(8), 24–30.

Zeuthen, F. (1930). *Problems of monopoly and economic warfare*. London: Routledge and Kegan Paul.

Chapter 12
Online Construction Dispute Negotiation

Sai On Cheung and Tak Wing Yiu

Abstract Disputes are common in construction projects and negotiating disputes is part of the daily routine of construction professionals. The advance in Information Technology (IT) has made tremendous impact on the way businesses operate. Making use of IT technology, a computerised construction dispute negotiation programme namely CoNegO (Construction Negotiation Online) is proposed. CoNegO utilises the Smartsettle™ software technology. With the built-in facilities of Smartsettle™, it is possible to conduct negotiation online, hence removing geographical barriers between negotiators. Smartsettle™ is developed on the concept of 'Even Swaps' in which negotiators are required to evaluate possible options available on the basis of their relative importance. As construction disputes are characterised by multiple factors and dimensions, the problem fits nicely with the 'trade-off' methodology that underpins Even Swaps. The use of CoNegO is illustrated by a simulated negotiation.

12.1 Introduction

The application of Information Technology (IT) has attracted world-wide attention. In construction, ample research has been conducted to investigate the applications of IT (Aouad and Price 1994; Aouad et al. 1996; Betts et al. 1991; CICA 1990; O'Brien and Al-Soufi 1994; Samuelson 1998; Shash and Al-Amir 1997).

S. O. Cheung (✉)
Construction Dispute Resolution Research Unit, Department of Civil and Architectural Engineering, City University of Hong Kong, Hong Kong, People's Republic of China
e-mail: saion.cheung@cityu.edu.hk

T. W. Yiu
Department of Civil and Environmental Engineering, Faculty of Engineering, The University of Auckland, Auckland, New Zealand
e-mail: k.yiu@auckland.ac.nz

S. O. Cheung (ed.), *Construction Dispute Research*,
DOI: 10.1007/978-3-319-04429-3_12,

Advancement in IT enables construction activities to be programmed and executed in a speedy and cost-effective manner. It is no longer regarded as an enhancement to traditional business, but an innovative agent. Ahmad et al. (1995) suggested that IT makes previously impossible things possible for the enterprises in the industry. The rapid development of software products has made the most impact. For example, 3-Dimensional Computer-Aided Drafting Tools such as AutoCAD and Integra, are indispensable planning and design tools for architects, engineers and contractors (Reinschmidt et al. 1991). In addition, Project Information Management System (PIMS) is now widely used to handle tasks such as construction programming, information storage and retrieval (Lloyd et al. 2001). Access to project information through the Internet is also well-documented (Huang et al. 1999; Lam and Chang 2001). As for the use of information technology in construction dispute resolution, reported studies include the use of: computer simulation for claim assessment during a mediation process (AbouRizk and Dozzi 1993); computer-supported conflict mitigation system (Pena-Mora et. al. 1993), computer agents to facilitate negotiation (Pena-Mora and Wang 1998), and projects for improving communication to help engineers to carry out negotiation tasks (Cutkosky and Tennenbaum 1996; Divita et. al. 1998; Fruchter 1996; Rezgui et. al. 1998; Roddis 1998; Schmitt 1998). Nonetheless, the use of online systems to facilitate construction dispute resolution remains few. In practice, dealing with construction disputes is in fact an important part of project managers' daily routine. Hence, with the effective and cost-saving dispute resolution process, they could easily settle the disputes without the intervention of third parties. This is also the reason why negotiation is always the preferred option other than mediation, arbitration and litigation among the various dispute resolution procedures, In fact, negotiation is the most commonly used dispute resolution procedure (Fisher and Ury 1986; Merna and Bower 1997). Due to the important role that negotiation plays in construction management, the use of on-line facilities to assist in negotiation is not only of academic interest but also invaluable to improving communication at the project operation level.

12.2 Current Development

Computer-based negotiation support system (NSS) and other group decision-support system (GDSS) products have been developed to deal with negotiations and decision making in response to the needs of industry. These systems are often used in group decision-making, which take place in an electronic meeting room environment, such as PLEXZSYS (Nunamaker et al. 1987). Bui and Shahun (1997) introduced the utility of a conflict resolution framework 'Evolutionary Systems Design' (EDS) by utilising a Negotiation Support System. Kersten and Noronha (1999) developed a negotiation support system known as InterNeg Project, assisting users to analyse decisions. This study discusses an on-line construction negotiation system, named 'CoNego' hereafter. The concept of CoNegO

(Construction Negotiation Online) is first introduced, followed by examining the SmartSettle^{TM} system (platform for CoNegO) and its associated online facilities. The development of CoNegO is then presented.

12.3 Underpinning Concepts

The aim of negotiation is to settle a dispute. In the negotiation process, proactive communication, exchange of ideas and prioritisation of issues are essential. Incidentally, the computing abilities of speedy communication, data accessibility and a common system make it ideal for the development of CoNegO (Ahmad et al. 1995). Figure 12.1 presents the conceptual framework for the development of CoNegO.

In CoNegO, the communication component is the internet. The data accessibility component manages the sharing of information by the negotiators. In negotiation, fact or evidence is often called upon to justify an argument. Hence, a well-organised set of project data is not only useful but essential. The common system component is concerned with tools that can be used to aid decision-making and help to reach a settlement in a more systematic manner. Commonly used tools are Knowledge-based Expert Systems and the Case-based Reasoning Approach (Li 1996).

To be a useful tool, CoNegO needs to provide a set of standard and rational principles to guide negotiators. This is vital as it is common that negotiation principles are often neglected during the negotiation process. SmartSettle^{TM} Program is negotiation software, which takes advantage of the power of the network to bring disputants to negotiate despite in different locations. It can take any tentative agreement and suggest alternative approaches that the parties can consider. It also makes use of a trade-off technique called Even Swaps (Cheung et al. 2002; Hammond et al. 1998, 1999) that provides a practical way of making trade-offs among a given set of objectives across a range of alternatives. The built-in online facility of SmartSettle^{TM} is also central to the CoNegO (ICAN 2000).

12.4 CoNegO

The advancement of IT has further removed geographical barriers to communication. SmartSettle^{TM}, via internet, enables online negotiation. SmartSettle^{TM} is the central component of CoNegO. It is a software with an interactive online facility for the negotiating parties. During negotiation, SmartSettle^{TM} elicits the case description, preference information and proposals from all parties. The primary objective of SmartSettle^{TM} is to help parties reach a settlement.

The contents of negotiation can easily be stored in the computer database for further retrieval and record. In an e-negotiation environment, two disputing parties

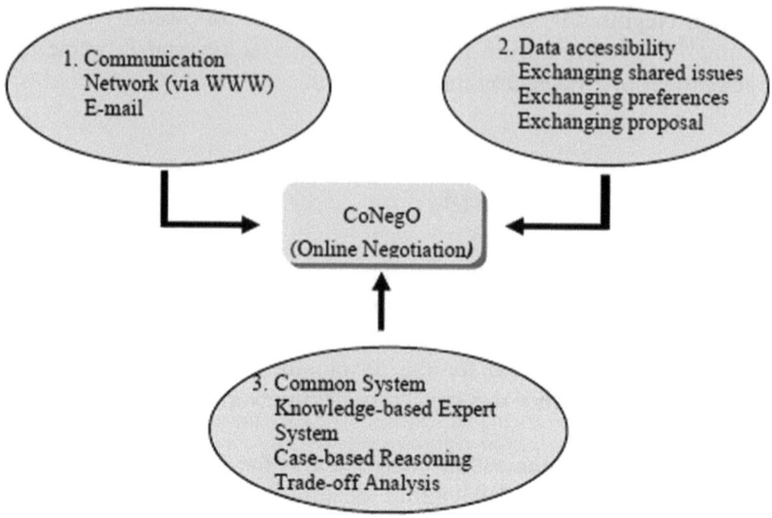

Fig. 12.1 Development framework of CoNegO

are communicating with each other by using a neutral server. Figure 12.2 shows the mechanism of online facilities of SmartSettle^{TM}. The server assists in the negotiation process by providing instantaneous responses from either party on accepting or exchanging proposals so that each party can acquire the updated information as the negotiation progresses. Furthermore, the server stores details of the case so that the users can extract information from the server instantaneously or through a subsequent continuation if the negotiation stopped midway or could not be concluded in one setting.

12.5 Illustration

To illustrate the use of CoNegO, a hypothetical case is utilised. The hypothetical case was concerned with a negotiation between contractor and client regarding the settlement of a dispute involving Extension of Time, Loss/Expenses and Cost of Acceleration. Extension of Time (EOT) refers to the additional time granted to the Main Contractor under the stipulated ground of the contract. Loss/Expenses (L/E) refers to the amount reimbursed to the Contractor due to the causes for which the employer is responsible. Cost of Acceleration refers to the additional cost reimbursable to the Contractor for catching up with qualifying delay. In this respect, one of the experts was selected from a consultant firm and the other was from a contractor firm. They were invited to participate and negotiate on the hypothetical case using CoNegO. Both experts have over 10 years of experience in dealing with construction claims and negotiations, and are referred to as 'negotiators' hereafter.

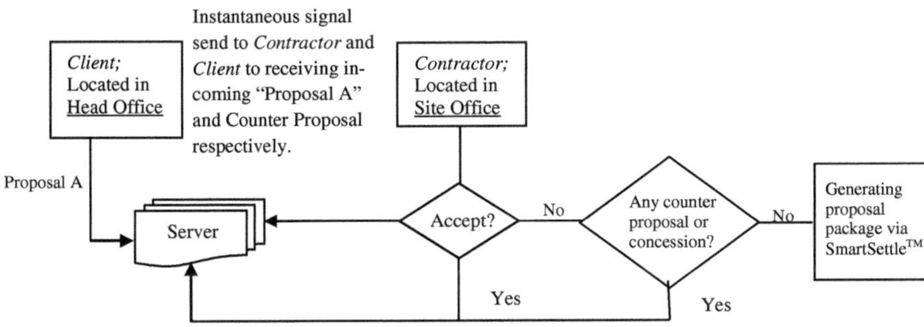

Fig. 12.2 Mechanism of online SmartSettle™ facilities

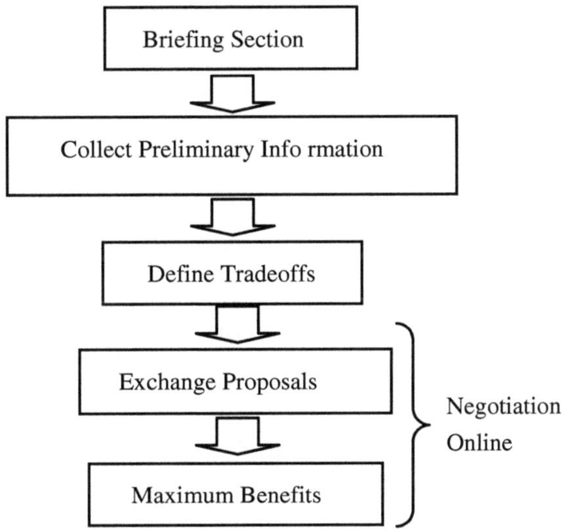

Fig. 12.3 Arrangements of the simulation

The simulation process was arranged as shown in Fig. 12.3. Negotiators were briefed with the working of CoNegO before actual simulation.

During the simulation, both negotiators were physically separated. Briefing was first given to the negotiators to introduce the SmartSettle™ program and explain the procedures involved in the simulated environment. These included a brief description of the hypothetical case and a fill-in Data In-take form. The Data In-take form collected preliminary information on the negotiators such as the case information, individual preferences and bargaining ranges of each issue.

The two negotiators first studied the hypothetical construction claim case. Having understood the circumstances of the case, the negotiators then formulated the bargaining ranges for each of the three issues to be negotiated. A bargaining range is a set of possible decision values for a particular issue (ICAN 2000). These ranges were then recorded in the Data In-take Form (D.I.F.). The form was designed to record their bargaining ranges in numeral values. Table 12.1 shows the

Table 12.1 Acceptable ranges of the negotiator

	Pessimistic value	Optimistic value
(a) *Client*		
E.O.T. (Unit: day)	40	30
L/E (Unit: $/day)	6,500	3,200
Acceleration cost pay to contractor (Unit: $/day)	13,000	7,000
(b) *Contractor*		
E.O.T. (Unit: day)	35	55
L/E (Unit: $/day)	6,000	7,000
Acceleration cost pay to contractor (Unit: $/day)	10,000	20,000

Table 12.2 Bargaining ranges with relative importance weights

Issue abbreviation	RI	Bargaining range	
		Worst	Best
(a) *Client side*			
1. EOT	30	40	30
2. L/E	40	6,500	3,200
3. AccCost	30	13,000	7,000
Total	100		
(b) *Contractor side*			
1. EOT	60	35	55
2. L/E	30	6,000	7,000
3. AccCost	10	10,000	20,000
Total	100		

Client and Contractor acceptable bargaining ranges. The pessimistic value represents the baseline of the negotiator for a particular issue which implies that no further concession will be offered beyond this value. While the optimistic value represents the value with the highest satisfaction for the negotiator.

Having familiarised themselves with the case and established the acceptable range for each issue, the negotiators were then required to assess the relative importance of the issues. The relative importance is an indication of how important one issue is relative the other. Basing on the information from Tables 12.1 and 12.2 shows the D.I.F. with relative importance weights included.

The next task was to define Tradeoffs by using the Even Swaps Method. As shown in Table 12.3, Swap 1, Swap 2 and Swap 3 were performed by the negotiators. The term 'Ref.' stands for reference alternative package. It is the value which the negotiators consider to be a possible final outcome. In going from the reference alternative package to Swap 1, the client side reasoned that a one day increase in EOT would sufficiently counter 100 decrease of L/E from 4000 to 3900. Hence, three equivalent alternatives are generated in this way.

Finally, negotiators were asked to provide satisfaction ratings for the range of acceptable values. By default, the satisfaction graphs are linear for all issues. In order to fine-tune the satisfaction value of each party, SmartSettle^{TM} allows

Table 12.3 Even swaps exercise

Issue abbr.	Ref.	Swap 1		Swap 2		Swap 3	
(a) *Client side*							
EOT	32	+1	31		32	−1	31
L/E	4,000	−100	3,900	−200	3,800		4,000
AccCost	9,500		9,500	−1,000	8,500	−500	9,000
(b) *Contractor side*							
EOT	50	+1	51		50	+1	51
L/E	6,800	+100	6,900	+200	7,000		6,800
AccCost	18,000		18,000	+1,500	19,500	+750	18,750

Table 12.4 Assessment of satisfaction rating

Issue abbreviation	Least preferred value	25 % satisfaction scale	50 % satisfaction scale	75 % satisfaction scale	Most preferred value
(a) *Client side*					
EOT	40	39	38	36	35
L/E	6,500	6,000	5,500	4,000	3,200
AccCost	13,000	12,000	11,000	9,000	7,000
(b) *Contractor side*					
EOT	35	37	39	45	55
L/E	6,000	6,100	6,400	6,600	7,000
AccCost	10,000	13,000	15,000	17,500	20,000

negotiators to plot satisfaction graphs. In this simulation, the data for these graphs are set in stages of a 25 % satisfaction scale. Table 12.4 summarises the results.

12.6 Negotiation online

Based on the data from the D.I.F., the rating and satisfaction graphs of each issue are generated. The satisfaction graph of the Client is shown in Fig. 12.4. The satisfaction graphs are linear by default. The negotiators can adjust these graphs in accordance with their bargaining range and issue values. The results of such mathematical function can accurately predict the level of satisfaction for each value of the related issue.

12.7 Processing Even Swaps Method in SmartSettle™

The Relative Importance (R.I.) weightings shown in Table 12.2 are the values subjectively assessed by the negotiating parties. These values may not accurately predict the relative importance of each issue. Based on the satisfaction scale and

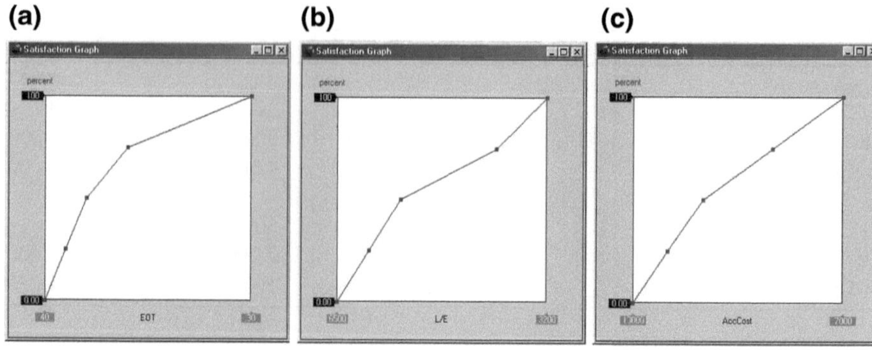

Fig. 12.4 Satisfaction graph from client side **a** E.O.T.; **b** Loss and expense; **c** Cost of acceleration

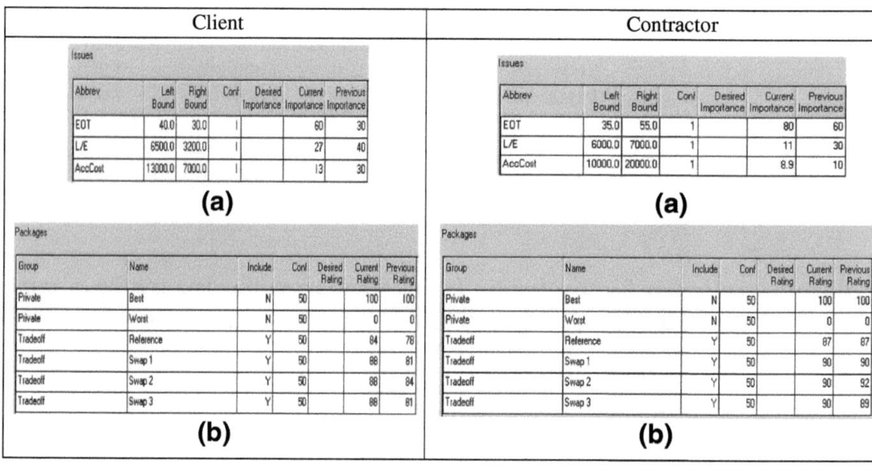

Client

Issues

Abbrev	Left Bound	Right Bound	Conf	Desired Importance	Current Importance	Previous Importance
EOT	40.0	30.0	1		60	30
L/E	6500.0	3200.0	1		27	40
AccCost	13000.0	7000.0	1		13	30

(a)

Packages

Group	Name	Include	Conf	Desired Rating	Current Rating	Previous Rating
Private	Best	N	50		100	100
Private	Worst	N	50		0	0
Tradeoff	Reference	Y	50		84	78
Tradeoff	Swap 1	Y	50		88	81
Tradeoff	Swap 2	Y	50		88	84
Tradeoff	Swap 3	Y	50		88	81

(b)

Contractor

Issues

Abbrev	Left Bound	Right Bound	Conf	Desired Importance	Current Importance	Previous Importance
EOT	35.0	55.0	1		80	60
L/E	6000.0	7000.0	1		11	30
AccCost	10000.0	20000.0	1		8.9	10

(a)

Packages

Group	Name	Include	Conf	Desired Rating	Current Rating	Previous Rating
Private	Best	N	50		100	100
Private	Worst	N	50		0	0
Tradeoff	Reference	Y	50		87	87
Tradeoff	Swap 1	Y	50		90	90
Tradeoff	Swap 2	Y	50		90	92
Tradeoff	Swap 3	Y	50		90	89

(b)

Fig. 12.5 Rating comparisons after Even Swaps (Client and contractor) **a** Issues; **b** Packages

the Even Swaps Method, R.I. can be defined in a rational way. By analysing the data in Table 12.3, Even Swaps Method can successfully be applied in such a way to make the three packages (e.g. Swap 1, Swap 2 and Swap 3) equivalent to the Reference and to each other in terms of satisfaction. The changes in rating and relative importance are tabulated in SmartSettle^{TM} and the screenshot of such tables are shown in Fig. 12.5. By comparing the current and previous R.I., it is found that both parties regard E.O.T. as the most important issues in this simulation.

In terms of the Algorithms adopted in SmartSettle^{TM} for analysing the preference of each party, alternatives which equivalent to the reference, are required to enable SmartSettle^{TM} to determine more precisely the total satisfaction levels by comparing the alternatives. In this hypothetical case, for example, on the Client side, the rating of the alternative (E.O.T. = 31 days, L/E = 3,900 and AccCost = 9,500) is equivalent to (E.O.T. = 32 days, L/E = 3,800 and AccCost = 8,500). The total satisfaction TS_j for each party j associated with equivalent alternative k will be equal to the sum over all decisions i of the weighted relative additional satisfaction functions $R_{ij}(V_{ijk})$ selected by that party. Thus, for each of the party j, such as the Client side (Cl) in this hypothetical case, the total satisfaction associated with the first alternative (k = 1) is:

$$TS_{Cl} = w_{1(Cl)} \times R_{1(Cl)}\left(E.O.T._{(Cl)1}\right) + w_{2(Cl)} \times R_{2(Cl)}\left(L/E_{(Cl)1}\right) + w_{3(Cl)} \times R_{3(Cl)}\left(AccCost_{(Cl)1}\right) + C_{Cl} \tag{12.1}$$

where C is an unknown scale adjustment constant. Thus, for the other alternatives (say k = 2 and 3), the satisfaction equations are:

$$TS_{Cl} = w_{1(Cl)} \times R_{2(Cl)}\left(E.O.T._{(Cl)2}\right) + w_{2(Cl)} \times R_{2(Cl)}\left(L/E_{(Cl)2}\right) + w_{3(Cl)} \times R_{3(Cl)}\left(AccCost_{(Cl)2}\right) + C_{Cl} \tag{12.2}$$

and

$$TS_{Cl} = w_{1(Cl)} \times R_{1(Cl)}\left(E.O.T._{(Cl)3}\right) + w_{2(Cl)} \times R_{2(Cl)}\left(L/E_{(Cl)3}\right) + w_{3(Cl)} \times R_{3(Cl)}\left(AccCost_{(Cl)3}\right) + C_{Cl} \tag{12.3}$$

In these three equations, for each party j, the weights w_{ij} are still unknown, as is the total additional satisfaction TS_j. We can also introduce two other equations defining the zero and 100 % levels of total satisfaction as:

$$0 = w_{1j} \times R_{1j}\left(E.O.T._{jn}\right) + w_{2j} \times R_{2j}\left(L/E_{jn}\right) + w_{3j} \times R_{3j}\left(AccCost_{jn}\right) + C_j \tag{12.4}$$

where the $E.O.T._{jn}$, L/E_{jn} and $AccCost_{jn}$ are the least preferred values in the identified bargaining ranges and

$$100 = w_{1j} \times R_{1j}\left(E.O.T._{jm}\right) + w_{2j} \times R_{2j}\left(L/E_{jm}\right) + w_{3j} \times R_{3j}\left(AccCost_{jm}\right) + C_j \tag{12.5}$$

where the $E.O.T._{jm}$, L/E_{jm} and $AccCost_{jm}$ are the most preferred values in the identified bargaining ranges. With the above equations, the unknown value can be solved. The calculation of total satisfaction functions can also be performed for the expanding decision variables.

After the preferences and relative importance were clearly defined, the two negotiators are ready to negotiate on-line. In this hypothetical case, a total of six

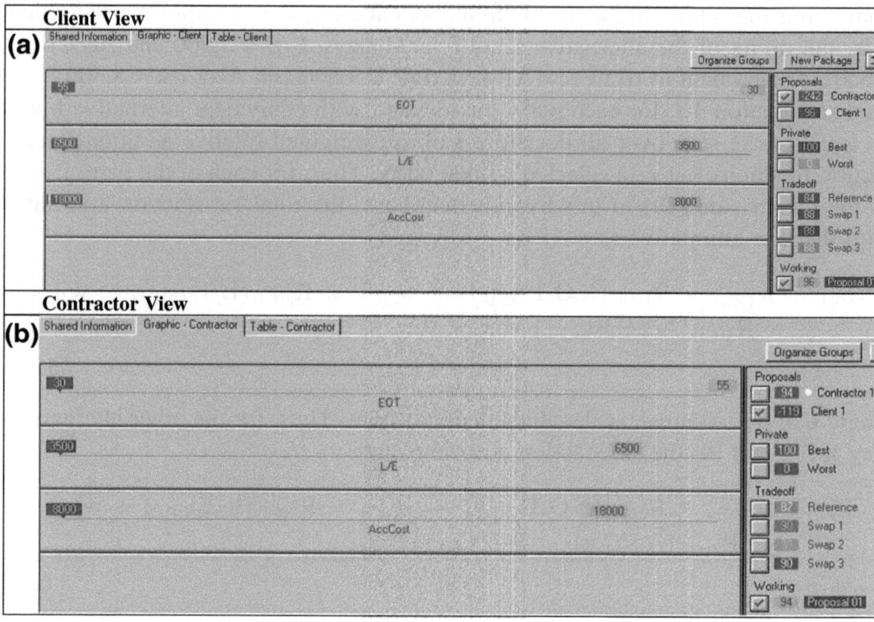

Fig. 12.6 First proposal

proposals were exchanged between the Contractor and Client negotiators. The screenshots of the initial proposal of both sides are shown in Fig. 12.6.

The 'flags' in Fig. 12.6 show the value of each issue. The given rating beside each proposal reflects the users' satisfaction in respect of that issue. The distance between the flags becomes closer every time the parties make a concession. When the flag on a particular issue is overlapped, agreement on that issue becomes possible.

In the first proposal, the satisfaction rating of the first proposal on the Client side and Contractor side is 96 and 94 respectively. On the other hand, the rating of the Contractor's proposal (assessed by the Client's satisfaction curve) is −242. The rating of Client's proposal, (assessed by the Contractor's satisfaction curve) is −119. Thus, in the first proposal, both parties cannot reach an agreement and further proposals are needed in this simulation.

On the second proposal, concession was offered by the parties. The Client's view is shown in Fig. 12.7. The ratings of the second proposal are reduced. The distant between the flags are now shortened as compared with the first proposal.

Further progress was made in the third proposal (Fig. 12.7 refers). In particular, agreement is reached for the acceleration cost issue due to the great concession on the Contractor side. Figure 12.8 shows that the flags on the AccCost issue are overlapped. No further negotiation on this issue was required. When comparing the two outstanding issues, E.O.T. appeared to be a barrier to reaching an agreement. With the concession on the Client side in the fourth proposal, a further

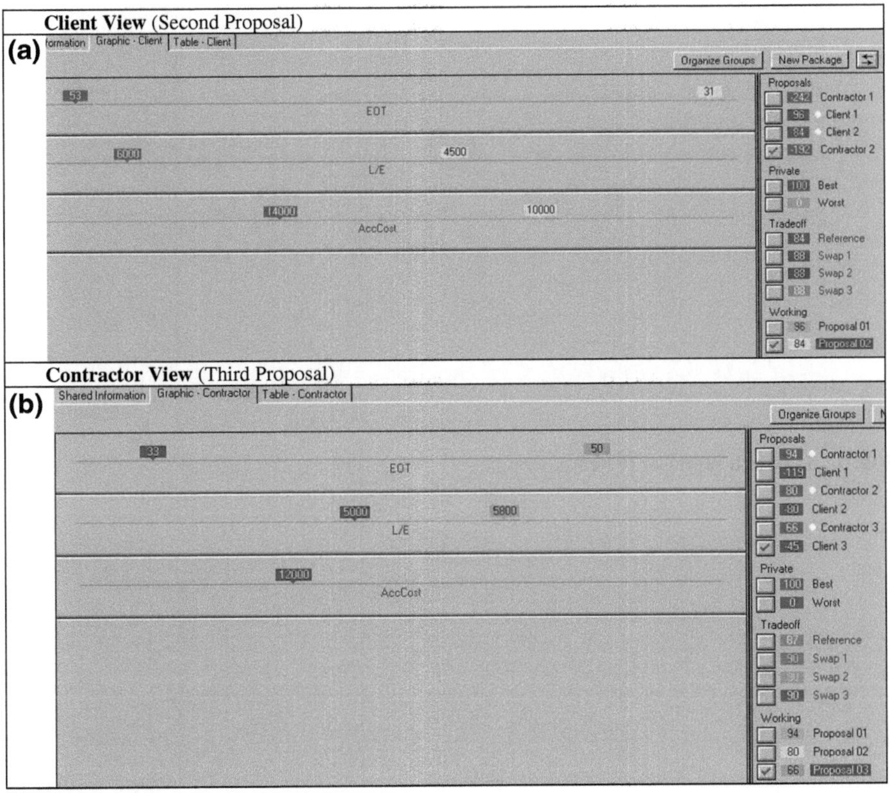

Fig. 12.7 Second and third proposal

step was made to reach an agreement on L/E (Fig. 12.8 refers). However, the difference with regard to E.O.T. remains large. Both parties must consider further concession in order to reach the agreement.

In the fifth proposal (Fig. 12.9 refers), both parties decided to offer a larger concession on this outstanding issue. However, such concession was not adequate to reach mutual agreement and a further exchange was made (sixth proposal, Fig. 12.10 refers). The simulation came to the end with mutual agreement regarding to the three issues of this dispute.

12.8 Maximise Benefits

After tentative agreement was reached, the negotiators can achieve further 'Improvement'. The "Improvement" function of SmartSettle^{TM} enables the search for an optimal package on the efficiency frontier that distributes benefits to all the negotiating parties according to their level of influence (ICAN 2000). The concept

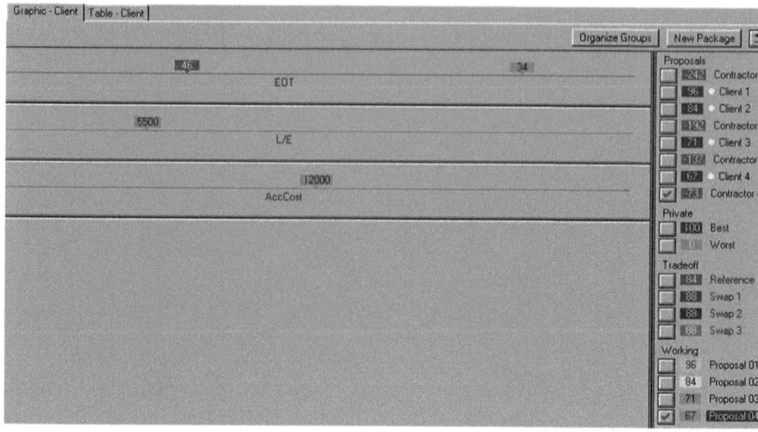

Fig. 12.8 Fourth proposal (Client's view)

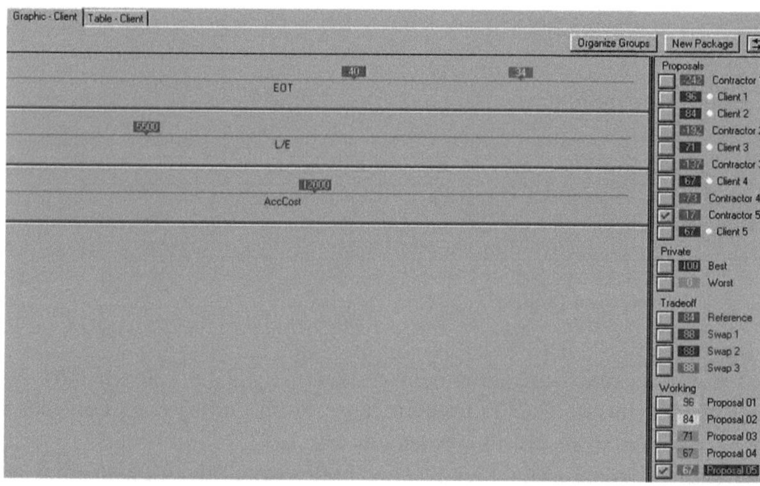

Fig. 12.9 Fifth proposal (Client's view)

of efficiency frontier can easily illustrate the benefit of CoNegO (Fig. 12.11 refers). Referring to the hypothetical case, the efficiency frontier represents the best possible outcomes for both parties. Based on the tentative agreement, SmartSettle™ attempts to divide benefits fairly to each party by generating an Improvement package that moving towards the efficiency frontier. The rating of such a package is certainly higher than the tentative solution made by both parties, who can consider the suggestion as the new settlement agreement.

The improved package generated has a higher satisfaction rating than the tentative agreement. It is found that the rating of the improved package is 68, which is higher than that of the tentative agreement (with satisfaction rating of 62).

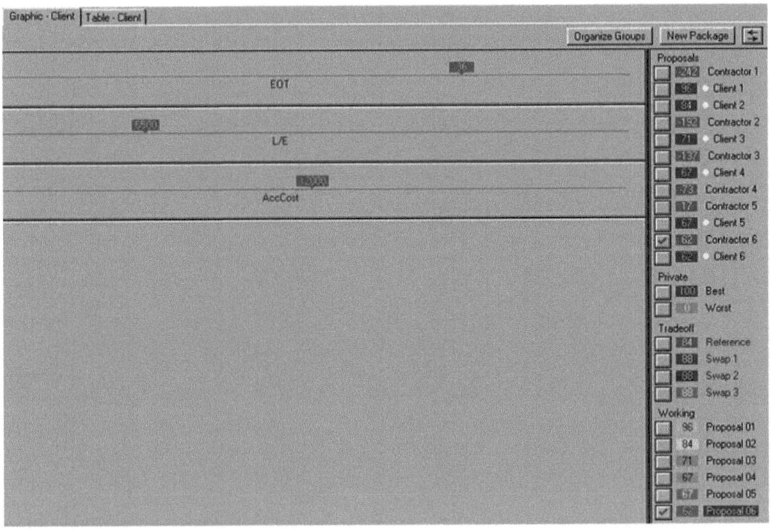

Fig. 12.10 Sixth proposal (Client's view)

Fig. 12.11 Efficiency frontier in CoNegO

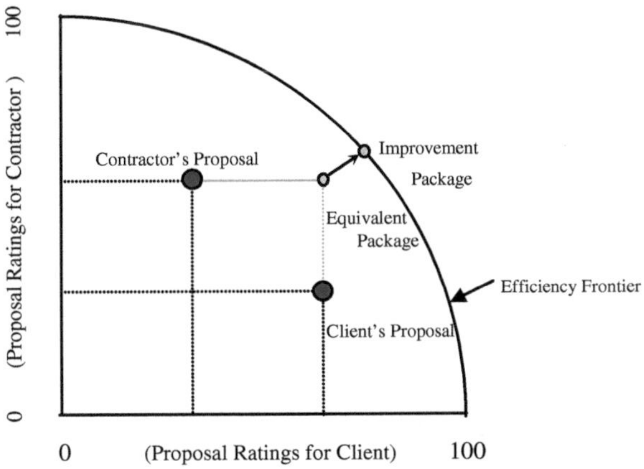

This is shown in Fig. 12.12. After several proposal submissions, negotiation and the generation of improvement, both parties have no hesitation in accepting the 'Improvement' package as the mutual agreement in this Simulation. Figure 12.13 summarises the proposals made by the negotiating parties.

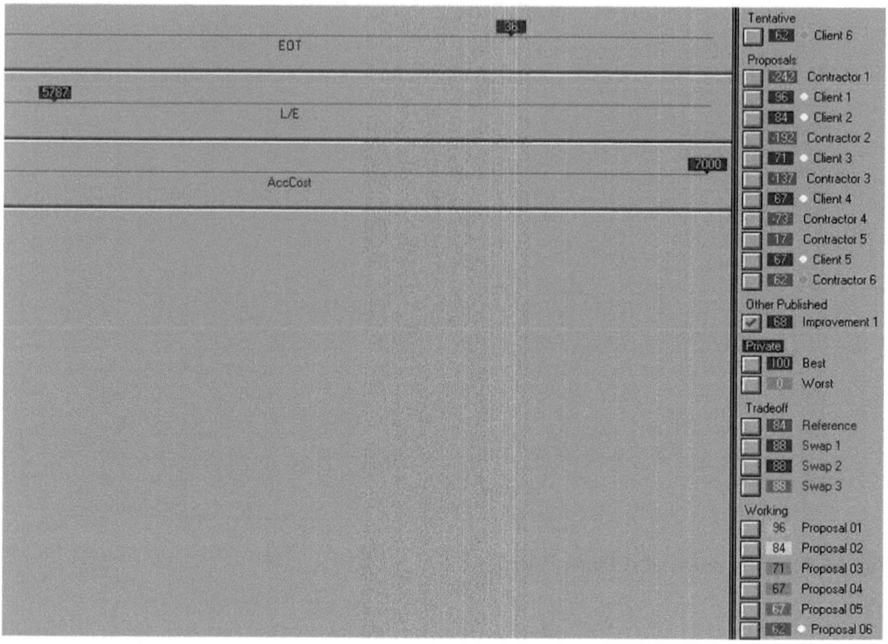

Fig. 12.12 Improvement of the client view

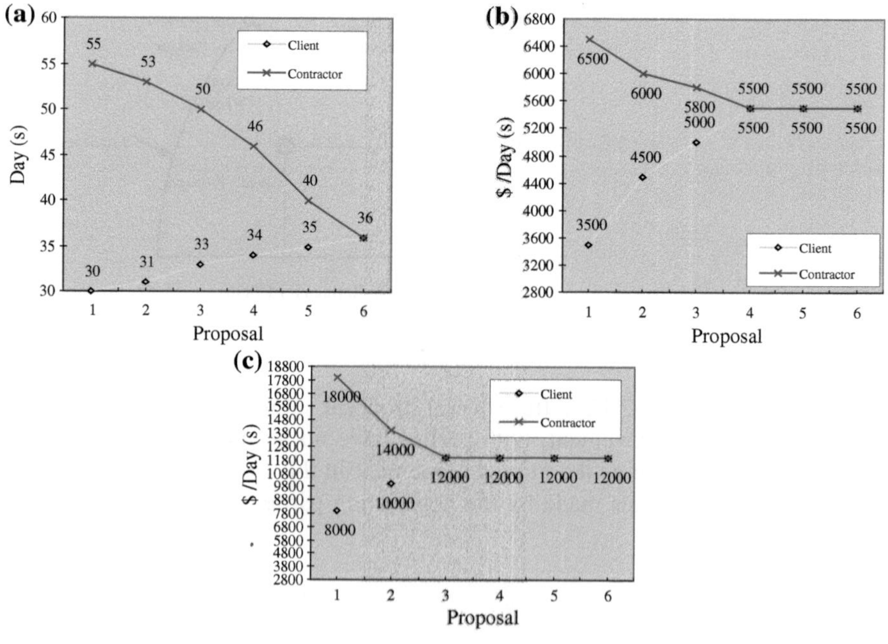

Fig. 12.13 Proposal trends **a** E.O.T.; **b** Loss and expenses; **c** Cost of acceleration

12.9 Discussion

The hypothetical case illustrates the working of CoNego, the main advantages of which include:

12.9.1 Enhanced Efficiency of Negotiation Preparation

CoNegO formalises the process involved in typical construction negotiation. The Internet connection of CoNegO enables the negotiation take place at a distance. Negotiators can exchange their offers, counteroffers data through a secure neutral server. Thus, the time for document presentation, negotiation meetings can be reduced. Furthermore, the use of the Data In-take Form improves the negotiation preparation stage which enables negotiators to list, define and eventually evaluate their alternatives on disputing issues.

12.9.2 Computing Facilitates of CoNegO

CoNegO, utilising the computing capacity and the communication strength of the Internet, provides a user-friendly and interactive environment. Through constructing the Satisfaction Graphs, negotiators can better understand their satisfactions on each issue and define trade-offs with a set of unique equivalent alternatives using the Even Swaps Method. CoNegO, by making use of the computing power to conduct trade-off, can devise satisfaction and suggest improvement. These tools can reduce negotiating time and cost.

12.9.3 Flexible Management and Involvement

In the preliminary stage of negotiation, it is normally started at the site level between the Contractor and the Client's agent. With CoNegO, senior staff can observe or even supervise the negotiation through access to the neutral server. This can lead the senior members to understand the development of the negotiation thus avoiding failure due to discontinuity of negotiators involved.

12.10 Chapter Summary

CoNegO is an Internet-based computerised construction negotiation support system. It is developed based on the SmartSettle™ program that embraces the Even Swaps method for trade-off analysis. Construction negotiation typically involves multiple issues, systematic prioritising and making trade-offs assist the formulation of a settlement package. CoNegO is an invaluable tool to complement the often subjective approach to negotiation. It is aimed primarily to provide a structured approach in construction negotiation. With the help of experts in the field, use of CoNegO was simulated with a hypothetical case.

Acknowledgments The content of this chapter has been published in Volume 130(6) of the Journal of Construction Engineering and Management and is used with the permission from ASCE.

References

AbouRizk, S. M., & Dozzi, S. P. (1993). Application of computer simulation in resolving construction disputes. *Journal of Construction Engineering and Management, 119*(2), 355–373.

Ahmad, I., Russell, J., & Abou-zeid, A. (1995). Information technology (IT) and integration in the construction industry. *Construction Management and Economic, 13*, 163–171.

Aouad, G., Alshawi, M., & Bee, S. (1996). Priority topics for construction information technology. *International Journal of Construction Information Technology, 4*(2), 45–66.

Aouad, G., & Price, A. (1994). Construction planning and information technology in the UK and US construction industries: A comparative study. *Construction Management and Economics, 12*, 97–106.

Betts, M., Cher, L., Mathur, K., & Ofori, G. (1991). Strategies for the construction sector in the information technology era. *Construction Management and Economics, 9*, 509–528.

Bui, T., & Shahun, M. F. (1997). Negotiation processes, evolutionary systems design and negotiator. available from http://www.vacets.org/vtic97/txbui3.htm.

Cheung, S. O., Yiu, T. W. & Suen, H. (2002). Construction negotiation: An even swaps APPROACH. In: *Proceedings of first international conference on construction in the 21st century*, 25–26 April, Miami, Florida, USA.

Construction Industry Computing Association (CICA). (1990). *Building on IT for the 1990s: A survey of the IT trends and needs in the construction industry*. London: Peat Marwick McLintock.

Cutkosky, M. R., & Tennenbaum, J. M. (1996). Collaborative engineering over the internet. *Communications of the ACM, 39*(9), 78–87.

Divita, E. L., Kunz, J. C., & Fischer, M. A. (1998). *Collaborative desktop engineering. AI in structural engineering, lecture notes in AI*. New York: Springer.

Fisher, R., & Ury, W. (1986). *Getting to yes: Negotiating agreement without giving in*. London: Hutchinson Business.

Fruchter, R. (1996). Conceptual, collaborative building design through shared graphics. IEEE Expert, Intelligent System & their Applications, *11*(3), 33–41.

Hammond, J. S., Keeney, R. L. & Raiffa, H. (1998). Even swaps: A rational method for making trade-offs. Harvard Business Review, March-April, 137–148.

Hammond, J. S., Keeney, R. L., & Raiffa, H. (1999). *Smart choices: A practical guide to making better decisions*. Boston: Harvard Business School Press.
Huang, Y. P., Huang, C. Y., Chan, S. S., & Lin, W. C. (1999). The world wide web and the databases for retaining wall design. *Advances in Engineering Software, 30*(9–11), 799–808.
ICAN System Inc. (2000). SmartSettle™ user guide for SmartSettle™ software beta build 789. Abbotsford, BC.
Kersten, G. E., & Noronha, S. J. (1999). WWW-based negotiation support: Design, implementation, and use. *Decision Support Systems, 25*, 135–154.
Lam, H. F. & Chang, T. Y. (2001). Internet-based construction project information management system (I. Construction). In: *Proceedings of international conference on construction*, (Vol. 1, pp. 450–461). Hong Kong.
Li, H. (1996). Case based reasoning for intelligent support of construction negotiation. *Information and Management, 30*, 231–238.
Lloyd, R. H., Boughton, G. S. & Siu, K. C. (2001). A project information management system, providing greater collaboration and improved communications on major construction works. In: *Proceedings of international conference on construction*, (Vol. 1, pp. 462–473). Hong Kong.
Merna, A., & Bower, D. (1997). *Dispute resolution in construction and infrastructure projects*. Hong Kong: Hong Kong Asia Law & Practice Publishing Ltd.
Nunamker, J. F., Applegate, L. M., & Konsynki, B. R. (1987). Facilitating group creativity: Experience with a group decision support system. Journal of Management Information Systems, *3*(4), 5–19 Spring.
O'Brien, M., & Al-Soufi, A. (1994). A survey of data communications in the UK construction industry. *Construction Management and Economics, 12*, 457–465.
Pena-Mora, F., Sriram, D., & Logcher, R. (1993). Design rationale for computer-supported conflict mitigation. *Journal of Computing in Civil Engineering, 9*(1), 57–72.
Pena-Mora, F., & Wang, C. Y. (1998). Computer-supported collaborative negotiation methodology. *Journal of Computing in Civil Engineering, 12*(2), 64–81.
Reinschmidt, K. F., Griffis, F. H., & Bronner, P. (1991). Integration of engineering, design and construction. *ASCE Journal of Construction Engineering and Management, 117*(4), 756–772.
Rezgui, Y., Cooper, G., & Brandon, P. (1998). Information management in a collaborative multifactor environment: The COMMIT approach. *Journal of Computing in Civil Engineering, ASCE, 12*(3), 136–143.
Roddis, W. M. K. (1998). *Knowledge-based assistants in collaborative engineering. AI in structural engineering, lecture notes in AI*, New York: Springer.
Samuelson, O. (1998). A study of the use of IT in the construction industry, The life cycle of construction IT innovations. In: *Proceedings CIB W78 conference*, Stockholm. 3–5 June.
Schmitt, G. (1998). *A new collaborative design environment for engineers and architect, AI in structural engineering, lecture notes in AI*, New York: Springer.
Shash, A. A., & Al-Amir, M. (1997). Information technology in contractors' firms in Saudi Arabia. *Construction Management and Economics, 15*, 187–200.

Chapter 13
Withdrawal as a Form of Construction Dispute Negotiation Failure

Sai On Cheung and Pui Ting Chow

Abstract Failing to reach a settlement in construction project dispute negotiation (CPDN) is not uncommon. One of the failing scenarios is withdrawal (WA), a situation where a negotiator loses interest in continuing with the discussion and leaves the negotiation table. This study aims to understand better withdrawal in construction dispute negotiation. With reference to a wage negotiation, the symptoms of withdrawal are discussed. It is hypothesised that WA is influenced by behavioural primers such as personality, motivation and cognition. A framework between WA and these behavioural primers is proposed and empirically tested. Emotion is found to be the most significant behavioural trigger of withdrawal.

13.1 Introduction

Economists assume negotiators are utility-driven and make decisions that maximises the utilities derivable (i.e. Pareto optimality) (Allred 2004; Bazerman and Chugh 2006; Thompson 2005). They interact to uncover each other's concerns, identify possible contract zones, make tradeoffs and devise settlement solutions. Construction disputes are content specific and appear to suit this rational model of principled negotiation (Yiu et al. 2008). Settling construction project dispute through negotiation is therefore the most sensible means of resolution. Cheung et al. (2009) developed taxonomies of dispute sources, negotiators' tactics and negotiation outcome. In analysing the interactions among these three groups of

S. O. Cheung (✉) · P. T. Chow
Construction Dispute Resolution Research Unit, Department of Civil and Architectural Engineering, City University of Hong Kong, Hong Kong, People's Republic of China
e-mail: saion.cheung@cityu.edu.hk

P. T. Chow
e-mail: ronnie.chow.pt@hotmail.com

S. O. Cheung (ed.), *Construction Dispute Research*,
DOI: 10.1007/978-3-319-04429-3_13,

taxonomies, seeking progress was found to be the most versatile tactic. Moreover, a narrow focus on rationality depersonalises negotiators and implies that the negotiation process is dictated by the desired optimal outcomes. In fact, failing to reach a negotiated settlement in construction project dispute negotiation (CPDN) is not uncommon (Cheung et al. 2006). Unveiling the reasons for failing negotiations are good lessons to be learnt. One of the failing scenarios is withdrawal (WA), a situation where a negotiator loses interest in continuing with the discussion and leaves the negotiation table. It is also noted that negotiators are subjected to influences such as personality, motivation and cognition. This study aims to identify the most influential behavioural primers on WA. With reference to the behavioural tradition of negotiation studies, a relationship framework between WA and behavioural primers is proposed. The relationship framework is then empirically tested. Emotion is found to be the significant trigger of withdrawal.

13.2 Withdrawal as a Form of Negotiation Failure

Models of negotiation often take "snapshot" views of the process. This approach does not consider the dynamics involved and often fails to unveil causes of negotiation failure. Tjosvold et al. (1999) suggested that: (1) failed negotiation often results in distrust among negotiating parties and uncertainty about the future functioning of the relationship; (2) unsatisfactory performance is manifested through withdrawing behaviour (O'Connor and Arnold 2001); and (3) withdrawing behaviour may develop as self-reinforcing cycles (Beersma and De Dreu 2005) (Fig. 13.1).

Watkins (1998) suggested that (1) negotiator's interest is sensitive to early events which are profoundly influential to events thereafter; (2) negotiator's withdrawal action impacts the negotiation in an irreversible way and create barriers to further negotiation; and (3) a slight loss of interest may have no apparent effect on the negotiation, but a greater loss may lead to disproportional consequence of complete withdrawal. The effort needed to put negotiation back on track once withdrawal happened is enormous and the phenomenon is analogous to the catastrophe flags proposed by Gilmore (1981). These flags are used to explain phenomena of catastrophic change. Anecdotal data from a steel benders wage negotiation in Hong Kong is used to illustrate the proposed conceptualisation of withdrawal in construction dispute negotiation.

13.3 Withdrawal Symptoms

In 2007, Hong Kong endured the longest bar-bender strike in postwar history. More than 800 workers participated in the strike lasting over 36 days from 6 August to 12 September (Table 13.1). A substantial daily wage cut from HK$1300

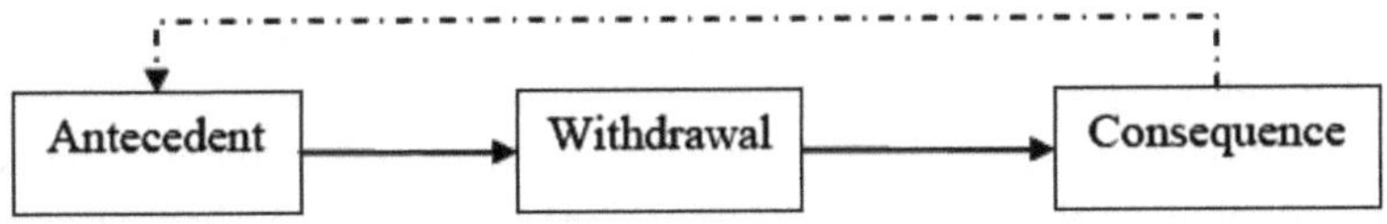

Precedent dysfunctional behaviors	Loss of interest to negotiate	Wasted resources and time
Divergent view	Reduction in socio-psychological attraction to organization	Organization regression
Slowing down prelude	Accumulation of minor dispute	Organization disintegration

Fig. 13.1 Withdrawal cycle

Table 13.1 Critical events in the negotiation between the Hong Kong Construction Industry Bar-Bending Workers' Union (CIB-BWU) and the Hong Kong Bar-Bending Contractors Association (BCA) in 2007

Date	Events
01 August	Supposed date of wages negotiation
08 August	Strike at a construction site in To Kwa Wan (Kowloon)
09 August	Sit-in protest outside the Chief Executive's official residence
10 August	Strike at a construction site in To Kwa Wan (Kowloon)
11 August	March from To Kwa Wan to Tsim Sha Tsui then to the Central Government Offices on Hong Kong Island
12 August	An un-announced meeting between CIB-BWU and BCA
12 August	BCA's offer: HK$850 backdated to 1 August, HK$950 started from 1 September, 8.5 work hours with 15-min break at work
	CIB-BWU's demand: HK$950 backdated to 1 August, 8 work hours
12 August	Failure to resolve divergent views within CIB-BWU and negotiation withdraw
13 August	Assembly in Kowloon and Cordon off two nearby streets
13 August	Extensive media report on the negotiation issues
23 August	Resumed negotiation
23 August	Bar-benders hold divergent views on protest
23 August	Negotiation slowed down
25 August	A number of bar-benders accepted an improved pay offer
28 August	Only one major developer agreed to pay wage of HK$950 a day and negotiation collapsed
31 August	Resumed negotiation
02 September	Negotiations paused
05 September	A minor strike
05 September	Negotiation slowed down
12 September	Final round of negotiation

to HK$800 happened in 1997. The wage has then been frozen, largely because of the low level of construction activities started in 1997 till 2007. The negotiation was punctuated with several crises including walk-outs of the delegations. Daily wage and working condition reviews had been held between the Hong Kong

Construction Industry Bar-bending Workers Union (CIB-BWU) and the Hong Kong Bar-Bending Contractors Association (BCA). Both parties agreed to adjust wages through negotiation in August annually. Nevertheless, the postponement of the wages negotiation meeting to November 2007 triggered the fear of continual stagnation on wages. The economy recovery in 2007 sparked off a wage raise demand. Bar-benders' initially demanded the daily wage to be raised to HK$950 per day and 8-h work day. The trade association only offered HK$850 per 8.5-h work day with a 15-min break at work. The strike affected approximately 60 private and 10 government projects. Although the two sides eventually reached an agreement, both endured heavy losses. Many workers went into heavy debt while they were on picket lines. In addition, there were huge liquidated damages imposed on the contractors for the irretrievable projects' delay. Through an anecdotal longitudinal analysis of the negotiation between the two parties from literature search and interviews with key personnel, several withdrawal symptoms were identified. Among them, preceding dysfunctional behaviours, divergent views and slowing down are found to be useful detectors of withdrawal.

13.3.1 Preceding Dysfunctional Behaviours

Firstly, ultimate withdrawal was found to be preceded by a series of dysfunctional behaviours. It manifests in the form of contribution reduction if leaving the organisation is not a viable option. Employees are concerned over the degree to which the organisation values their contribution and cares about their well-being. As such, the employer-employee relationship is strengthened through exchange of reward for increased efforts by the employees. In fact, the delay in wages negotiation which usually took place on 1 August negated this reciprocating expectation (Rousseau 1995). The bar-benders showed increased dissatisfaction toward the wages package. The psychological impact of dissatisfaction was translated into reduced employee morale, tardiness, unnecessary absenteeism and lateness which exacerbated both task and relationship conflicts within the work place (Duffy et al. 2000). Members of the CIB-BWU refused to work overtime as a means to press for concession. This served as an ultimatum before withdrawal. These withdrawing behaviours if not corrected are likely to cause further damages. The number of employee displaying withdrawing behaviour increased and ultimately led to collective withdrawal.

13.3.2 Divergent View

Secondly, divergent demand among the fractions within the CIB-BWU was also a major factor that had led to withdrawal. Negotiators are subjected to various sources of tension, especially when they are being evaluated by constituents and

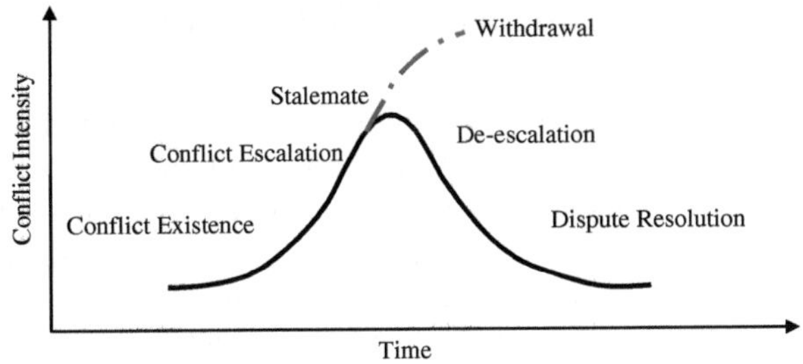

Fig. 13.2 Conflict intensity during negotiation

bounded by the respective mandates. The newly formed Construction Site Workers General Union (CSWGU) held major different views on the wages issues with CIB-BWU. The lack of unanimously perceived organisational support from unions had led to distressing behaviour of the bar-benders. One of the psychological consequences of distress was deterioration of perceived general functioning (MacBride et al. 1981). In other words, bar-benders experienced dissonance of union rivalry followed by diverging supports. The bar-benders were unable to identity a common authority. Consequently, bar-benders were left detached and de-motivated to continue with the negotiation. Withdrawal provided a leeway to reduce the unpleasant feelings of the perceived disproportional input to outcome (Adams 1965). Siegel and Lane's (1982) discrepancy theory offers another explanation on how the dissonance on perceived and received outcomes would lead to withdrawal. The lack of authority thus accountability resulted in lowing respect and recognition of the union by the bar-benders. The negotiators turned to withdrawal and left the negotiation table.

13.3.3 Slowing Down

Thirdly, slowing down of the negotiation process was found to be a preemption of withdrawal. Insufficient progress reinforced the perception that the wage issue was a zero-sum game. Position bargaining tactics were therefore used to uphold their demand. The hard negotiating rhetoric reduced the parties' flexibility to search for mutual interests. The involvement of wide media coverage further provoked confrontation and position hardening. With an impasse in mind, it is difficult for parties to contemplate a settlement. The polarised mentality of self and other further hardened the parties' stance. When negotiator judgment falls systematically short of rationality, they would perform sub-optimally by turning into aggressive mode. Making concession would become extremely unlikely. The situation becomes stalemate (Fig. 13.2). The longer the time of negotiation, the more difficult to have retrospective wages adjustment, protracted conflicts can create a pool of future unresolved issues, frustration and resentment. Hence, failure to adequately address and deal with conflicts would create stalemate and impasse. The onslaught of negative emotion was one of the significant obstacles to wages negotiation.

13.4 Consequences of Withdrawal

The relationship of the withdrawing behaviour and withdrawal can be illustrated in a behaviour-response framework. Withdrawing behaviour can be gauged as the level of interest of a negotiator to continue with a negotiation. The lower the interest, the more likely that withdrawal will occur. It can thus be outlined as a bi-polar withdrawal-continuation continuum. At one pole, the negotiator continues with the negotiation with a positive attitude in considering proposals from the perspective of the other side of the negotiation table. At the other end of the continuum, if a negotiator loses interest to continue, he/she would withdraw. Negotiators must have some goals realised as part of the settlement. That means the negotiator is looking for tangible, intangible or both types of return. Locke and Latham (1990) operationalised goals attainment as a performance standard of negotiation. In this regard, a failing negotiation may be viewed as substandard performance of the negotiator. This works against withdrawal unless an alternative means is considered as more appropriate than continuing with the negotiation. Furthermore withdrawal would turn the efforts expanded in the previous rounds of negotiation futile (O'Connor and Arnold 2001). Most significantly, withdrawal would cause extra effort to resume the negotiation. Payoff restructuring, organisational unity and information asymmetry are associated with withdrawal.

13.4.1 Payoff Restructuring

A negotiation involves seeking an outcome that offers more value than individuals' best alternative to a negotiated agreement (BATNA) (Fisher and Ury 1983). The options available are the possible agreements upon which negotiators might possibly engage with. Nevertheless, major projects entail hundreds of issue and a multitude of implicit and explicit interests, resulting in substantially complex negotiations among construction contracting parties. The process consists of the formation of social contract, in which individuals voluntarily subjugate their self-interest to benefits provided by the contract (Muroaro and Kujala 2007; Williamson 1985). In other words, individuals refrain from opportunistic practices because of the contract. Nonetheless, once the negotiating parties fail to reach a settlement and withdraw, there are costs of withdrawal imposed on both parties. Very often this would mean a reduction in the resources available that could have shared for a negotiated settlement. It leads to increase in self-interest seeking activities, whereby pay-off structure and incentive are adjusted for subsequent dealings (Williamson 1985). The failure in previous dealing inevitably induces uncertainty. A negotiator can perceive a credible offer unacceptable due to this uncertainty. The pessimism of potential failure in a resumed negotiation is therefore matched with request for more. In fact, negotiation can be metaphorically represented by a dance in the joint utility space (Raiffa 1982).

Once negotiation stalls, inertia of deadlock grows. A withdrawn party would have to give an attractive re-opening offer in order to reinstate the negotiation dance. As a result, reaching an agreement in a withdrawn negotiation often requires extra time and concessions than if there has been no withdrawal.

13.4.2 Organisational Unity

Construction has been labeled as an outsourcing industry. Networks of organisations produce the customised built facilities to satisfy clients' needs by using resources from multiple functional groups. Nevertheless, project team, in the form of a temporary organisation does not operate in isolation from its members. A highly performing individual member alone may not necessarily lead to efficient and effective team performance. Nonetheless, a dysfunction one would undoubtedly worsen the interdependent organisational network. In fact, individuals are bounded by multiple human concerns and cognitive backgrounds, thus they do not necessarily conform to the rational ideology (Muroaro and Kujala 2007). It is not uncommon that negotiator finds it remote to receive unanimous direction from the interest groups. Diverging views are dysfunctional for group decision because it prohibits a unified response (Staw et al. 1981). Diverging views are found to be prominent antecedences and consequences of withdrawal. The decision to trigger greater persistence or contribute to a reduction of effort depends not only on how a negotiator perceives the outcomes but also the degree of confidence to achieve them (e.g. Leiter 1991). In other words, individuals would respond to potential failure by scaling back their efforts as they have little faith that persistence would pay off (e.g. Bandura and Cervone 1983). Indeed, within one camp, some may have relatively optimistic expectations regarding the ultimate outcome even they fail to make progress, while others may retreat in the face of potential failure with increasing effort (O'Connor and Arnold 2001). These divergent views cause inflexibility. Negotiation representatives often responds to such diverging views by standing firm instead of crafting creative solution (Tetlock 1999). In particular, withdrawal is often a result of divergent views within one camp (Jex and Bliese 1999).

13.4.3 Information Asymmetry

In a multi-issue negotiation, preferences of the negotiating parties are rarely known (Fassina 2004). Thus, a negotiator may need to prevent his/her counterpart from harboring on unrealistic and false expectation (Lax and Sebenius 1991). A withdrawn party is often required to disclose additional information in order to re-start the negotiation (Williamson 1985). The resumption is only possible if the counterpart is willing to take risks that previous withdrawal will not be repeated. Two forms of information matter here: (i) outcome information that concerns the amount of profit, knowledge of results or the final settlement of the negotiation, (ii) process information that addresses individuals' preferences, strategies, interests, environment and perceptions (Thompson and DeHarpport 1994). High level of information sharing would facilitate logrolling. Nevertheless, with this valuable information in resuming a withdrawn negotiation, a counterpart can readjust his/her offer, aspirations (Whyte and Sebenius 1997). Disclose of outcome information is associated with high joint gain in a cooperative context (O'Connor and Arnold 2001). Day and Klein (1987) found that high level of information asymmetry is associated with opportunistic negotiation behaviour. The potential pitfall of disclosing too much information is that it reveals the bottom line of the withdrawn party. If the risk of opportunism in a particular relationship is sufficiently high, considerable resources must be directed to control and monitor rather than other productive means in furthering the negotiation (Wathne and Heide 2000). In principle, any form of opportunistic behaviour has the risk of restricting value creation and redistribution (Wathne and Heide 2000). Information asymmetry can exist in a variety of settings, and it would arise when information is incomplete. A negotiating counterpart may possess more transaction information and use it to his/her advantage. In general, information asymmetry limits one's ability to detect the other's opportunistic motives (Williamson 1975).

13.5 Withdrawing Behaviour in Construction Dispute Negotiation

With breakdown as an ultimate state of WA, negotiator's degree of loss of interest in continuing with the negotiation is translated as the states of WA (Chow and Cheung 2008). Figure 13.3 presents the mapping between states of WA and negotiation behaviour. The lower the interest in continuing, the higher the state of withdrawal and the more likely that breakdown will occur. At one end of the interest scale, negotiators having a strong desire to settle will positively approach the negotiations and enthusiastically consider every proposal on the table. At the other end of the scale, negotiators completely losing interest in continuing will leave the negotiations which make previous efforts futile (O'Connor and Arnold 2001). The change from "continuation" to "breakdown" determines the fate of a negotiation. More significantly, extra efforts are needed to resume a withdrawn negotiation as noted in the deadlock of many international disputes (Brooks 2008; Chow et al. 2008a, b; Fisher 2001; Persson 1994).

Prescriptive negotiation models treat negotiators as rational utility maximisers and a negotiated settlement derives higher overall utility (Thompson 2005). This proposition suggests that continuation of negotiation is more beneficial than withdrawing from it. However, negotiation behaviour may depart from the utility driven framework (Bazerman and Chugh 2006). For example, perceived inability

Fig. 13.3 State of withdrawal and negotiating behaviour (adapted from Chow et al. 2008a, b; Zeeman 1977)

Interest to Continue	Negotiating Behaviour	State of Withdrawal
Low	Breakdown	High
	Irrational argument	
↓	Rational discussion	↓
	Concession	
Strong	Apologies	Low

to reach a negotiated agreement can trigger breakdown. If nothing happens to revitalise the desire, negotiators will drop out and leave the negotiation table. Pruitt and Kim (2004) offered a plausible explanation by linking WA with the behaviour of the negotiators. For example, emotion has been found to affect negotiators' judgment when challenged by counterparts (Van Kleef et al. 2006). With reference to Fig. 13.3, negotiations normally start with rational discussion. Depending on the behavioural characteristics of the negotiators, the discussions may continue with concessions or irrational argument. In extreme cases, negotiations may end with either apologies of conceding parties or breakdowns of contending parties (O'Connor and Arnold 2001).

13.6 Behavioural Primers of Withdrawal

WA is often manifested by sudden attitudinal change (Persson 1994; Sokolova and Szpakowicz 2007; Van der Maas et al. 2003). Reverting negotiation to the ante-WA position has been found to be effort-laden (Cheung et al. 2009; Van der Maas et al. 2003). Pinkley and Northcraft (1994) explained that behavioural primers are subconscious frames upon which an individual prioritises issues. Schweitzer and DeChurch (2001) advocate three behavioural primers influence the dynamics of negotiations. Bazerman and Chugh (2006) suggested that negotiators can reach agreements more readily if their subconscious frames match. Along this line of conceptualisation, three behavioural primers: (1) motivation (task vs. relation), (2) cognition (emotion vs. rationality), and (3) personality (competition vs. cooperation), are included in a WA-behavioural primers relationship framework (Pinkley and Northcraft 1994; Ross and Ward 1995). Thompson (2005) suggests that negotiators who are task-, emotion-, or competition-oriented achieve lower joint outcomes than those who are relation-, rationality-or cooperation-focused because of their inability to find integrative trade-offs (Kashiwagi 2004). The projections of negotiation behaviour under each of the three behavioural primers are discussed seriatim.

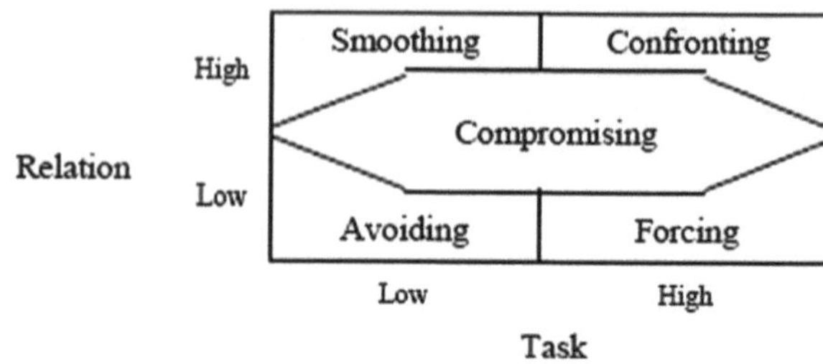

Fig. 13.4 Motivation primer with task subscale versus relation subscale (adapted from Kerzer 2006)

13.6.1 *Motivation Primer (Task Subscale Vs. Relation Subscale)*

Motivation primer (MO) examines the influence of aspirations on negotiation behaviour (Thompson 2005). Aspirations can be task or relation oriented which are not mutually exclusive (Bono et al. 2002; Simons and Peterson 2000). Negotiators with high aspirations on task subscale focus on the outcome. They take negotiations as jobs to be completed and define desired outcomes (Yiu et al. 2008). On one hand, task-oriented negotiators thus place stronger emphasis on the substance of the dispute and are aggressive in pursuing their desired outcomes. On the other hand, negotiators who are relation-concerned are more willing to compromise in order to maintain relationships. Their altruistic behaviour is more salient in situations where the interests are at odds than where the interests align. In this regard, Pruitt and Kim (2004) suggested that negotiators who have high degrees of concern for people downplay disagreement more than negotiators who are concerned about outcomes do. In construction, Bresnen and Marshall (2000) studied MO in construction project partnerships and suggested that contracting parties who value relationship look for goodwill to reinforce short-term collaboration and build long-term trust (Arditi and Yasamis 1998). Figure 13.4 presents the projections of negotiation behaviour under the influence of MO (Kerzner 2006). Negotiators who are low task-oriented are more likely to reconcile or concede to the situation whereas high task-oriented negotiators force their way to accomplish the task. Relationship-oriented negotiators are more willing to accommodate in order to maintain a good relationship with their counterparts. Nevertheless, negotiators who worry too much on relationship allow unnecessary concessions. Negotiators focusing on both task and relationship are assertive and mindful to avoid direct confrontation.

13.6.2 *Cognition Primer (Emotion Subscale Vs. Rationality Subscale)*

Cognition primer (CO) governs how negotiators interpret and present a dispute (De Dreu 2006). Negotiators construct mental maps according to their cognitive representations. Past experience plays a major role (Druckman et al. 2009).

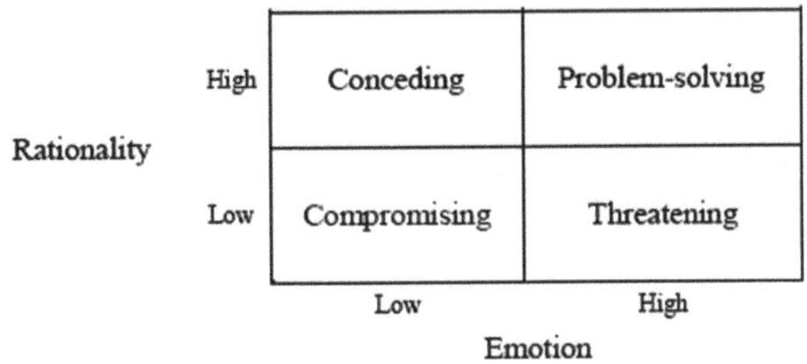

Fig. 13.5 Cognition primer with emotion subscale versus rationality subscale (adapted from Bazerman and Chugh 2006; De Dreu 2006; Druckman et al. 2009; Van Kleef et al. 2006)

Together with the information available, judgment is made (Bazerman and Chugh 2006). The responses of emotional negotiators are very direct (Van Kleef et al. 2006). Negative emotions stimulate retaliatory responses whereas positive emotions induce mutual understanding. On the other hand, rational negotiators evaluate and select alternatives objectively (De Dreu 2006). They persuade their counterparts by logical reasoning. Figure 13.5 shows the projections of negotiation behaviour under the influence of CO. Unemotional negotiators keep their nerves during negotiation and pay less attention to others' emotional cues. Emotional negotiators however, are very sensitive to others' emotional calls. A problem-solving attitude is expected if negotiators are high in both emotion and rationality subscales.

13.6.3 Personality Primer (Competition Subscale Vs. Cooperation Subscale)

Personality primer (PE) refers to the personal attributes of the negotiators such as trait and aptitude (Bono et al. 2002); Lewicki et al. (2010) use the term individual differences to identify personal characteristics of negotiators that affect their behaviour (Thompson 2005). PE manifests natural behavioural inclinations of negotiators during negotiation (O'Connor and Arnold 2001). Competitive negotiators are more risk seeking and more likely to argue for a better deal. They treat negotiations as distributive. Cooperative negotiators employ integrative strategies more readily. Figure 13.6 gives the projections of negotiation behaviour under the influence of PE (Blake and Mouton 1994; Rahim and Magner 1994; Thomas and Kilmann 1974). On one hand, competitive negotiators believe that pressure statements can stare the counterparts down and therefore often adopt high-handed approaches. On the other hand, negotiators who are low in the competition subscale are likely to respond passively (De Dreu 2006; Rahim and Magner 1994). They stay away from the argument and in extreme situations flee the negotiations altogether. Zeeman (1977) postulated that this attack-retreat phase transition behaviour can come out all of a sudden and its impact on the negotiation outcome is catastrophic. The attitude of cooperative negotiators is principally value creating. Highly cooperative negotiators are accommodating. They are willing to

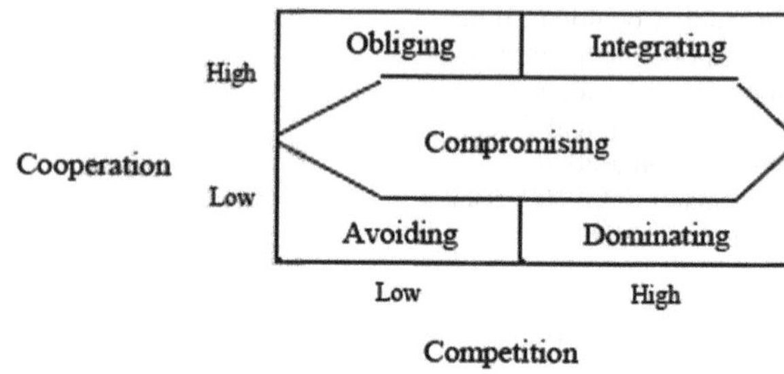

Fig. 13.6 Personality primer with competition subscale versus cooperation subscale (adapted from Blake and Mouton 1994; Rahim and Magner 1994; Thomas and Kilmann 1974)

explore integrative potentials and agree to ways that create greater value (Lewicki et al. 2010; Pruitt and Kim 2004). Settlements are thus more likely. The attitude of negotiators who are low in the cooperation subscale is somewhat one of being unconcerned (Allred 2004; Lewicki et al. 2010). These negotiators have low regard for each other's interests and priorities. Compromising negotiators are intermediate in both the competition and cooperation subscales. They are likely to suggest solutions that partially satisfy both parties. Therefore, compromising might just be splitting the difference, exchanging concessions, or seeking quick middle-ground solutions. Collaborative negotiators are high in both subscales. They work with their counterparts to find a solution that can satisfy the concerns of both. Collaboration might take the form of exploring disagreement in order to learn from each other's perspective (Bagozzi and Heatherton 1994).

13.7 A Relationship Framework Between Withdrawal and Behavioural Primers

Based on the afore-described projections of negotiation behaviour summarised in Figs. 13.4, 13.5 and 13.6, a withdrawal-behavioural primers relationship framework for construction project dispute negotiation (CPDN) and the three behavioural primers is proposed. Figure 13.7 presents the relationship framework underpinned by the following propositions:

H1 Task-motivated negotiators are more likely to withdraw in CPDN
H2 Relationship-motivated negotiators are less likely to withdraw in CPDN
H3 Cognitively emotional negotiators are more likely to withdraw in CPDN
H4 Cognitively rational negotiators are less likely to withdraw in CPDN
H5 Competitive negotiators are more likely to withdraw in CPDN
H6 Cooperative negotiators are less likely to withdraw in CPDN
H7 The three behavioural primers affect withdrawal in CPDN in varying degrees

The relationship framework is then analysed by structural equation modeling (SEM). SEM has been used to examine the inter-relationship among variables of

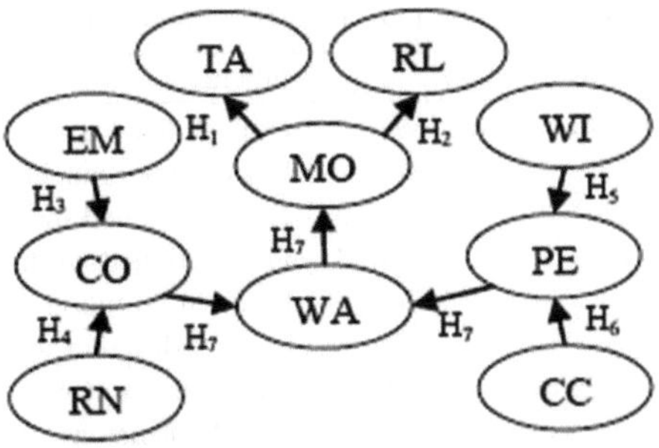

Note WA: Withdrawal; MO: Motivational; TA: Task; RL: Relationship; CO: Cognitive; EM: Emotion; RN: Rational; PE: Personality; WI: Competition; CC: Cooperation

Fig. 13.7 Schematic diagram of the relationship framework between withdrawal and the three behavioural primers

emotion in negotiation (De Dreu 2006), negotiators' performance (Overbeck et al. 2010) and construction partnering (Wong et al. 2009). The use of SEM allows the examination of relationships between WA (dependent variable) and behavioural primers (independent variables) in a holistic manner (Hair et al. 1998; Sharma 1996). In terms of model development, SEM combines confirmatory and exploratory purposes (Molenaar et al. 2000). In addition, the tested relationships are free from measurement errors that have been estimated and removed (Ullman and Bentler 2004). Thus, SEM is particularly useful in evaluating the proposed relationship framework. WA, the three behavioural primers and their respective subscales form a second-order factor SEM model in the relationship framework. They are latent variables and are approximated by observable indicators (also known as manifest and/or measurable variables) (Byrne 2010). Six indicators of WA are modified from the Maslach Burnout Inventory (Maslach et al. 2001) to suit the specific context of CPDN (Chow and Cheung 2008). For MO, TA uses six indicators to assess negotiators' level of aspiration to resolve the disputes (Kerzner 2006; Simons and Peterson 2000). This study adopts another six items measuring RL as the negotiator level of concern relationship (Bono et al. 2002; Simons and Peterson 2000). For CO, there are six indicators each for EM and RN. They measure the cognitive map negotiators constructed during the negotiation (Butler and Chinowsky 2006; De Dreu 2006; Druckman et al. 2009; Fraser and Hipel 1984; Van Kleef et al. 2006). Referring to Dual Concern Grid by Blake and Mouton (1994), Conflict Resolution Inventory by Rahim and Magner (1994), and Conflict Handling Style by Thomas and Kilmann (1974), WI and CC are constructed. Table 13.2 summarises the measurement statements used.

The arrows in the SEM model represent possible influences of one unit on another. The unit can be an indicator or a variable as the case may be (Hair et al. 1998; Sharma 1996). Straight arrows hypothesise causations between variables. The variables emanating arrows are the cause and the variables arrows pointing to are the effect. Curved and double-headed arrows indicate that variables may correlate with each other but there is no assumed causation between variables

Table 13.2 The indicators of withdrawal and three behavioural primers

Coding		Indictor
WA ($\alpha = 0.64$)	WA_1	I felt drained from the negotiation
	WA_2	I felt burned out in negotiating the dispute
	WA_3	I felt frustrated ineffectively dealing with the negotiation
	WA_4	I felt fatigued working on the trivial things of the negotiation
	WA_5	I was in no mood to resolve the dispute
	WA_6	I was indifferent to the negotiation
MO ($\alpha = 0.71$)	TA_1	I was determined to manage the dispute
	TA_2	I made decisions on matters of the dispute on my own
	TA_3	I exercised the entitled rights in the negotiation
	TA_4	I pointed out the disagreements in the negotiation
TA ($\alpha = 0.86$)	TA_5	I emphasised the material issues of the dispute
	TA_6	I concentrated on the outcomes of the negotiation
RL ($\alpha = 0.87$)	RL_1	I made concessions in order to maintain a good relationship with the counterpart
	RL_2	I downplayed the level of disagreement to preserve the relationship
	RL_3	I reinforced collaboration by making concessions
	RL_4	I looked for goodwill in the negotiation
	RL_5	I accommodated the counterpart's requests to build a relationship
	RL_6	I showed respect for the relationship with my counterpart
CO ($\alpha = 0.68$)	EM_1	I conveyed emotion messages on how I felt about the dispute to my counterpart
	EM_2	I cared about the emotions elicited by my counterpart
EM ($\alpha = 0.81$)	EM_3	I expressed negative emotions as deterrents for the adverse behaviour of my counterpart
	EM_4	I showed positive emotions to elicit favourable behaviour from my counterpart
	EM_5	I used emotional expressions as influencing tactics
RL ($\alpha = 0.79$)	EM_6	I expressed standoff when disagreement occurred
	RN_1	I evaluated the alternatives to negotiated agreement according to established principles
	RN_2	I persuaded the counterpart to accept my proposal by means of logic and reasoning
	RN_3	I established ground rules for the negotiation
	RN_4	I acquired all the relevant knowledge for decision-making
	RN_5	I organised the negotiation in a systematic way
	RN_6*	I was cognitively overloaded by the complexity of the negotiation
PE ($\alpha = 0.74$)	WI_1	I suggested creative solutions in favour of my own benefit
	WI_2	I pressed for concessions from the counterpart
	WI_3	I repeated my offer to assert my will
WI ($\alpha = 0.89$)	WI_4	I pointed out the mistakes made by the counterpart
	WI_5	I insisted on my own position by using power
	WI_6	I asked the counterpart for the reasons for his/her position

(continued)

Table 13.2 (continued)

Coding		Indictor
CC (α = 0.85)	CC_1	I acknowledged the concerns of the counterpart
	CC_2	I entertained the concerns of the counterpart
	CC_3	I promoted a mutual understanding of the needs and priorities
	CC_4	I notified both similarities and differences
	CC_5	I shared my views with the counterpart and encouraged him/her to do the same
	CC_6	I suggested integrative solutions and sought the support of the counterpart

Note: All scales are anchored with a 7-point Likert-scale (1 = Strongly Disagree; 7 = Strongly Agree), WA: Withdrawal; MO: Motivation; TA: Task; RL: Relation; CO: Cognition; EM: Emotion; RN: Rationality; PE: Personality; WI: Competition; CC: Cooperation; *: reverse score; α: Cronbach's alpha

(Hair et al. 1998). It is important to note that there is terminology in SEM which has multiple meanings in other fields. The specification of the second-order factor model posits that the indicators estimate their first-order factors. The respective underlying subscales influence the three behavioural primers. The behavioural primers, in turn, affect WA. Thus, the SEM models are sets of linear equations to be resolved by multiple regression methods holistically with maximum likelihood estimation applied (Wong et al. 2009). For example, task and relation subscales are the first-order factors of the motivation primer that is a second-order factor. They are, in turn, respectively measured by their indicators (TA_1 to TA_6 for task subscale and RL_1 to RL_6 for relation subscale).

The present study applied a two-step analysis to assess the SEM models (Hair et al. 1998; Jöreskog and Sörbom 1996). The first step involves assessing the internal consistency and inter-item relationships of the four latent constructs (WA and the three behavioural primers) and their respective subscales in the measurement SEM model. Cronbach's alphas are used to test the internal consistency. A higher value in Cronbach's alphas indicates a greater level of internal consistency. Cronbach's alphas that lie between 0.6 and 0.7 are regarded as "sufficient" and Cronbach's alphas greater than 0.7 are regarded as "good" (Sharma 1996). Pearson correlation testing is used to assess the inter-item relationships. Pearson correlation (both positive and negative) ranged from 0.0 to 0.2, 0.2 to 0.4, 0.4 to 0.7 and 0.7 to 1.0 are considered as "trivial", "small", "moderate", and "strong" respectively. The indicators and the respective error terms are connected to correct the measurement errors of the indicators and to form the measurement SEM model. The initial SEM model links the dependent and independent variables in the measurement SEM model. It is subjected to examination in the second step (Hair et al. 1998). The resulting set of Goodness-of-fit indices shows the statistical fit of the model (Table 13.3). The relationship framework is refined by applying the adjustments proposed by the SEM modification indexes (Arbuckle and Wothke 1999). The refined model is then re-tested. Further refinements are performed until the goodness-of-fit (GOF) indices are satisfied to obtain the final SEM model.

Table 13.3 Goodness of fit indices (GOF) and their threshold values

GOF	GOF range	Threshold	SEM model	
			Initial	Final
x^2/DF	0 or above	1.00–3.00	1.48	1.59
GFI	0 no fit to 1 perfect fit	0.76 or above	0.73	0.80
TLI	0 no fit to 1 perfect fit	0.70 or above	0.69	0.78
CFI	0 no fit to 1 perfect fit	0.73 or above	0.71	0.77
RMSEA	0 perfect fit to 1 no fit	0.1 or below	0.09	0.08

Note: GOF: goodness of fit indices; x^2 /DF: Chi square/degree of freedom, GFI: goodness of fit index; TLI: Tucker-Lewis index; CFI: comparative fit index; RMSEA: root mean square error of approximation

In addition, bootstrapping analysis is conducted to provide more reliable analysis results when the data set is small. Bootstrapping also allows the testing of the significance of parameter estimates from data (Janz and Prasarnphanich 2009). It is used to test the significance of path coefficients with 1000 replications. To avoid the risk of sample non-normality, bootstrapping analysis is applied to augment the reliability of the results of small sample sized SEM analysis (Kline 1998; Ozorhon et al. 2008; Paiva et al. 2008; Sadler-Smith and Smith 2004). The bases and working of bootstrapping analysis can be found in the work of Bollen and Stine (1992), Carmel et al. (1993) and Cheung and Lau (2008). The statistical analyses as mentioned were conducted by Statistical Package for Social Science (SPSS) Version 17.0 and Analysis of Moment Structures (AMOS) Version 17.0 with maximum likelihood estimation.

13.8 Methodology and Data Collection

A questionnaire is designed to include the measurement statements of the indicators listed in Table 13.2 for data collection. With reference to a recently completed CPDN, the respondents were asked to indicate the extent to which they agreed on the indicators in describing what had happened on a Likert scale of 1–7. For example, the respondents were asked to indicate how often they experienced WA in the specified CPDN. The study was conducted in Hong Kong and the targeted respondents were Hong Kong-based construction professionals including architects, engineers, quantity surveyors, and project managers with experience in CPDN. They were randomly selected from the construction firms listed in the directories and professional institutions, for example, the Hong Kong Contractor Association (HKCA), the Hong Kong Institute of Architects (HKIA), the Hong Kong Institute of Surveyors (HKIS) and the Hong Kong Institution of Engineers (HKIE). A total of 230 shortlisted respondents were contacted. Those who agreed to participate in the study were sent the data collection questionnaire. A total of

Table 13.4 Summary of the profile of respondents by types of profession and company

Response	Type of profession						
Type of company	Architect	Surveyor	Project manager	Engineer	Lawyer	Project coordinator	Grand total
Government	2	9	1	2	–	8	22
Main-contractor	1	8	6	13	–	3	31
Sub-contractor	–	–	4	7	–	2	13
Developer	9	3	2	–	–	–	14
Consultant	2	10	–	9	1	1	23
Grand total	14	30	13	31	1	14	103

103 questionnaires was returned, giving the response rate of about 40 %, comparable to the typical response rate achieved in construction research studies (Butler and Chinowsky 2006; Chan et al. 2001). 77 % of the returned questionnaires were completed by respondents having more than 5 years of experience. The negotiated cases were mainly related to building, civil and building services engineering works. Table 13.4 has summarised the profile of respondents by types of profession and company.

13.9 Results

13.9.1 Step 1: Checking the Construct Reliabilities, Inter-Relationships, and Model Validity of Withdrawal and the Three Behavioural Primers in the Measurement SEM Model

The construct internal consistency assesses the reliabilities of the measurement SEM models. In this study, four base measurement SEMs were constructed: one for WA and three for behavioural primers. WA, the three behavioural primers, and the six respective subscales achieved Cronbach's alpha values above 0.7, suggesting that the indicators significantly related to the respective constructs (Hair et al. 1998). Hence, the indicators as well as their groupings were retained in the relationship framework. These four base measurement SEMs were then linked to form the initial SEM of the proposed relationship framework (Fig. 13.8). The interrelationships of the variables and indicators were then considered holistically. As previously discussed, the overall fitness of the initial SEM is assessed by the GOF indices and subsequent modifications are used to improve the model fitness. The initial SEM of the relationship framework achieved marginally satisfactory GOF indices (Table 13.3).

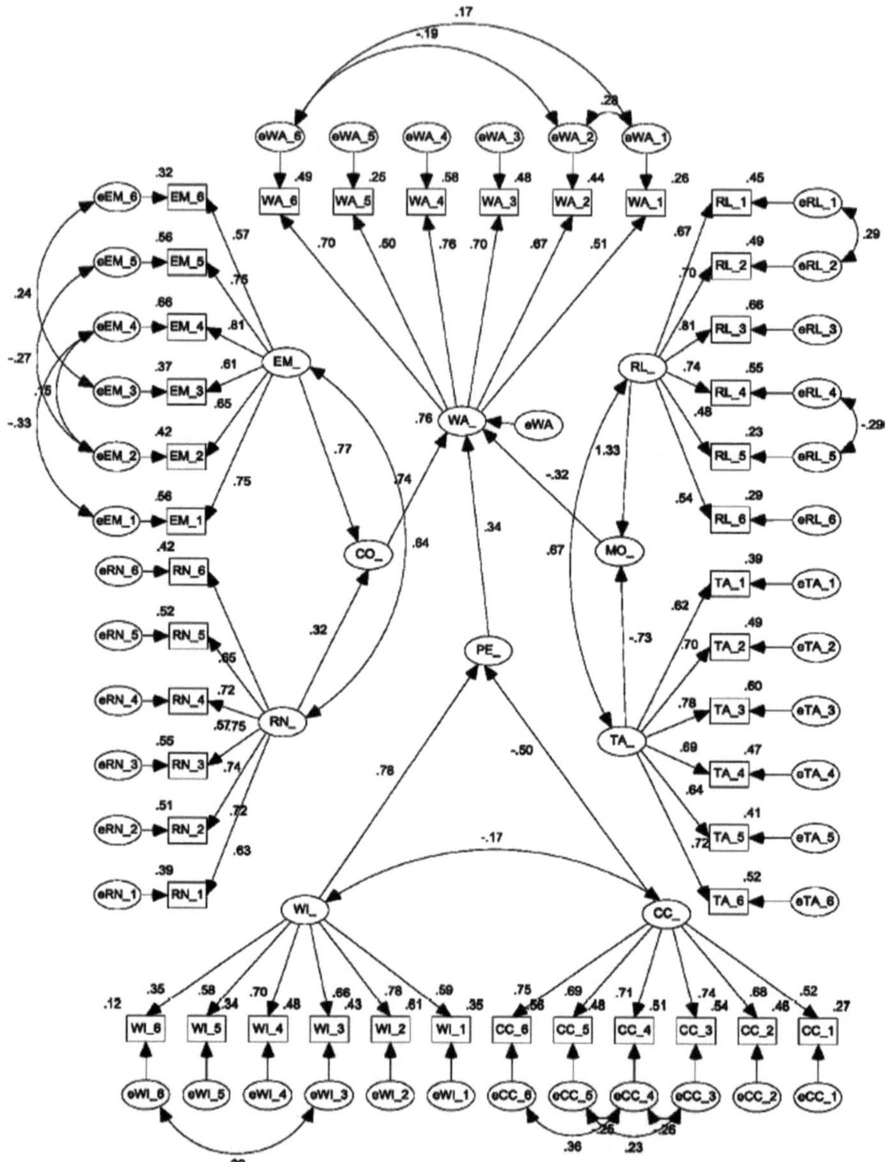

Note WA: Withdrawal; MO: Motivational; TA: Task; RL: Relationship; CO: Cognitive; EM: Emotion; RN: Rational; PE: Personality; WI: Competition; CC: Cooperation

Fig. 13.8 Stage 2: Final SEM model of the relationship framework between withdrawal and the three behavioural primers

13.9.2 Step 2: Checking the Model fit of the Final SEM Model of the Relationship Framework

The relationship framework contains 42 indicators. With this number it is not uncommon to achieve lower Goodness of Fit as there is a high level of random error (Bagozzi and Heatherton 1994; Rahim and Magner 1994). The analysis uses the covariance matrix for the 42 indicators, and the maximum likelihood estimation estimates the parameters. In the final SEM, each of the 42 indicators loads on only its associated latent variable and correlate with the error term. With all the Goodness of fit (GOF) indices falling within the acceptable range, the final SEM is accepted (Table 13.3 and Fig. 13.8). It is also noted that a sample size of 100 (preferably 200) is recommended for SEM analyses to avoid risk of sample non-normality (Hair et al. 1998; Jöreskog and Sörbom 1996; Kline 1998). In addition Hair et al. (1998) recommended that the ratio between the sample size and the number of free parameters should be 5:1 under normal distribution theory. Otherwise, the estimated regression weights of both latent and independent indicators may become statistically insignificant with high standard errors. Notwithstanding, Hair et al. (1998) suggested that even a small sample size of 50 might provide valid results in SEM analyses. For example, Mohamed (2003) employed SEM analysis with 44 data sets to investigate the implications of risk in the performance of the international joint venture construction projects. The SEM-based studies by Paiva et al. (2008) and Sadler-Smith and Smith (2004) also used less than 100 data sets. In this regard, bootstrapping is employed as a remedial method to augment the reliability of SEM analysis results (Ozorhon et al. 2008; Sadler-Smith and Smith 2004). The results of bootstrapping in the study indicate that the parameter estimates obtained from the SEM analysis are statistically significant (Molenaar et al. 2000; Ozorhon et al. 2008; Paiva et al. 2008; Wong et al. 2009). In sum, to avoid the risk of sample non-normality, bootstrapping analysis is applied to augment the reliability of the results of small sample sized SEM analysis (Kline 1998; Ozorhon et al. 2008; Paiva et al. 2008; Sadler-Smith and Smith 2004). The bases and detailed working of the method are beyond the scope of this paper but can be found in the work of Bollen and Stine (1992), Carmel et al. (1993) and Cheung and Lau (2008).

13.10 Discussion

The measurement SEM models and the initial SEM model of the relationship framework were firstly tested. After a series of modifications, the final SEM model achieved satisfactory Goodness of Fit indices. Figure 13.8 presents the final SEM model and Table 13.5 highlights the statistical inferences on the inter-relationships among the variables. The following gives some of the key observations. For the constructs of TA and RL, both are significantly related to MO (path [a] and path [b]).

Table 13.5 Summary of the testing of the propositions

Path		Related hypothesis	β	Sig.	Hypothesis supported
[a]	TA → MO	H_1	−0.788	0.006	Marginally
[b]	RL → MO	H_2	1.328	0.000	Yes
[c]	EM → CO	H_3	0.765	0.000	Yes
[d]	RN → CO	H_4	0.320	0.120	No
[e]	WI → PE	H_5	0.784	0.000	Yes
[f]	CC → PE	H_6	−0.499	0.115	No
[g]	MO → WA	H_7	−0.322	0.000	Yes
[h]	CO → WA	H_7	0.739	0.000	Yes
[j]	PE → WA	H_7	0.338	0.003	Yes

Note: WA: Withdrawal; MO: Motivation; TA: Task; RL: Relation; CO: Cognition; EM: Emotion; RN: Rationality; PE: Personality; WI: Competition; CC: Cooperation; β: Standardised regression weight

However, in terms of effect on WA, relation-focused negotiators are less likely to withdraw (path [b]–[g]) than task-oriented negotiators (path [a]–[g]). For CO, only EM is significant (path [c]). An emotional negotiator is more likely to withdraw (path [c]–[h]). In fact, CO shows the greatest influence on WA (path [h]). Likewise, competitive negotiators are more likely to withdraw in relation to PE (path [e]–[j])). Accordingly, of the seven hypotheses H1, H2, H3, H5 and H7 are statistically supported while H4 and H6 are not. Qualitatively, the respondents are in the opinion that task-, emotion-and competition-oriented negotiators are more likely to withdraw. The counter force against ultimate WA is when the negotiators are relation-oriented. WA is also less likely for cooperative and rational negotiators, but with less significant results.

This study has focused on withdrawal in negotiation, an area that is almost uncharted in construction project dispute negotiation study. The danger of ultimate WA looms when negotiators display forcing (high task, low relation), threatening (highly emotional) and competing (highly competitive) behaviour. This behaviour can be strategically used by the negotiator to press for concessions and are often found to be useful when the counterpart is cooperative or relation-minded (Cheung et al. 2009; Fulmer and Barry 2004). WA as a strategy would lure compromise by a counterpart who wishes to avoid impasse (O'Connor and Arnold 2001; Wallihan 1998). Negotiators thus need to be mindful of the strategic use of WA. Emotional WA is more problematic. The finding of this study suggests that emotion is the most likely cause of WA among the subscales of behavioural primers. Emotional and cooperative negotiators who also care about relation are likely to give into the pressure of the counterparts and make unwarranted concessions (Overbeck et al. 2010). They are likely to be the ultimate losers. The caveat against being overly emotional perhaps is the bringing back of rationality. The use of external consultants and the use of reality testing before responding may bring the rationality required (Yiu and Cheung 2005).

One interesting finding of the study is that construction project disputes are content specific and subject to the governance of the contract. Pragmatic negotiators are less likely to withdraw. This study highlights the criticality of keeping emotion under control in CPDN. The decision-making approach of Bazerman and Chugh (2006) may offer insight to explain why rationality subscale is statistically insignificant in the relationship framework. Fulmer and Barry (2004) summarised that decision-making approach to negotiation emphasises situational determinants. Thus, given a particular situation or context, individual negotiators are expected to react in more or less the same ways within the confines of bounded rationality and to be subjected to similar kinds of biases and decision-making errors. From this perspective, rationality subscale in cognition primer explains little variance in withdrawal. Bazerman and Chugh (2006) added that even if RN does impact WA, it is of limited interest because they are demographic or dispositional, and thus there is little that negotiators can do to change them. This is an observation that echoes the findings of this study. Nonetheless, not all researchers subscribe to this point of view. A number of scholars have continued to focus on the impact of PE. Recent research has explored other factors such as social value orientation (De Dreu 2006; Olekalns and Smith 2003). This study can be viewed as an extension to these studies in incorporating PE as a behavioural primer influencing WA. It is found that competitive negotiators are more likely to withdraw than cooperative negotiators. On this note ElShenawy (2010) raises an intriguing question "does negotiation training improve negotiators' performance?" Specifically, it was found that training enhances the skills of negotiators. This includes the better use of principled negotiation as well as the ability to suppress subconscious PE (Sadler-Smith and Smith 2004). This echoes well the key findings of this study. EM in relation to CO is found to be the most critical contributor to WA in CPDN (Butler and Chinowsky 2006). Good negotiators are those who are good at the technical aspects of the dispute as well as being emotionally stable (Wong et al. 2009).

13.11 Chapter Summary

Withdrawal refers to the situation where a negotiator loses interest in continuing with the negotiation-an area that is under researched in construction project dispute negotiation. WA is the penultimate state before the negotiation reaches failure. Under the rational school, negotiators are utility maximisers and are willing to create value through tradeoffs. Nevertheless, WA is likely to occur if negotiating behaviour is left uncontrolled. The behavioural tradition of negotiation study aptly explains this phenomenon. In this study, WA in CPDN is posited as a behavioural response influenced by three behavioural primers: MO, CO, and PE. A withdrawal-behavioural primers relationship framework is proposed. With data collected in Hong Kong, the relationship framework was investigated by the technique of SEM. The readers are reminded of the cultural context of the study with regard to the source of data. In addition, bootstrapping analysis has been used to augment the reliability of the results due to the relatively small number of data sets used. A WA-behavioural primers relationship framework underpinned by several propositions is proposed. At the primer level, the findings suggested that task-,emotion-,and competition-oriented negotiators are more likely to withdraw. The counter force against WA is when the negotiators are relation-oriented. Construction project disputes are content specific and resolution options are governed by contract. They fit neatly with the rational model of principled negotiation. This study timely reminds one of the influence of behaviour on the success or otherwise of CDPN. Emotion is found to be the most detrimental trigger of withdrawal. Recent study has highlighted the possibility of containing emotion through structured negotiation training.

Acknowledgments Special thanks to Miss Ming Sum Chiang for collecting data for the study. Part of the content of this chapter has been published in Volume 137(12) of the Journal of Construction Engineering and Management and is used with the permission from ASCE.

References

Adams, J. S. (1965). Inequity in social exchange. *Advances in Experimental Social Psychology, 62*(3), 335–343.

Allred, K. G. (2004). Distinguishing best and strategic practices: A framework for managing the dilemma between claiming and creating value. *Negotiation Journal, 16*(4), 387–397.

Arbuckle, J. L., & Wothke, E. (1999). *AMOS 4.0 user's guide*, Spss Inc. Small Waters Corporation, Chicago.

Arditi, D., & Yasamis, F. (1998). Incentive/disincentive contracts: Perceptions of owners and contractors. *Journal of Construction Engineering and Management, 124*(5), 361–373.

Bagozzi, R. P., & Heatherton, T. F. (1994). A general approach to representing multifaceted personality constructs: Application to state self-esteem. *Structural Equation Modeling, 1*(1), 35–67.

Bandura, A., & Cervone, D. (1983). Self-evaluative and self-efficacy mechanisms governing the motivational effects of goal systems. *Journal of Personality Social Psychology, 45*(5), 1017–1028.

Bazerman, M. H., & Chugh, D. (2006). Bounded awareness: Focusing failures in negotiation. In L. Thompson (Ed.), *Negotiation theory and research* (pp. 7–26). New York: Psychological Press.

Beersma, B., & De Dreu, C. K. W. (2005). Conflict's consequences: Effect of social motives on post negotiation creative and convergent group functioning and performance. *Journal of Personality and Social Psychology, 89*(3), 358–374.

Blake, R. R., & Mouton, J. S. (1994). *The managerial grid.* Houston: Gulf.

Bollen, K. A., & Stine, R. A. (1992). Bootstrapping goodness-of-fit measures in structural equation models. In K. A. Bollen & J. S. Long (Eds.), *Testing structural equation models* (pp. 111–135). Thousand Oaks: Sage Publication.

Bono, J. E., Boles, T. L., Judge, T. A., & Lauver, K. J. (2002). The role of personality in task and relationship conflict. *Journal of Personality, 70*(3), 311–343.

Bresnen, M., & Marshall, N. (2000). Partnering in construction: A critical review of issues, problems and dilemmas. *Construction Management & Economics, 18*(2), 229–237.

Brooks, S. P. (2008). Enforcing a turning point and imposing a deal: An analysis of the darfur abuja negotiations of 2006. *International Negotiation, 13*(3), 413–440.

Butler, C. J., & Chinowsky, P. S. (2006). Emotional intelligent and leadership behaviour in construction executives. *Journal of Management in Engineering, 22*(3), 119–125.
Byrne, B. M. (2010). *Structural equation modeling with AMOS* (2nd ed.). New York: Routledge.
Carmel, E., Herniter, B. C., & Nunamaker, J. F. (1993). Labour-management contract negotiation in an electronic meeting room: A case study. *Group Decision and Negotiation, 2*(1), 27–60.
Chan, A. P. C., Ho, D. C. K., & Tam, C. M. (2001). Design and build project success factors: Multivariate analysis. *Journal of Construction Engineering and Management, 127*(2), 10–93.
Cheung, S. O., Chow, P. T., & Yiu, T. W. (2009). Contingent use of negotiator tactics in construction dispute negotiation. *Journal of Construction Engineering and Management, 135*(6), 466–476.
Cheung, G. W., & Lau, R. S. (2008). Testing mediation and suppression effects of latent variables: Bootstrapping with structural equation models. *Organizational Research Methods, 11*(2), 296–325.
Cheung, S. O., Yiu, T. W., & Yeung, S. F. (2006). A study of styles and outcomes in construction dispute negotiation. *Journal of Construction Engineering and Management, 132*(8), 806–814.
Chow, P. T., & Cheung, S. O. (2008). Developing a conceptual framework of catastrophic withdrawal behaviour in construction dispute, *Proceedings of the BEAR 2008 conference* (pp. 659–669) Sri Lanka.
Chow, P. T., Cheung, S. O., & Chiang, M. S. (2008a). Consequences of withdrawal in negotiation, *Proceedings of 11th East Asia-Pacific conference on structural engineering and construction (EASEC-11), building a sustainable environment*, Nov 19–21, 2008, Contract and Organization C05_08, EASEC-11-699, (pp. 942–943). Taipei.
Chow, P. T., Cheung, S. O., & Chiang, M. S. (2008b). Withdrawal signals in construction dispute negotiation, *Proceedings of CIB W112 International Conference on Securing High Performance through Cultural Awareness and Dispute Avoidance.* Shanghai.
Day, G. S., & Klein, S. (1987). Cooperative behaviour in vertical markets: The influence of transactions costs and competitive strategies. In M. J. Houston (Ed.), *Review of marketing* (pp. 39–66). Chicago: American Marketing Association.
De Dreu, C. K. W. (2006). Rational self-interest and other-orientation in organizational behavior: A critical appraisal and extension of Meglino and Korsgaard (2004). *Journal of Applied Psychology, 91*(6), 1245–1252.
Druckman, D., Olekalns, M., & Smith, P. S. (2009). Interpretive filters: Social cognition and the impact of turning points in negotiation. *Negotiation Journal, 25*(1), 13–40.
Duffy, M. K., Shaw, J. D., & Stark, E. M. (2000). Performance and satisfaction in conflicted interdependent groups: When and how does self-esteem make a difference? *The Academy of Management Journal, 43*(4), 772–782.
ElShenawy, E. (2010). Does negotiation training improve negotiators' performance? *Journal of European Industrial Training, 34*(3), 192–210.
Fassina, N. E. (2004). Constraining a principal's choice—outcome versus behavior contingent agency contracts in representative negotiations. *Negotiation Journal, 1*, 435–459.
Fisher, R. J. (2001). Cyprus: The failure of mediation and the escalation of an identity-based conflict to an adversarial impasse. *Journal of Peace Research, 38*(3), 307–326.
Fisher, R., & Ury, W. (1983). *Getting to yes*. New York: Houghton Mifflin.
Fraser, M. N., & Hipel, K. W. (Eds.). (1984). *Conflict analysis: Models and resolutions*. New York: North-Holland.
Fulmer, S. I., & Barry, B. (2004). The smart negotiator: Cognitive ability and emotional intelligence in negotiation. *International Journal of Conflict Management, 15*(3), 245–272.
Gilmore, R. (1981). *Catastrophe flags, in catastrophe theory for scientists and engineers*. New York: Wiley.
Hair, J. F., Anderson, R. E., & Tatham, R. L. (1998). Multivariate data analysis (5th ed.) (pp. 88–92). Upper Saddle River: Prentice Hall.
Janz, B. D., & Prasarnphanich, P. (2009). Freedom to cooperate: Gaining clarity into knowledge integration in information systems development teams. *IEEE Transactions on Engineering Management, 56*(4), 621–635.

Jex, S. M., & Bliese, P. D. (1999). Efficacy beliefs as a moderator of the impact of work-related stressors: A multilevel study. *Journal of Applied Psychology, 84*(4), 349–361.

Jöreskog, K. G., & Sörbom, D. (1996). *LISREL 8 user's reference guide*. Chicago: Scientific Software.

Kashiwagi, D. T. (2004). *Best value procurement: How to use information systems to minimize risk, increase performance, and predict future success* (2nd ed.). Performance Based Studies Research Group, Arizona State University, Tempe.

Kerzner, H. (Ed.). (2006). *Project management: A systems approach to planning, scheduling, and controlling* (9th ed.). Berea: Wiley.

Kline, R. B. (1998). *Structural equation modeling*. New York: The Guilford Press.

Lax, D. A., & Sebenius, J. K. (1991). Negotiating through an agent. *Journal of Conflict Resolution, 35*(3), 474–493.

Leiter, M. P. (1991). Coping patterns as predictors of burnout: the function of control and escapist coping patters. *Journal of Organizational Behavior, 12*(2), 123–144.

Lewicki, R. J., Barry, B., & Saunders, D M. (2010). Individual differences. In R. J. Lewicki, B. Barry, & D. M. Saunders (Eds.), *Negotiation* (Vol. 5, pp. 363–393) (6th ed.). Boston: McGraw Hill.

Locke, E. A., & Latham, G. P. (1990). *A theory of goal setting and task performance*. Englewood Cliffs: Prentice Hall.

MacBride, A., Lancee, W., & Freeman, S. J. J. (1981). The psychosocial impact of a labour dispute. *Journal of Occupational Psychology, 54*(2), 125–133.

Maslach, C., Jackson, S. E., & Leiter, M. P. (2001). Job burnout. *Annual Review of Psychology, 52*, 397–422.

Mohamed, S. (2003). Performance in international construction joint ventures: Modeling perspective. *Journal of Construction Engineering and Management, 129*(6), 619–626.

Molenaar, K., Washington, S., & Diekmann, J. (2000). Structural equation model of construction contract dispute potential. *Journal of Construction Engineering and Management, 126*(4), 268–277.

Muroaro, J., & Kujala, J. (2007). Project negotiation analysis. *International Journal of Project Management, 25*(7), 722–733.

O'Connor, M. K., & Arnold, J. A. (2001). Distributive spirals: Negotiation impasses and the moderating role of disputant self-efficacy. *Organizational Behaviour and Human Decision Processes, 84*(1), 148–176.

Olekalns, M., & Smith, P. L. (2003). Social motives in negotiation: The relationships between dyad composition, negotiation processes and outcomes. *International Journal of Conflict Management, 14*(3/4), 233–243.

Overbeck, J. R., Neale, M. A., & Govan, C. L. (2010). I feel, therefore you act: Intrapersonal and interpersonal effects of emotion on negotiation as a function of social power. *Organizational Behaviour and Human Decision Processes, 112*(2), 126–139.

Ozorhon, B., Arditi, D., Dikmen, I., & Birgonul, M. T. (2008). Implications of culture in the performance of international construction joint ventures. *Journal of Construction Engineering and Management, 134*(5), 361–370.

Paiva, E. L., Roth, A., & Fensterseifer, J. E. (2008). Organizational knowledge and the manufacturing strategy process: A resource-based view analysis. *Journal of Operations Management, 26*(1), 115–132.

Persson, S. (1994). Deadlocks in international negotiations. *Cooperation and Conflict, 29*(3), 211–244.

Pinkley, R. L., & Northcraft, G. B. (1994). Conflict frames of reference: Implications for dispute processes and outcomes. *Academy of Management Review, 37*(1), 193–205.

Pruitt, D. G., & Kim, S. H. (2004). *Social conflict: Escalation, stalemate and Settlement* (3rd ed.). New York: McGraw Hill.

Rahim, M. A., & Magner, N. R. (1994). Convergent and discriminant validity of the Rahim organizational conflict inventory–II. *Psychological Reports, 74*(1), 35–38.

Raiffa, H. (1982). *The art and science of negotiation*. Cambridge: Belknap Press.

Ross, L., & Ward, A. (1995). Psychological barriers to dispute resolution. In M. P. Zanna (Ed.), *Advances in experimental social psychology* (Vol. 27, pp. 255–297) San Diego: Academic Press.

Rousseau, D. M. (1995). *Psychological contracts in organization*. Thousand Oaks: Sage Publication.

Sadler-Smith, E., & Smith, P. J. (2004). Strategies for accommodating individuals' styles and preferences in flexible learning programmes. *British Journal of Educational Technology, 35*(4), 395–412.

Schweitzer, M. E., & DeChurch, L. A. (2001). Linking frame in negotiations: gains, losses and conflict frame adoption. *International Journal of Conflict Management, 12*(2), 100–113.

Sharma, S. (1996). *Applied multivariate techniques* (pp. 116–123). New York: Wiley.

Siegel, L., & Lane, I. M. (1982). *Personnel and organizational psychology*. Homewood: Irwin.

Simons, T. L., & Peterson, R. S. (2000). Task conflict and relationship conflict in top management teams: The pivotal role of intragroup trust. *Journal of Applied Psychology, 85*(1), 102–111.

Sokolova, M., & Szpakowicz, S. (2007). Strategies and language trends in learning success and failure of negotiation. *Group Decision and Negotiation, 16*(5), 469–484.

Staw, B. M., Sandelands, L. E., & Dutton, J. E. (1981). Threat rigidity effects in organizational behaviour: A multilevel analysis. *Administrative Science Quarterly, 26*(4), 501–524.

Tetlock, P. E. (1999). Coping with trade-offs: Psychological constraints and political implications. In S. Lupia, M. McCubbins, & S. Popkin (Eds.), *Political reasoning and choice*. Berkeley: University of California Press.

Thomas, K. W., & Kilmann, R. H. (1974). *Thomas-Kilmann conflict mode instrument*. Tuxedo: XICOM Inc.

Thompson, L. L. (Ed.). (2005). *The mind and heart of the negotiator* (3rd ed.). Upper Saddle River: Prentice Hall.

Thompson, L., & DeHarpport, T. (1994). Social judgement, feedback, and interpersonal learning in negotiation. *Organizational Behaviour and Human Decision Processes, 58*(3), 327–345.

Tjosvold, D., Morishima, M., & Belsheim, J. (1999). Complaint handling on the shop floor: Cooperative relationships and open-minded strategies. *International Journal of Conflict Management, 10*(1), 45–68.

Ullman, J., & Bentler, P. M. (2004). Structural equation modelling. In H. Melissa, & A. Bryman (Eds.), *Handbook of data analysis*, London: Sage.

Van der Maas, H. L. J., Kolstein, R., & van der Pligt, J. (2003). Sudden transitions in attitudes. *Sociological Methods & Research, 32*(2), 125–152.

Van Kleef, G. A., De Dreu, C. K. W., Pietroni, D., & Manstead, A. S. R. (2006). Power and emotion in negotiation: Power moderates the interpersonal effects of anger and happiness on concession making. *European Journal of Social Psychology, 36*, 557–581.

Wallihan, J. (1998). Negotiating to avoid agreement. *Negotiation Journal, 14*(3), 257–268.

Wathne, K. H., & Heide, J. B. (2000). Opportunism in interfirm relationships: Forms, outcomes, and solutions. *Journal of Marketing, 64*(4), 36–51.

Watkins, M. (1998). Building momentum in negotiations: Time-related costs and action-forcing events. *Negotiation Journal, 14*(3), 241–256.

Whyte, G., & Sebenius, J. K. (1997). The effect of multiple anchors on anchoring in individual and group judgment. *Organizational Behaviour and Human Decision Processes, 69*(1), 75–85.

Williamson, O. (1975). *Markets and hierarchies: Analysis and antitrust implications*. New York: The Free Press.

Williamson, O. (1985). *The economic institution of capitalism: Firms, markets, relational contracting*. New York: Free Press.

Wong, P. S. P., Cheung, S. O., & Fan, K. L. (2009). Examining the relationship between organizational learning styles and project performance. *Journal of Construction Engineering and Management, 135*(6), 497–507.

Yiu, T. W., & Cheung, S. O. (2005). A study of construction mediator tactics: Part II The contingent use of tactics. *Building and Environment, 42*(2), 762–769.
Yiu, T. W., Cheung, S. O., & Chow, P. T. (2008). Logistic regression modeling of construction negotiation outcomes. *IEEE Transactions on Engineering Management, 55*(3), 468–471.
Zeeman, E. C. (1977). *Catastrophe theory: Selected paper, 1972-1977*. Boston: Addison-Wesley Publishing Company.

Chapter 14
Mediating and Moderating Effect of Tension on Withdrawal: Commitment Relationship in Construction Dispute Negotiation

Pui Ting Chow and Sai On Cheung

Abstract The success or failure of a negotiation depends on the commitment of the negotiators for a settlement. Withdrawal refers to a situation in which a construction negotiator loses the interest to continue with a negotiation. A withdrawing negotiator is likely to abandon a negotiation. It is proposed that the higher the commitment of the negotiators, the less likely they will lose interest and hence a greater chance to achieve negotiated settlement. Furthermore, feeling tensed or relaxed is having a bearing on the cognitive reasoning of a negotiator. A certain level of tension helps negotiators stay focused on the disputing issues and engenders commitment. Tension is affecting commitment thus withdrawal. The roles and impact of tension on the withdrawal-commitment relationship are investigated in this study. It is hypothesised that: (1) tension mediates the withdrawal-commitment relationship, and (2) tension moderates the withdrawal-commitment relationship. With data collected from experienced construction dispute negotiators, regression analyses were conducted to test the hypotheses. Tensioned negotiators are generally more committed to a negotiated settlement than their low-tensioned counterparts. However, if the withdrawing tendency reaches its threshold value, the loss of commitment of high-tensioned negotiators is much quicker than their low-tensioned counterparts. This reminds managers that optimal level of tension can mobilise human resources to the betterment of a negotiated settlement, but excessive level tension can raise the state of withdrawal of the negotiators and lower commitment. In this regard, management may adjust the tension level by varying the settlement targets as well as changing the memberships of the negotiation team.

P. T. Chow (✉) · S. O. Cheung
Construction Dispute Resolution Research Unit, Department of Civil and Architectural Engineering, City University of Hong Kong, Hong Kong, People's Republic of China
e-mail: ronnie.chow.pt@hotmail.com

S. O. Cheung
e-mail: saion.cheung@cityu.edu.hk

S. O. Cheung (ed.), *Construction Dispute Research*,
DOI: 10.1007/978-3-319-04429-3_14,

14.1 Introduction

For construction organisations, one way to enhance their competitive edge is to reduce non-productive cost such as those used in handling dispute (Cheung et al. 2000). It is well accepted that negotiation is the most cost-effective means to resolve dispute. However, not every negotiation ends with settlement. Sometimes, a dispute is just non-negotiable due to irreconcilable divergence in interest of the disputing parties. In other instances, a negotiation fails simply because one or more of the negotiators have lost interest to continue; a situation identified as withdrawal (Chow and Cheung 2008). Losing interest to continue (hereafter as withdrawal) has a negative effect on the form of commitment (hereafter as commitment) that is manifested by a negotiator's continuing effort to invest in a relationship as well as an acceptance of joint goals and values. Thus, a negotiated settlement is only possible if the negotiators are committed to achieving it. The conceptual link between withdrawal and commitment has been studied in meta-analytic and causation studies (Mobley et al. 1979; Steel and Ovalle 1984). Commitment has been examined as both an antecedent and a consequence of withdrawal (Black 2008; Cullen et al. 2003; Mowday et al. 1984). Mathieu and Zajac (1990) suggested that a low withdrawing negotiator is more likely to engage in "extra-role" behaviours resulting in inspirational commitment. A negotiator with high level of commitment, in turn, has a better chance to derive common goals with the counterpart and is less likely to withdraw (O'Reilly and Chatman 1986). On the contrary, low commitment indicates that a negotiator perceives the value of maintaining the relationship with the counterpart to be low and thus displays high level of withdrawing behaviour (Mathieu and Zajac 1990). Moreover, three observations suggest that there may be intervening variable in the withdrawal-commitment relationship. First, a high withdrawing negotiator without pressure discourages his counterpart from resolving the dispute by delaying his response through inaction and isolation (Deutsch et al. 2006; Simonson and Staw 1992). Second, a high withdrawing negotiator under great tension is likely to make unwarranted pre-mature make-or-break decision. He is too keen to get out of the negotiation. His strategies are therefore either denial or aggression (Deutsch et al. 2006; Lee et al. 2006; Watson et al. 1992). Third, a low withdrawing negotiator who is able to suppress the effect of tension is in a better position to initiate collaborative responses. Notable manifestations include compassionate, empathetic and committal negotiating behaviour (Deutsch et al. 2006).

Negotiating construction dispute is stressful and the tension arising there-from influences the way a negotiator values and prioritises options (Fryer 2004; Schwarz and Clore 2007). Certain level of tension arising from the need to obtain desired result may urge a negotiator to stay focused on getting the dispute settled (Nordqvist et al. 2004). However, a stressful negotiator may view a looming negotiation as threat. He may hastily reach a suboptimal deal and, in extreme situation, even walk off without any conscientious attempt for a settlement (O'Connor and Arnold 2001). In this regard, it is proposed that the level of withdrawal (independent control variable) under the influence of tension (independent mediating and moderating variable) predicts the level of commitment (dependent variable). Research has pointed explicitly to the important roles of tension in a withdrawal-commitment relationship (Morgan and Hunt 1994; Sommer et al. 1996). A better understanding of its roles shall reduce withdrawal and thereby maintain the chance of having a negotiated settlement. The contribution of this study is the holistic treatment of tension, withdrawal, and commitment in construction dispute negotiation. It is hypothesised that tension both mediates and moderates the withdrawal-commitment relationship. The mediating and moderating effects of tension in the withdrawal-commitment relationship are first discussed. Then, tension, withdrawal, commitment and their attributes in construction dispute negotiation are elaborated seriatim.

14.2 Tension as a Mediator

Negotiators are supposedly committed to achieve optimal outcomes. However, the drives for economic return is countered by the call for psychological relax. Negotiators may fail to unfreeze energy and get motivated to step up their efforts if the tension level is too low (Deutsch et al. 2006). Moreover, too much tension would lead to psychological collapse because of a negotiator's inability to cope with the stress (O'Connor and Arnold 2001; Yiu and Cheung 2007). Tension as a mediator in the withdrawal-commitment relationship represents the generative mechanism through which withdrawal is able to influence commitment (Baron and Kenny 1986). In general, high-tensioned negotiators have a more negative attitude toward commitment than their low-tensioned counterparts (Deutsch et al. 2006; Lee et al. 2006). It is assumed that the relationships among withdrawal, tension, and commitment vary across project-specific parameters (e.g. contract procurement method, project type, etc.). In this connection, the withdrawal-tension (withdrawal → tension) and tension-commitment (tension → commitment) relationships are further discussed.

14.2.1 Withdrawal-Tension Relationship (Withdrawal → Tension)

Withdrawal is manifested by a reduction in a negotiator's attention to or interest in a negotiation. Blau (1985) first defined three categories of withdrawal; unavoidable, stable periodic and increasing chronic. Roznowski and Hanisch (1990) classified withdrawal as either excusable or inexcusable. The decision to terminate a negotiation could be described as a sequence of cognitive stages whereby the potential withdrawer feels dissatisfied with his prevailing status in the negotiation (Mobley et al. 1979). Each successive step represents an increased and cumulative propensity to withdraw. A withdrawing negotiator first decreases performance, displays a bad attitude, refuses working to potential and broadcasts limitations in achieving a settlement. Then, he usually expresses some forms of escape like displaying unfavorable negotiation behaviour, arriving late, leaving early or complete absence from project meetings with the aim of minimising the time to be spent on the negotiating task (Hanisch and Hulin 1991). It is followed by an "*intention to search for alternatives*" and an "*active search and evaluation of alternatives*" suggesting his intention to remove himself from both the situation and his assumed role (Hanisch and Hulin 1991; Mobley et al. 1979). Finally, breakdown is resulted. Several studies have found that withdrawal is positively associated with tension (Bhanugopan and Fish 2006). According to Sheridan and Abelson (1983), a negotiator's progression through different stages of withdrawal is mediated by his perception of the anticipated severity of tension. Tension exacerbates the 'avoiding' effect of withdrawal and their resonance causes a negotiator to shirk his duty or even to walk away from the negotiation table in order to stay away from the stressful environment. Thus, increasing withdrawal and progressive dysfunctional negotiators' behaviours are exemplified by tension (Chow and Cheung 2008).

14.2.2 Tension-Commitment Relationship (Tension → Commitment)

Optimum level of tension of a negotiation would mobilise psychological resources to the achievement of a negotiated settlement (Deutsch et al. 2006). However, excessive tension lowers a negotiator's commitment (Jaros et al. 1993). Anderson and Weitz (1992) observed that asymmetries in commitment probably are the consequences of previous tense negotiation experiences (Mowday et al. 1984; Simonson and Staw 1992). Tension evokes the illusion that negotiators can act irrationally and uneconomically (Lempereur and Colson 2010). A tense negotiator is more willing to abandon a relationship and less willing to reciprocate compromises made by his committed counterpart. Furthermore, the unilateral commitment to a negotiated settlement made by the counterpart invites the practice of opportunism in a stressful environment (Delerue-Vidot 2006; Gundlach et al. 1995). Tangible progress to a resolution is possible only if both negotiators are committed to a proposal (MacFarlane et al. 2003). Mutually committed and recognised relationships serve to reinforce exchange and prevent opportunistic exploitation (Cook and Wall 1980). Mathieu and Zajac (1990) found that reduced commitment is one of the major outcomes of reduced attachment to a social group, role conflict and role ambiguity which are the attributes of tension. Commitment is the driver behind any means to achieve a negotiated settlement (Ring and Van de Ven 1994). A committed negotiator can rationally effectuate negotiated outcomes

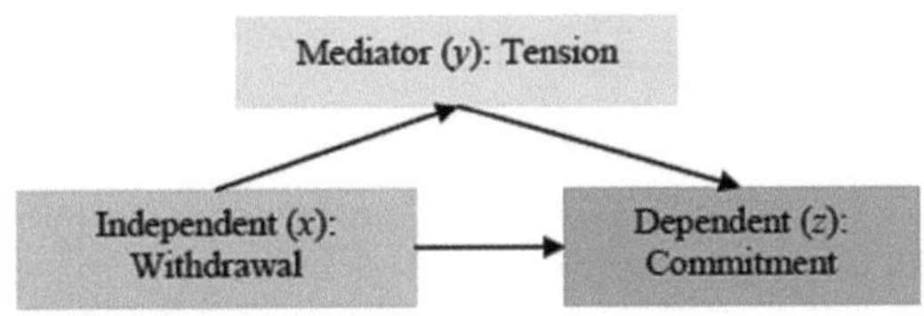

(1) withdrawal (independent variable *x*) affects tension (mediator *y*);

(2) withdrawal (independent variable *x*) affects commitment (dependent variable *z*);

(3) tension (mediator *y*) affects commitment (dependent variable *z*) in controlling withdrawal (independent variable *x*); and

(4)the effect of withdrawal (independent variable *x*) on commitment (dependent variable *z*) in the second condition is smaller than that of the third one.

H_1: Tension mediates the withdrawal-commitment relationship.

Fig. 14.1 The conceptual diagram of mediating effect

to match the precedent-based settlement which emphasises its consistency and certainty (MacFarlane et al. 2003)

For a mediation relationship, a complete mediation model has the form $x \rightarrow y \rightarrow z$, where x is the antecedent (i.e. withdrawal); y is the mediator (i.e. tension); and z is the consequence (i.e. commitment) (James and Brett 1984). Tension is a mediator in the withdrawal-commitment relationship, if the following conditions are met (Fig. 14.1).

14.3 Tension as a Moderator

Tension as a moderator in the withdrawal-commitment relationship partitions withdrawal into subgroups that establish its domains of maximal effectiveness in regard to commitment (Baron and Kenny 1986). Review of negotiation literatures offers mixed predictions on the moderating effect of tension on the withdrawal-commitment relationship. A high-tensioned negotiator is unlikely to reach an agreement particularly when he experiences high level of withdrawal. High withdrawal reflects a sense of anger and frustration leading to a negative expectation on the negotiation outcome. Even if a negotiated settlement is ultimately reached, the desire for revenge, non-compliance or creation of future dispute lingers (Saraydar 1971). As such, a high withdrawing negotiator is less committed in a tensed negotiation. It is proposed that tension positively affects withdrawal especially when it is high. In other words, the withdrawal-commitment relationship will be

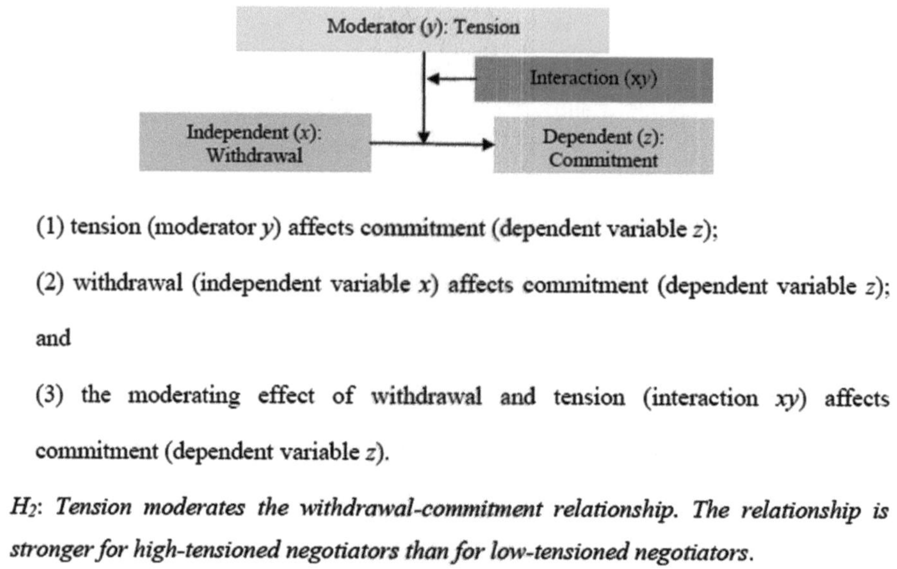

Fig. 14.2 The conceptual diagram of moderating effect

stronger for high-tensioned rather than low-tensioned negotiators. It is expected that high-tensioned negotiators are having higher negative expectation from the negotiation than their low-tensioned counterparts. If one's level of withdrawal increases from low to high, it will have a stronger impact on the level of commitment for a high-tensioned negotiator than a low-tensioned one. Accordingly, the slope of regression line of the withdrawal-commitment relationship will be steeper for high-tensioned negotiators than the low-tensioned. Therefore, a significant interaction effect between withdrawal and tension on commitment is predicted. The relation between withdrawal and commitment is thus moderated by tension (Sheridan and Abelson 1983). The dependent variable z (i.e. commitment) is a probabilistic function of x (i.e. withdrawal), y (i.e. tension) and their interaction xy (i.e. withdrawal*tension). Tension is a moderator in the withdrawal-commitment relationship if the following conditions are met (Fig. 14.2).

14.4 Variables and Measures

14.4.1 Tension

The measurement scale of tension has been developed in the light of role theory. The attributes of tension used in the study are role overload, role conflict and role ambiguity as shown in Table 14.1 (Caplan and Jones 1975; Kahn et al. 1964; Singh 1998; Weatherly and Tansik 1992). Role overload is defined as a lack of

Table 14.1 List of observable variables of tension (Beehr et al. 2000; Caplan and Jones 1975; Kahn et al. 1964, pp. 21–23; Singh 1998; Weatherly and Tansik 1992)

Code	List of observable variables
TN_RO_01	I had insufficient manpower and materials to handle the negotiation task
TN_RO_02	I was not given enough time to do what was expected of me in negotiating the dispute
TN_RO_03	I had too many negotiation tasks to handle
TN_RO_04	I could not work efficiently because I was interfered by the complexity of the negotiation task
TN_RO_05	I often experienced a marked increase of work load during the dispute negotiation process
TN_RC_01	I had to make decisions which could not satisfy my counterpart
TN_RC_02	I had to closely work with my counterpart who operated quite differently
TN_RC_03	I received incompatible requests from my counterpart
TN_RC_04	I had to make decisions from my counterpart's point of view
TN_RC_05	I needed to make decision on unnecessary thing in the negotiation
TN_RA_01	I felt uncertain about how much authority I had in negotiating the dispute
TN_RA_02	I had unclear goals and objectives for the negotiation task
TN_RA_03	I would not divide the negotiation task properly
TN_RA_04	I did not know exactly what was expected of me in the negotiation
TN_RA_05	I was unclear about the purposes of the negotiation

Note: Anchor and reverse scored
TN: Tension; *RO*: Role overload; *RC*: Role conflict; *RA*: Role ambiguity

adequate resources required to materialise the role expectations or demands (Brumels and Beach 2008; Singh et al. 1996). It occurs if a negotiator is demanded to accomplish a task with insufficient resources or limited capabilities. According to Jones et al. (1995), role overload in construction dispute negotiation can be quantitative and qualitative. Quantitative overload happens when a negotiator is asked to complete a negotiation task, however the resources given may be limited, or there are too many associated works. Qualitative overload describes a situation where a negotiator may not have adequate skill and/or ability to complete a negotiation task, even if more resources are made available. Role conflict occurs when a negotiator faces incompatible demands arising from concurrent occurrence of two or more events (Onyemah 2008). It is described as a feeling of being torn in multiple directions. A negotiator with role conflict is unable to find ways to reconcile the demand from the respective roles. He gets frustrated when his roles are conflicting and as a result he is unable to complete the task in a satisfactory manner (Brumels and Beach 2008). Role ambiguity is defined as a feeling that a negotiator finds oneself absence or lack of adequate information available to fulfill his role satisfactorily (Kahn et al. 1964; Singh et al. 1996). It occurs when expectations for a particular negotiation position are vague, unclear, or ill defined. These contradictory roles and responsibilities are often associated with poorly defined requirements, haphazard performance, and inconsistent evaluations (Hardy and Conway 1988). In such circumstances, a negotiator is often uncertain about the settlement options (Onyemah 2008).

Table 14.2 List of observable variables of withdrawal (Beck and Martin 1995; Blau 1985; Chow and Cheung 2008; Hanisch and Hulin 1991; Mobley et al. 1979)

Code	List of observable variables
WB_EE_01	I was emotionally drained as a result of the negotiation
WB_EE_02	My energy was used up at the end of the negotiation
WB_EE_03	I felt fatigued when I got up in the morning and had to face the negotiation another day
WB_EE_04	I was burned out as a result of the negotiation
WB_EE_05	I felt frustrated by the negotiation
WB_PA_01	I dealt very effectively with the negotiation[a]
WB_PA_02	I felt I had positively influenced my counterpart through the negotiation[a]
WB_PA_03	I could easily create a relaxed atmosphere with my counterpart[a]
WB_PA_04	I felt exhilarated after working closely with my counterpart[a]
WB_PA_05	I had accomplished many worthwhile things in the negotiation[a]
WB_DP_01	I felt I treat my counterpart as an impersonal 'object'
WB_DP_02	I had become more callous toward my counterpart since I participated in the negotiation
WB_DP_03	I worried that the negotiation had hardened me emotionally
WB_DP_04	I didn't really care what happened to my counterpart
WB_DP_05	I felt my counterpart had blamed me for some of his/her own problems

[a] Anchor and reverse scored
WB: Withdrawal; *EE*: Emotional exhaustion; *PA*: Reduced personal accomplishment; *DP*: Depersonalisation

14.4.2 Withdrawal

In this study, withdrawal has been operationalised as emotional exhaustion, reduced personal accomplishment, and depersonalisation by aggregating responses to items related to "*thinking of withdraw*"; "*desirability of withdraw*" and "*likelihood of withdraw*" (Hanisch and Hulin 1991) (Table 14.2). Emotional exhaustion describes the adverse reactions of a negotiator to the tedium has in a negotiation (Maslach et al. 2001). It is suggested that as emotional resources are depleted or drained, a negotiator is no longer able to control his temper. Densten (2001) further characterised emotional exhaustion as a phenomenon of lack of energy. Reduced personal accomplishment refers to the tendency of a negotiator evaluating oneself negatively, particularly when compared with the counterpart. It leads to low level of motivation and self-esteem at the negotiation table (Densten 2001). Thus, a negotiator easily feels unhappy and dissatisfied with his performance in the negotiation (Maslach et al. 2001). This decline in one's feeling of competence reinforces pessimistic attitude and runs against commitment. *Depersonalisation* refers to the unfeeling and callous responses by a negotiator toward his counterpart. Densten (2001) elaborated depersonalisation as detachment, emotional callousness and development of negative and cynical attitude toward the counterpart. Such negative reaction is partly linked to the experience of emotion exhaustion. Emotion exhaustion and depersonalisation are thus somewhat related (Maslach et al. 2001).

Table 14.3 List of observable variables of commitment (Fiss 1983; MacFarlane et al. 2003; Meyer and Allen 1991; Mowday et al. 1984; Ring and Van de Ven 1994; Simonson and Staw 1992)

Code	List of observable variables
CM_AC_01	I was pleased to spend my leisure time to deal with the negotiation
CM_AC_02	The project team is like a family and I felt like part of it
CM_AC_03	I did feel I was emotionally attached to my counterpart
CM_AC_04	This project had a great deal of personal meaning to me
CM_AC_05	I felt a strong sense of belonging to my project team
CM_CC_01	Too much of my career life would be disrupted if I decided to call for breakdown from the negotiation
CM_CC_02	It was too costly for me to call for breakdown from the negotiation
CM_CC_03	Staying with the project team was a matter of necessity as much as desire
CM_CC_04	I felt that I would have few other contracting partners left to choose if I called for breakdown from the negotiation
CM_CC_05	There were only limited choices of alternative package of resolution, if I called for breakdown from the negotiation
CM_NC_01	I thought that project team these days often changes their counterparts[a]
CM_NC_02	I believed that members of a project team must always show their contribution to the resolution of disputes
CM_NC_03	Calling breakdown from the negotiation was an unethical symbol to me
CM_NC_04	I did not feel it would be right to leave my original position even if I got another offer for a better role elsewhere
CM_NC_05	I believed in the value of remaining loyal to my project team in resolving the dispute

[a] Anchor and reverse scored
CM: Commitment; *AC*: Affective commitment; *CC*: Continual commitment; *NC*: Normative commitment

14.4.3 Commitment

Behavioural approach has focused on identifying conditions under which commitment are manifested (O'Reilly and Chatman 1986). Despite the different forms of conceptualisation, attributes of commitment is expressed in three general themes; (i) affective attachment to the project team, (ii) perceived costs associated with leaving the negotiation table, and (iii) obligation to remain with the project team where each negotiator has a commitment profile reflecting the degree of desire, need, and obligation to a negotiated settlement (Meyer and Allen 1991). In this study, commitment is categorised as affective, continual and normative (Table 14.3). Affective commitment refers to a negotiator's emotional attachment to, identification with, and involvement in a project team. A negotiator with strong affective commitment contributes to a negotiation because he wants to do so. Also, affective commitment reflects a desire to maintain a membership in a project team, that develops largely as a result of project experiences and more importantly the feeling of comfort and personal competence created (Meyer and Allen 1991). Continual commitment refers to an awareness of the costs associated with leaving the project team. A negotiator who is primarily linked to a project team due to continual commitment stays on with the negotiation because he needs to do so (Meyer and Allen 1991). Continual commitment reflects the degree to which a negotiator experiences a sense of being locked in a place because of the high costs of leaving and termination (Beck and Martin 1995; Jaros et al. 1993). Common antecedents of continual commitment are increasing side bets or investments, and initiating alternative proposals (Meyer and Allen 1991). Normative commitment reflects a feeling of obligation to continue. A negotiator with a high level of normative commitment believes that he ought to remain with the negotiation (Meyer and Allen 1991). Wiener (1982) defined normative commitment as the "*totality of internalised normative pressures to act in a way which meets organisational goals and interests*", and suggests that a negotiator exhibits these behaviours solely because he '*believe(s) it is the "right" and moral thing to do*'. Normative commitment differs from continual commitment because it does not necessarily vary with individual calculation of inducement or sunk cost (Jaros et al. 1993; Mathieu and Zajac 1990; Mowday et al. 1984). Wiener (1982) further proposed that it is the culture of a project team that frames normative commitment.

14.5 Methodology

14.5.1 Participants

The prospective participants of the study were construction practitioners who (1) were practicing and (2) had construction dispute negotiation experiences. The sample was developed based on the company directories and the member directories of construction professional institutes such as the Hong Kong Institute of Architects (HKIA) and the Hong Kong Institute of Surveyors (HKIS) (Far East Trade 2003). Participants were contacted either by phone, fax or email. After the prospective participants accept the invitation, the researchers then send them the questionnaire by fax or email.

14.5.2 Measures

In this study, the self-reported control variables include participants' gender (dummy coding: female = 0, male = 1), project organisation (dummy coding: client = 0, contractor = 1), working experiences (in years) and project sum (in HK$), etc. Other demographic data like professions and type of membership in the professional institutions were collected but were not further analysed in the study. Respondents were asked to provide their degree of agreement of the statements listed in Tables 14.1, 14.2 and 14.3. 7-point Likert-scale was used for the

Table 14.4 Models of mediation-and-moderation regression analyses

Regression equations	Model
$y = \beta_0 + \beta_a a + \beta_b b + \beta_x x + \varepsilon$	(1)
$z = \beta_0 + \beta_a a + \beta_b b + \beta_y y + \varepsilon$	(2)
$z = \beta_0 + \beta_a a + \beta_b b + \beta_x x + \varepsilon$	(3)
$z = \beta_0 + \beta_a a + \beta_b b + \beta_x x + \beta_y y + \varepsilon$	(4)
$z = \beta_0 + \beta_a a + \beta_b b + \beta_x x + \beta_y y + \beta_{xy} xy + \varepsilon$	(5)

β_i: coefficient of i; ε: error term; a: demographics (i.e. gender); b: negotiation-related variables (i.e. project organisation, working experiences, project sum and duration); x: independent variable; y: mediator in Eqs. 1, 3 and 4 or moderator in Eqs. 2, 3 and 5; z: dependent variable; xy: interaction; a and b are control variables

measurements of tension, withdrawal and commitment where "1" indicates strongly disagree and "7" indicates strongly agree (Cooper and Schindler 2000). The scores of tension, withdrawal and commitment were calculated with anchored items reverse scored. All the data analyses were performed by SPSS version 17.0.

14.5.3 Reliability

Cronbach's alpha, average inter-item correlation and item total correlation of reliability measurement are employed in the study. They are used to assess the degree of internal consistency of items (Hair et al. 1998). Cronbach's alphas of all factors are greater than 0.7 (varies from the lowest: 0.7 of TN_RC to the highest: 0.9 of WB_EE) suggesting that the factors are internally consistent as well as the data set is reliable for further statistical analyses.

14.5.4 Mediation-and-Moderation Regression

Mediation-and-moderation regression analyses are used to determine the extent to which tension as a mediator and a moderator plays in the withdrawal-commitment relationship (James and Brett 1984; O'Connor et al. 2005). In this regard, 5 regression equations are developed (Table 14.4). Demographic data and project particulars were added as control variables in steps 1 and 2 respectively in each of the regression. To meet the four conditions of mediation effect, (1) the F values of models 1, 3 and 4 should be significant; (2) β_x and β_y of models 1, 3 and 4 should be significant; and (3) the values of β_x of model 3 should be much greater than that of model 4. To meet the three conditions of moderation effect, (1) the F values of models 2, 3 and 5 should be significant; and (2) β_x, β_y and β_{xy} of models 2, 3 and 5 should be significant.

14.6 Results and Findings

The data collection questionnaire was either emailed (157) or faxed (356) to the sample. Hundred thirty usable responses were obtained (25.3 % return rate). The sample has an average experience of 12.19 years (SD = 10.58 years). By profession, the respondents are made up of architects (9 %), surveyors (60 %), project managers (5 %), engineers (25 %) or lawyer (1 %). Seventy nine percentage of them worked for client (e.g. private developer, government bodies, consultant, etc.) and the remaining of them worked for contractor (e.g. main contractor, sub-contractor, etc.). They were involved in building (54 %), civil (23 %), building services (11 %), or maintenance (12 %) projects. In each of the multiple regression models, demographic data (i.e. gender) and negotiation-related variables (e.g. project sum and working experiences) were first added in steps 1 and 2 respectively. Gender and negotiation-related variables were not all significantly related to commitment and were not further discussed. Table 14.5 gives the general statistics of the three dimensions.

14.6.1 Tension as a Mediator

To investigate tension as a mediator, the procedures mentioned earlier in the moderation-and-mediation regression analyses were employed (i.e. Models 1, 3 and 4 in Table 14.6). First, withdrawal affects tension (mediator) ($F = 3.87$, $p = 0.001$). Second, withdrawal affects commitment ($F = 2.46$, $p = 0.028$). Third, tension does affect commitment in the presence of withdrawal ($\beta_y = -0.22$, $p = 0.017$; $F = 3.04$, $p = 0.006$). Finally, when the first three conditions are confirmed, then, the effect of the independent variable (withdrawal) on the dependent variable (commitment) in the third model ($\beta_x = -0.19$, $p = 0.005$) is greater than in the fourth one ($\beta_x = -0.12$, $p = 0.087$). The results met the requirements. Tension is thus a mediator in the withdrawal-commitment relationship and Hypothesis 1 is supported.

14.6.2 Tension as a Moderator

Withdrawal, tension and their interaction were entered in steps 3, 4 and 5 respectively in Model 5. The third and fourth variables produced two main effects on commitment (Table 14.6). Supporting the notion that withdrawal is related to commitment, withdrawal explained 5.8 % of variance in commitment (ΔF (1, 123) = 7.99, $p = 0.005$) in Step 3 of Model 5. Withdrawal produces a negative effect on commitment. Step 4 of Model 5 revealed that tension also explained 4.1 % of variance in commitment (ΔF (1, 122) = 5.91, $p = 0.017$). Tension negatively

Table 14.5 Descriptive, correlation and Crobach's alpha of the study variables

I	II	μ	SD	EE	PA	DP	AC	CC	NC	RO	RC	RA	WB	CM	TN
a	Gender	0.64	0.48	0.02	−0.01	0.05	−0.06	−0.07	−0.25**	0.08	0.06	−0.07	0.03	−0.16	0.02
b_1	Comp.	0.21	0.41	0.01	0.13	−0.13	−0.01	−0.10	0.04	0.10	0.04	0.05	0.00	−0.04	0.08
b_2	Exp.	12.19	10.58	−0.18*	−0.29**	−0.14	0.09	0.21*	−0.07	0.00	−0.05	−0.41**	−0.25**	0.11	−0.23*
b_3	Sum	281	196	−0.05	−0.07	0.00	−0.01	0.03	0.09	0.04	−0.02	−0.09	−0.05	0.05	−0.04
b_4	Dur.	2.72	1.16	0.14	0.07	0.02	−0.01	0.05	0.04	−0.02	−0.07	0.13	0.10	0.04	0.03
	EE	3.97	1.32	**0.92**	0.71**	0.29**	−0.06	−0.31**	−0.02	0.42**	0.42**	0.49**	0.87**	−0.18*	0.58**
	PA	3.45	1.05	−6.33**	**0.79**	0.31**	−0.12	−0.51**	0.07	0.44**	0.37**	0.56**	0.85**	−0.27**	0.61**
	DP	2.99	1.07	−7.77**	−4.19	**0.78**	−0.30**	−0.30**	−0.17	0.06	0.16	0.24**	0.65**	−0.35**	0.21*
	AC	4.37	1.06				**0.85**	0.36**	0.41**	0.04	−0.13	−0.03	−0.19*	0.79**	−0.04
	CC	4.69	1.13				2.95*	**0.85**	0.24**	−0.31**	−0.34**	−0.27**	−0.46**	0.75**	−0.39**
	NC	3.74	0.94				−6.51**	−8.41**	**0.77**	0.11	−0.19*	0.25**	−0.05	0.70**	0.10
	RO	4.13	0.91							**0.77**	0.50**	0.44**	0.39**	−0.09	0.83**
	RC	4.42	0.77							4.00**	**0.70**	0.24**	0.41**	−0.30**	0.69**
	RA	3.44	1.12							−7.15**	−9.35**	**0.85**	0.55**	−0.04	0.79**
y	WB	3.47	0.91										**0.88**	−0.33**	0.59**
x	CM	4.26	0.78											**0.85**	−0.16
z	TN	4.00	0.72												**0.83**

Bold diagonal: Crobach's alpha; *Upper-diagonal*: Correlation (df 130); *Lower-diagonal*: Pairwise sample *t*-statistic; significant (2-tailed) at the level of ** 0.05, * 0.01 level;

gender: female = 0, male = 1; *Comp*: Client = 0, Contractor = 1; *Exp.*: total construction dispute negotiation experience in year; *Sum*: project sum in million; *Dur*: project duration in year; *TN*: Tension; *WB*: Withdrawal; *CM*: Commitment; *I*: Variable notation corresponding to the regression analysis; *II*: Variable code; *RO*: Role overload; *RC*: Role conflict; *RA*: Role ambiguity; *EE*: Emotional exhaustion; *PA*: Reduced personal accomplishment; *DP*: Depersonalisation; *AC*: Affective commitment; *CC*: Continual commitment; *NC*: Normative commitment

Table 14.6 Statistic results of mediation-and-moderation regression analyses

Step	β_i	Model									
		1		2		3		4		5	
1	*a*	0.07	(0.14)	−0.28**	(0.14)	−0.29	(0.14)	−0.27	(0.14)	−0.26	(0.14)
2	*b*1	−0.03	(0.16)	−0.03	(0.16)	0.01	(0.17)	0.01	(0.16)	−0.04	(0.16)
	*b*2	0.00	(0.01)	0.01	(0.01)	0.01	(0.08)	0.01	(0.01)	0.00	(0.01)
	*b*3	0.00	(0.00)	0.00	(0.00)	0.00	(0.00)	0.00	(0.00)	0.00	(0.00)
	*b*4	−0.09	(0.07)	−0.02	(0.07)	0.03	(0.07)	0.01	(0.07)	0.05	(0.07)
3	*x*	0.30***	(0.07)			−0.19***	(0.07)	−0.12*	(0.07)	−1.04***	(0.30)
4	*y*			−0.29**	(0.09)			−0.22**	(0.09)	−0.87***	(0.23)
5	*xy*									0.20***	(0.07)
F		3.87***		3.00***		2.46**		3.04***		4.05***	
R^2		0.16		0.13		0.11		0.15		0.21	
Adj. R^2		0.12		0.09		0.06		0.10		0.16	
ΔR^2								0.04**		0.06***	
dfs		6	123	6	123	6	123	7	122	8	121

β_i: Variable, *a* and *b* control variables where *a* gender, *b*: project organisation; working experiences; project sum and duration; *x*: independent variable-withdrawal; *y*: mediator and moderator-tension; *xy*: interaction

* $p < 0.1$,

** $p < 0.05$

*** $p < 0.01$

β_i: unstandardised regression coefficients with the standard errors in parentheses

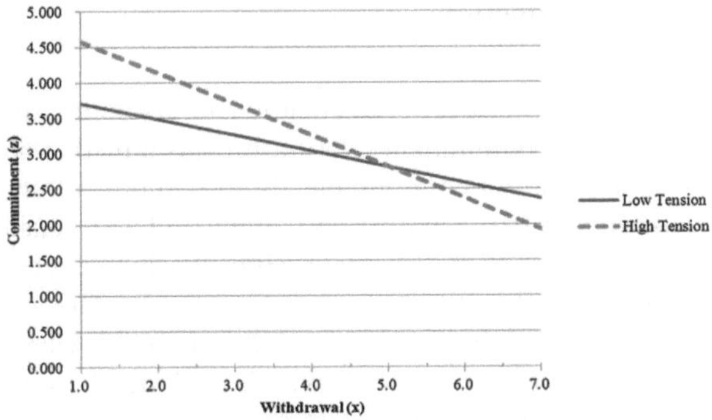

Fig. 14.3 The interaction effect of withdrawal (*x*) and tension (*y*) on commitment (*z*)

influences commitment, in the presence of withdrawal. The product of the two (withdrawal*tension) (*xy*) was significant ($\Delta R^2 = 0.063$, ΔF (1, 121) = 9.63, $p = 0.002$). The significant interaction effect on commitment was further investigated by examining the differences between high-tensioned and low-tensioned negotiators. The demarcation between the high- and low-tension negotiators is the median of the tension scores. Separate regression lines for the high- and low-tensioned groups are shown in Fig. 14.3. Low-tensioned negotiators had a negative but non-significant relationship in the withdrawal-commitment relationship ($\beta_L = -0.224, t = 0.207, p = 0.233$). However, for the high-tensioned group, the regression line had a significant and negative relationship ($\beta_H = -0.441$, $t = 4.035$, $p = 0.000$).

The regression lines for high- (Eq. 14.1) and low-tensioned participants (Eq. 14.2) can be expressed as follow:

$$Z_{\mathrm{H}} = 5.022 - 0.441\, x_{\mathrm{H}}, \tag{14.1}$$

$$Z_{\mathrm{L}} = 3.933 - 0.224\, x_{\mathrm{L}}, \tag{14.2}$$

x and *z* are identified by solving the above two equations. The high- and low-tensioned regression lines intersect at the point with withdrawal score of 5.018 and commitment score of 2.810 (7-point Likert scale: 1—Low; 7— High). High-tensioned participants had higher commitment than low-tensioned participants. When the negotiators had the withdrawal score higher than 5.018, high-tensioned negotiators had lower commitment than the low-tensioned. Moreover, the average commitment for high-tensioned negotiators ($\mu_H = 4.500$, $SD = 0.486$) was higher than that of low-tensioned negotiators ($\mu_L = 4.023$, $SD = 0.569$) (F (1, 128) = 22.518, $p = 0.000$), while most high-tensioned negotiators have higher withdrawing behaviour ($\mu_H = 4.157$, $SD = 0.450$; $\mu_L = 2.801$, $SD = 0.650$; F (1, 128) = 4.315, $p = 0.040$). There were five negotiators having withdrawal score higher than 5.018, three of them were high-tensioned negotiators and two of them were low-tensioned. Most of the high-tensioned negotiators have higher commitment than their low-tensioned counterparts except those with withdrawal score higher than 5.018. Thus, tension moderates the withdrawal-commitment relationship and Hypothesis 2 is supported.

14.7 Discussion

This present study proposes that tension is both a mediator and a moderator in the withdrawal-commitment relationship. As a mediator, tension exemplifies the effect of withdrawal on commitment. Both withdrawal and tension negatively influence commitment. As a moderator, high- and low-tensioned negotiators display different patterns of commitment. High-tensioned negotiators, generally, have higher commitment than low-tensioned negotiators. However, this pattern reverses when the state of withdrawal increases. In this study, when the withdrawal score reached 5.018 in a scale of 1 (low withdrawal) to 7 (high withdrawal), high-tensioned negotiators have lower commitment than their low-tensioned counterparts. In the regression analyses, the regression lines of low- and high- tension in the withdrawal-commitment relationship intersect at the withdrawal score of 5.018 and commitment score 2.810 (Both figures are with reference to a 7-point Likert scale of 1—Low to 7—High). This reference point suggests possible directions on managerial plan to gauge commitment in practice. When withdrawal is low, asserting pressure may be a good way to boost commitment. Once withdrawal reaches its threshold value, providing a relaxing negotiating environment may well promote commitment instead. There is a high price for negotiation failure and reverting a withdrawn negotiation is extremely resource laden (Cheung and Chow 2011). Once stalemate surfaces, resurgent measures such as recognising achievement attained provide the necessary conducive and intrinsic support to ease a tense environment. Low withdrawing negotiators develop and internalise tension to keep the project team motivated toward a negotiated settlement, while high withdrawing negotiators are motivated by satisfaction in accomplishing the task and are particularly interested in equity.

The withdrawal-commitment relationship is significantly negative for high-tensioned negotiators but not for low-tensioned negotiators. Thus, for the same increase in withdrawal, high-tensioned negotiators will have significantly greater decrease in commitment than low-tensioned negotiators. When tension is high, higher withdrawal does lead to lower commitment. In such situation, the call for psychological relax would devalue the utilities of the possible settlement options. These findings support the notion that certain level of tension mobilises human resources and keeps the negotiators stay focused on the disputing issues (Deutsch et al. 2006). Nevertheless, tension is a two-edged sword. On one hand, it drives focus on getting the dispute settled. On the other hand, increasing level of tension induces abscondment—a strong form of withdrawal (Cullen et al. 2003; Mathieu and Zajac 1990). Withdrawal is influenced by tension and has temporal and cross-situational stability. As tension increases, a withdrawing negotiator may quickly adjust his commitment, and also his “frame of reference” in evaluating his negotiation tasks, mostly pessimistic. In this regard, withdrawal hampers the interest to continue with the negotiation. Results of this study provide insight for construction practitioners, especially the measures of tension, withdrawal and commitment that have not been operationalised in previous studies.

The construction industry has advocated the use of alternative dispute resolution techniques like mediation and adjudication instead of adversarial processes such as arbitration and litigation. Moreover, negotiation remains the most cost effective means to resolve dispute. In fact, successful negotiation yields greater overall economic payoff. In search for success factors for achieving negotiated settlements, negotiation research has been furthered to examine the process and outcomes related issues such as withdrawal and tension. The desire to continue a negotiation is strongly related to the behavioural factors of the negotiators. High withdrawing negotiators are more prone to choose adversarial means in resolving dispute and less committed to a negotiated settlement. Certain level of tension can empower potential responses and enhance evaluations. Tensioned negotiators are keen to search for rapid recognition that drives commitment. However, too much of the tension arising from the negotiation will intensify the withdrawal tendency. The urge for a result would lead to compromises on the expected outcomes. Even if a negotiated settlement is ultimately reached, the desire for revenge, non-compliance or creation of future dispute lingers. Excessive tension in this regard would cause failure of the negotiation in extreme circumstances.

The interpretation of the research findings is restricted. First, it is mindful that self-reported measures are not longitudinal, thus the cause-and-effect relationship could be further enhanced and supplemented by qualitative analyses with greater case information. Second, the findings are subjected to the method variance such as geographical region and sample distribution. The data set is from Hong Kong and an uneven distribution of sample professions in the survey. The findings should be read in the light of this characteristic. Further research is thus needed to explore the antecedents and different situational variables of tension and withdrawal and to examine longitudinal data to see whether these findings can be replicated in different samples, occupations, and cultures.

14.8 Chapter Summary

Negotiation is the most cost-effective way to resolve dispute. Having a negotiated settlement reflects well on the negotiators. The psychological state of a negotiator underpins his behaviours and these behaviours govern the success or otherwise of a negotiation. One of these psychological states is commitment that has triggered researches to study the factors that contribute to its development, maintenance and enhancement. The attitudinal factors, i.e. withdrawal and tension, are pivotal to commitment. In this study, it is proposed that the effects of withdrawal and tension on commitment are more complex than revealed by previous studies. It is proposed that tension both mediates and moderates the withdrawal-commitment relationship. A questionnaire study was used to collect empirical data to examine the proposition. The results support the hypothesis that tension is both a meditator and a moderator in the withdrawal-commitment relationship. Tensioned negotiators are generally more committed to a negotiated settlement than their low-tensioned counterparts. However, if the tension is excessive, the loss of commitment is much quicker than the low-tensioned counterparts. It reminds managers that even tension can mobilise human resources to the betterment of a negotiated settlement, but too much of the tension would raise the state of withdrawal of the negotiators and in terms lowers commitment. In this regard, management may adjust the tension level by varying the settlement targets as well as changing the memberships of the negotiation.

Acknowledgments Special thanks to Mr. Fung Kong for collecting data for the study. The content of this chapter has been published in Volume 138(10) of the Journal of Construction Engineering and Management and is used with the permission from ASCE.

References

Anderson, E., & Weitz, B. (1992). The use of pledges to build and sustain commitment in distribution channels. *Journal of Marketing Research, 29*(1), 18–34.

Baron, R. M., & Kenny, D. A. (1986). The moderator–mediator variable distinction in social psychological research: Conceptual, strategic, and statistical considerations. *Journal of Personality and Social Psychology, 51*(6), 1173–1182.

Beck, T., & Martin, S. (1995). Trying to look bad at work: Methods and motives for management poor impressions in organisations. *Academy of Management Journal, 38*, 174–199.

Beehr, T. A., Jex, S. M., Stacy, B. A., & Murray, M. A. (2000). Work stress and co-worker support as predictors of individual strains and performance. *Journal of Organizational Behavior, 21*(4), 391–405.

Bhanugopan, R., & Fish, A. (2006). An empirical investigation of job burnout among expatriates. *Personnel Review, 35*(4), 449–468.

Black, G. S. (2008). Trust and commitment: reciprocal and multidimensional concepts in distribution relationships. *SAM Advanced Management Journal, 73*(1), 46–55.

Blau, G. (1985). Relationship of extrinsic, intrinsic, and demographic predictors to various types of withdrawal behaviors. *Journal of Applied Psychology, 70*, 442–450.

Brumels, K., & Beach, A. (2008). Professional role complexity and job satisfaction of collegiate certified athletic trainers. *Journal of Athletic Training, 43*(4), 373–378.

Caplan, R. D., & Jones, K. W. (1975). Effects of workload, role ambiguity, and Type A personality on anxiety, depression, and heart rate. *Journal of Applied Psychology, 60*, 713–719.

Cheung, S. O., & Chow, P. T. (2011). Withdrawal in construction project dispute negotiation. *Journal of Construction Engineering and Management, 137*(12), 1071–1079.

Cheung, S. O., Tam, C. M., Ndekugri, I., & Harris, F. C. (2000). Factors affecting clients' project dispute resolution satisfaction in Hong Kong. *Construction Management & Economics, 18*(3), 281–294.

Chow, P. T., & Cheung, S. O. (2008). Developing a conceptual framework of catastrophic withdrawal behavior in construction dispute. In *Proceedings of the BEAR 2008 Conference, Sri Lanka* (pp. 659–669).

Cook, J., & Wall, T. (1980). New work attitude measures of trust, organizational commitment and personal need non-fulfilment. *Journal of Occupational Psychology, 53*(1), 39–52.

Cooper, D. R., & Schindler, P. S. (Eds.). (2000). *Business Research Method.* Boston: McGraw Hill.

Cullen, J. B., Parboteeah, K. P., & Victor, B. (2003). The effects of ethical climates on organizational commitment: A double-study analysis. *Journal of Business Ethics, 46*(2), 127–141.

Delerue-Vidot, H. (2006). Opportunism and unilateral commitment: The moderating effect of relational capital. *Management Decision, 44*(6), 737–751.

Densten, I. (2001). Re-thinking burnout. *Journal of Organizational Behavior, 22*(8), 833–847.

Deutsch, M., Coleman, P. T., & Marcus, E. C. (Eds.). (2006). *The Handbook of Conflict Resolution: Theory and Practice*. San Francisco: Jossey-Bass Publishers.

Far East Trade (Ed.). (2003). *Hong Kong builder directory*. Hong Kong: Far East Trade Press (35).

Fiss, O. M. (1983). Against settlement. *Yale Law Journal, 93*, 1073–1090.

Fryer, B. G. (Ed.). (2004). *The practice of construction management: people and business performance*. Oxford: Blackwell.

Gundlach, G. T., Achrol, R. S., & Mentzer, J. T. (1995). The structure of commitment in exchange. *Journal of Marketing, 59*(1), 78–92.

Hair, J. F., Anderson, R. E., Tatham, R. L. & Black, W. C. (Eds.). (1998). *Multivariate data analysis* (5th ed., pp. 88–92). Upper Saddle River: Prentice Hall.

Hanisch, K. A., & Hulin, C. L. (1991). General attitudes and organizational withdrawal: An evaluation of a causal model. *Journal of Vocational Behavior, 39*, 110–128.

Hardy, M. E., & Conway, M. E. (Eds.). (1998). *Role theory: Perspectives for health professionals* (2nd ed.). Norwalk: Appleton & Lange.

James, L. R., & Brett, J. M. (1984). Mediators, moderators, and tests for mediation. *Journal of Applied Psychology, 69*(2), 322–333.

Jaros, S. J., Jermier, J. M., Koehler, J. W., & Sincich, T. (1993). Effects of continuance, affective, and moral commitment on the withdrawal process: An evaluation of eight structural equation models. *Academy of Management Journal, 36*(5), 951–995.

Jones, B., Flynn, D. M. & Kelloway, E. K. (1995). Perceptions of support from the organizations in relation to work stress, satisfaction and commitment. In S. L. Sauter, & L. R. Murphy (Eds.), *Organizational Risk Factors for Job Stress, American Psychological Association, Washington, DC* (pp. 41–52).

Kahn, R. L., Wolfs, D. M., Quinn, R. P., Snoek, J. D. & Rosenthal, R. A. (Eds.). (1964). *Organizational stress: Studies in role conflict and ambiguity*. New York: Wiley.

Lee, K. H., Yang, G. & Graham, J. L. (2006). Tension and trust in international business negotiations: American executives negotiating with Chinese executives. *Journal of International Business Studies, 37*(5), 623–641.

Lempereur, A. & Colson, A. (2010). Acknowledging emotions before problem-solving. In M. Pekar (Eds.), *The first move: A negotiator's companion* (Vol. 6, pp. 153–176). Hoboken: Wiley

Macfarlane, J., Manwaring, J., Zweibel, E. & Hamilton, J. W. (Eds.). (2003). *Dispute resolution: Readings and case studies*. Toronto: Emond Montgomery.

Maslach, C., Schaufeli, W. B., & Leiter, M. P. (2001). Job burnout. *Annual Review of Psychology, 52*, 397–422.

Mathieu, J. E., & Zajac, D. M. (1990). A review and meta-analysis of the antecedents, correlates, and consequences of organizational commitment. *Psychological Bulletin, 108*(2), 171–194.

Meyer, J. P., & Allen, N. J. (1991). A three-component conceptualization of organizational commitment. *Human Resource Management Review, 1*(1), 61–90.

Mobley, W. H., Griffeth, R. W., Hand, H. H., & Meglino, B. M. (1979). Review and conceptual analysis of the employee turnover process. *Psychological Bulletin, 86*(3), 493–522.

Morgan, R. M., & Hunt, S. D. (1994). The commitment-trust theory of relationship marketing. *Journal of Marketing, 58*(3), 20–38.

Mowday, R. T., Koberg, C. S., & McArthr, A. W. (1984). Psychology of the withdrawal process: A cross-validational test of Mobley's intermediate linkages model of turnover in two samples. *Academy of Management Journal, 27*(1), 79–94.

Nordqvist, S., Hovmark, S., & Zika-Viktorsson, A. (2004). Perceived time pressure and social processes in project teams. *International Journal of Project Management, 22*(6), 463–468.
O'Connor, M. K., & Arnold, J. A. (2001). Distributive spirals: Negotiation impasses and the moderating role of disputant self-efficacy. *Organizational Behavior and Human Decision Processes, 84*(1), 148–176.
O'Connor, M. K., Arnold, J. A. And Burris, E. R. (2005). Negotiators' bargaining histories and their effects on future negotiation performance. *Journal of Applied Psychology, 90*(2), 350–362.
O'Reilly, C. A., & Chatman, F. (1986). Organizational commitment and psychological attachment: The effects of compliance, identification, and internalization on pro-social behavior. *Journal of Applied Psychology, 71*(3), 492–499.
Onyemah, V. (2008). Role ambiguity, role conflict, and performance: Empirical evidence of an inverted-U relationship. *Journal of Personal Selling & Sales Management, 28*, 299–313.
Ring, P. S., & Van de Ven, A. H. (1994). Developmental process of cooperative interorganizational relationships. *Academy of Management Review, 19*(1), 90–118.
Roznowski, M., & Hanisch, K. (1990). Building systematic heterogeneity into work attitudes and behavior measures. *Journal of Vocational Behavior, 36*, 361–375.
Saraydar, E. (1971). Uncertainty, bargaining, and the strategy of commitment. *Journal of Economics, 31*(3/4), 309–322.
Schwarz, N., & Clore, G. L. (2007). Feelings and phenomenal experiences. In K. Kruglanski & E. T. Higgins (Eds.), *Social psychology: Handbook of basic principles* (2nd ed., pp. 385–407). New York: Guilford Press.
Sheridan, J. E., & Abelson, M. A. (1983). Cusp catastrophe model of employee turnover. *Academy of Management Journal, 26*(3), 418–436.
Simonson, I., & Staw, B. M. (1992). De-escalation strategies: A comparison of techniques for reducing commitment to losing courses of action. *Journal of Applied Psychology, 77*(4), 419–426.
Singh, J. (1998). Striking a balance in boundary-spanning positions: An investigation of some unconventional influences of role stressors and job characteristics on job outcomes of salespeople. *Journal of Marketing, 62*(3), 69–86.
Singh, J., Verbeke, W., & Rhoads, G. K. (1996). Do organizational practices matter in role stress processes? A study of direct and moderating effects for marketing-oriented boundary spanners. *Journal of Marketing, 60*, 69–91.
Sommer, S. M., Bae, S. H., & Luthans, F. (1996). Organizational commitment across cultures: The impact of antecedents on Korean employees. *Human Relations, 49*(7), 977–993.
Steel, R. P., & Ovalle, N. K. (1984). A review and meta-analysis of research on the relationship between behavioral intention and employee turnover. *Journal of Applied Psychology, 36*, 577–600.
Watson, D., Clark, L. A., Mclntyre, C. W., & Hamaker, S. (1992). Affect, personality, and social activity. *Journal of Personality and Social Psychology, 63*, 1011–1025.
Weatherly, K., & Tansik, D. A. (1992). Tactics used by customer-contract worker: Effect of roles stress, boundary spanning and control. *International Journal of Service Industry Management, 4*(1), 4–17.
Wiener, Y. (1982). Commitment in organizations: A normative view. *Academy of Management Review, 7*, 418–428.
Yiu, T. W., & Cheung, S. O. (2007). Behavioral transition: A framework for the construction conflict–tension relationship. *IEEE Transactions on Engineering Management, 54*(3), 498–505.

Chapter 15
Application of Bandura's Self-Efficacy Theory to Examining the Choice of Tactics in Construction Dispute Negotiation

Tak Wing Yiu and Sai On Cheung

Abstract This study investigates the confidence of negotiators in their own ability to successfully use tactics to achieve desired outcomes—a concept defined as negotiation-efficacy that underpins Bandura's self-efficacy theory. A questionnaire survey was used to measure the frequency of and confidence with which negotiators used negotiating tactics and the achievement of negotiation outcomes. With the collected data, confidence indices were created to reflect the strength of negotiation-efficacy for each negotiating tactic. Relationships of negotiation-efficacy and the achievement of negotiation outcomes were then examined by multiple regression analyses. The findings show that the strength of negotiation-efficacy is significantly related to the achievement of certain negotiation outcomes. In general, for negotiators who have negotiation-efficacy in executing distributive (integrative) tactics, negative (positive) negotiation outcomes are likely.

15.1 Introduction

In the construction industry, most project participants consider negotiation as the most effective way of solving disputes or claims (Steen and MacPherson 2000) because it helps to maintain harmonious relationships when one party is in conflict with others. However, not all negotiations end with settlement (Smith 1992; Ren et al. 2002). To polish one's negotiation skill, every negotiation is a learning

T. W. Yiu (✉)
Faculty of Engineering, Department of Civil and Environmental Engineering, The University of Auckland, Auckland, New Zealand
e-mail: k.yiu@auckland.ac.nz

S. O. Cheung
Construction Dispute Resolution Research Unit, Department of Civil and Architectural Engineering, City University of Hong Kong, Hong Kong, People's Republic of China
e-mail: saion.cheung@cityu.edu.hk

S. O. Cheung (ed.), *Construction Dispute Research*,
DOI: 10.1007/978-3-319-04429-3_15,

process. What has occurred previously should be used as a standard of assessment to choose what to aim for and what to do next (Ren et al. 2002; Gulliver 1979). The best example of this form of learning is the choice of negotiating tactics. Negotiators are likely to develop their own 'toolkits' of successful negotiating tactics, resulting in a high degree of confidence in using that particular set of tactics. This is one of the essential ways of assessing the choice of tactics in negotiation, because well selected and practised negotiating tactics prompt the counterpart to understand mutual interests, share information, generate alternatives/options and reach agreement (Walton and Mckersie 1965).

In the field of psychology, previous researches have also suggested that personality (Barry and Friedman 1998), social motivation (De Dreu and Van Kleef 2004), psychological biases (Bazerman 1994; Mnookin et al. 2000), and gender affect the choice of negotiating tactics (Eagly and Crowley 1986). Typically, negotiating tactics have been classified as contingent or non-contingent (Walton and McKersie 1993). Those tactics that can be applied to particular situations are categorised as contingent, whereas tactics that can be used in most situations are defined as non-contingent. In the construction industry, the contingent use of negotiating tactics was recently investigated by Cheung et al. (2009), who suggested that the achievability of negotiation outcomes through the use of certain tactics depends on the dispute source (Cheung et al. 2009).

This study aims to extend the research of Cheung et al. (2009) by investigating the relationships between the level of negotiators' confidence in choosing their negotiating tactics and the achievement of negotiation outcomes. The study is based on the notion that the more confident negotiators are in choosing their tactics, the more likely they will use those tactics, and also the greater are the chances of achieving their desired negotiation outcomes (Sullivan et al. 2006). This is underpinned by the well-established self-efficacy theory of Bandura (1977), which suggests that '*people who have confidence in their capabilities with respect to a specific task anticipate a successful performance, focus their thoughts on how they can succeed, and persist in the face of difficulty while people will avoid tasks for which they have a low level of self-efficacy*' (Bandura and Cervone 1983, 1986; Cervone and Peake 1986; Sullivan et al. 2006). This study applies the self-efficacy theory to construction dispute negotiation. It begins by briefly elaborating on Baudura's theory and discussing its applicability to construction negotiations. Next, the following research questions are addressed. (1) *How frequently do negotiators use certain negotiating tactics?* (2) *How confident are they in their ability to use those tactics?* (3) *What are the related negotiation outcomes?* Finally, the relationships between negotiation-efficacy and the achievement of negotiation outcomes are examined with multiple regression analyses.

The findings of this study will help construction academics and practitioners to understand how to achieve successful negotiation. Specifically, the study makes two important contributions to the study of construction dispute negotiation. First, the findings close a gap in the behavioural study of construction negotiation. The concept of self-efficacy as an important variable that directs the choice of tactics has not been researched in construction dispute negotiation. Second, the study offers a means of measuring the self-efficacy of using negotiating tactics for construction negotiators. It will then be possible to detect negotiator overconfidence, which has been proven to reduce concessionary behaviour and negotiation success (Einhorn and Hogarth 1978; Neale and Bazerman 1985). It can also serve as an instrument with which to test reality and prevent negotiators from underestimating (or overestimating) the entire negotiation situation (Lichtenstein et al. 1982). The industry will be benefitted by implementing these recommendations to make the negotiation process fairer and more efficient.

15.2 Self-Efficacy Theory

Self-efficacy theory was developed by Bandura (1977, 1986) and suggests that an individual has level of confidence in his or her ability to perform a certain task. This is a social-cognitive approach to describe task-specific self-confidence and how the level of an individuals' confidence will influence what he or she does. Bandura (1997, 1982) described four categories of experience that may be involved in the development of self-efficacy: enactive mastery (personal attainments), vicarious experience (modeling), verbal persuasion, and physiological arousal. An individual's cognitive appraisal and integration of these experiences will determine self-efficacy (Bandura 1982; Gist and Mitchell 1992). Most importantly, self-efficacy affects the choices that individuals make about how to spend their time (Sullivan et al. 2006), because they would avoid tasks that they have a low level of self-efficacy and prefer those for which they have a higher level (Bandura 1997; Sullivan et al. 2006). Likewise, negotiators have varying levels of self-efficacy for certain tactics, and they are more likely to use those tactics that they have high level of self-efficacy (Sullivan et al. 2006). A similar concept can be found in the work of Loucks-Atkinson and Mannell (2007), who applied the self-efficacy theory to measure people's confidence in their ability to successfully use negotiating strategies to overcome constraints. This is described as negotiation-efficacy (NE) in the field of leisure sciences.

Drawing inspiration from Loucks-Atkinson and Mannell's (2007) study, the definition of NE was slightly modified to suit the context of this study. It is defined here as '*the negotiator's confidence in his or her ability to successfully use tactics to conduct a negotiation to produce desired outcomes*' (Bandura 1997; Loucks-Atkinson and Mannell 2007; Sullivan et al. 2006). This definition expands the scope of self-efficacy theory that has been successfully applied in the fields of psychology (Bouffard-Bouchard 2001; Sullivan et al. 2006), leisure science (Loucks-Atkinson and Mannell 2007), and organisational behaviour (O'Connor and Arnold 2001), to which the study of negotiation-efficacy in construction dispute negotiation can be analogically referenced. Furthermore, the concept of NE has been successfully applied to the choice of tactics and performance (Loucks-Atkinson and Mannell 2007; O'Connor and Arnold 2001; Stevens and Gist 1997; Sullivan et al. 2006).

15.3 Method

A questionnaire was designed to collect data. The respondents were asked to provide their background information and the particulars of one of their most recently completed negotiations. The remaining three sections address the three research questions based on the respondents' replies.

15.3.1 How Frequently do Negotiators use Certain Negotiating Tactics?

This question measured the frequency of using various negotiating tactics (hereafter, the frequency variable). A literature review was first conducted to identify a list of negotiating tactics (see Table 15.1). The study adopts one of the major schools of thought in negotiating tactics (Walton and McKersie 1965). Negotiating tactics are classified into two types: distributive and integrative. Distributive tactics are individualistic, used to gain concessions from the other party, and concerned with getting resources on only one side of the negotiation (Giordano et al. 2007; Pruitt 1981; Sullivan et al. 2006). Integrative tactics are concerned with tradeoffs in fulfilling the interests of all of the negotiators, allowing creative solutions and maximising joint benefits (Giordano et al. 2007; Pruitt 1981; Weingart et al. 1996). The frequency of using the two types of negotiating tactics was measured on a 10-point scale from 1 for 'least frequent' to 10 for 'most frequent'.

15.3.2 How Confident are Negotiators in Their Ability to use Their Tactics?

The negotiation-efficacy described above (hereafter, the negotiation-efficacy variable) was measured in this part of the questionnaire. The study adopted the methodology developed by Bandura (1977) to rate the respondents' confidence in using each negotiating tactic successfully in a recently completed negotiation. This methodology involved the use of a 100-point confidence scale devised from probability rating, ranging from 0 = 'no confidence' to 100 for 'full confidence' (see Fig. 15.1). The scale has been widely applied in a number of previous studies (Brown et al. 1998; Gutkind et al. 2001; Loucks-Atkinson and Mannell 2007; Sullivan et al. 2006).

Furthermore, the 'strength' of negotiation-efficacy can be detected by combining the frequency and negotiation-efficacy variables to form a hybrid variable. This variable is described as 'confidence index' of using negotiating tactics and was developed based on the observation of Bandura (1997) and Sullivan et al. (2006)

Table 15.1 List of negotiating tactics

Negotiating Tactics	
Distributive negotiating tactics	
D1	I expressed my anger to my opponent[a]
D2	I tried to control the pace of a negotiation[a]
D3	I tried to object to an issue that was unfavorable to me[a]
D4	I used my power to change the perceptions of my opponent[a]
D5	I was silent when there was unacceptable offer[a]
D6	I escalated my arguments, threats and assertions of my needs[a]
D7	I maximised the information received and minimised the information given[a,b]
D8	I used my authority to make decisions in my favor[a,c]
D9	I attempted to increase time pressure by indicating the negotiation deadline[a,d]
D10	I accepted the other parties' point of view but not their offer[a,d]
D11	I attributed bad faith to my opponent when there was disagreement[a,d]
D12	I argued in support of my own position[a,d]
D13	I tried to hide my bottom line[e,f]
Integrative negotiating tactics	
I1	I tried to get chances for caucus and break[g]
I2	I began with easy issues on common ground[g]
I3	I tried to identify the core issue and clarify where each party stood[g]
I4	I brainstormed various options based on the interests of all parties[g]
I5	I attempted to exchange concession with my opponent[a]
I6	I asked for a time-out whenever deadlock was engaged[a]
I7	I tried to understand the situation from my opponent's point of view[a]
I8	I appeared patient during the negotiation[a]
I9	I provided others with more information[a]
I10	I gave counter-proposals to my opponent[a,f]
I11	I suggested a range of options or trade-offs across issues[d,h]
I12	I used information exchange as a mechanism for establishing trust[d]
I13	I attempted to reveal and acknowledge personal feelings[i]

[a] Churchman (1993)
[b] Wilson and Putnam (1993)
[c] Rahim (1983)
[d] Olekalns et al. (1996)
[e] Volkema and Fleury (2002)
[f] Barry and Friedman (1998)
[g] Bordone & Todd (2005)
[h] Thompson (1990)
[i] Schawarz and Peutsch (2001)

Rate the degree of confidence by recording a number from 0 to 100 using the following scale.										
0	10	20	30	40	50	60	70	80	90	100
uncertain					moderately certain					certain

Fig. 15.1 Negotiation-efficacy assessment—the 100-point probability scale (Bandura 1977, 1997)

Table 15.2 List of negotiation outcomes

Negotiator outcomes	
O1	Mutually beneficial solution for both parties was achieved[a]
O2	The relationship between parties was harmonious and the possibility of dealing with each other in the future increased[a]
O3	The time for solving problems was reduced[b]
O4	The solution satisfied the goals and needs of both parties[c]
O5	Less future disagreements are likely[c]
O6	Information exchange was increased[d]
O7	Trust was developed between parties[e]
O8	The level of conflict was reduced[e]
O9	A conflict-laden environment was eliminated[e]
O10	The quality of communication was enhanced[f]
O11	Innovation and creativity in the organisation were stimulated[f]
O12	Organisational decision making was improved[f]
O13	The opponents' needs could not be clearly defined[c]
O14	A higher level of ongoing conflict was experienced[e,g]
O15	Task conflict was turned into relationship conflict[e,g]
O16	Conflicts arose due to lack of basic information[e,g]
O17	Solution development was unlikely to be achieved, resulting in wasted resources[e,g]
O18	Stalemate, deadlock or impasse occurred[f]
O19	Organisational commitment and loyalty were affected[f]
O20	Information exchange was minimised and even false information was used[d]

[a] Schawarz and Peutsch (2001)
[b] Sheppard et al. (1989)
[c] Rahim et al. (2000)
[d] Pruitt and Lewis (1975)
[e] Friedman et al. (2000)
[f] Rahim (2001)
[g] Pruitt (1991)

that negotiators tend to frequently use certain negotiating tactics in which they have high confidence. As the frequency variable and the negotiation-efficacy variable are defined on scales of 1–10 and 0–100 respectively, the confidence index of using negotiating tactics (which is frequency x negotiation-efficacy) is defined on a hybrid scale of 0–1,000.

15.3.3 What are the Related Negotiation Outcomes?

A list of negotiation outcomes was identified from the literature (see Table 15.2). In the questionnaire, respondents were asked to report the degree to which they achieved the desired negotiation outcomes on a 7-point Likert scale from 1 for 'not achieved' to 7 'highly achieved'.

15.4 Sampling

The targeted respondents were construction industry professionals such as engineers, surveyors and project managers from the government, private developers, contractor firms and consultants. They were the key people in their project teams, who often stood on the front line to negotiate disputes or claims in construction projects. The sample was randomly selected from the Hong Kong Builder Directory and registers maintained by the Hong Kong Institution of Engineers (HKIE), the Hong Kong Institute of Surveyors (HKIS) and the Hong Kong Contractor Association (HKCA). The targeted respondents were first contacted. If they agreed to participate in the study, questionnaire was then sent either by post, fax, or email as preferred by them.

15.5 Results and Discussions

15.5.1 The Data

A total of 180 questionnaires were sent by fax or email and 101 were returned, giving a response rate of 56 %. Six of the returned questionnaires were not completed and were thus discarded, leaving 95 valid questionnaires for data analysis. In terms of organisation type, the dataset comprised respondents from the government (17 %), private developers (15 %), consultants (37 %) and contractor firms (31 %). Fifty percent of the respondents had more than 10 years of experience in construction negotiation.

15.5.2 Data Reliability

Cronbach's Alpha (α) was employed to assess the consistency of results across items within the scales used on the questionnaire (Rao and Schmidt 1998). It can determine whether an item is free from measurement error and identify inconsistent items (Funk et al. 2007). Cronbach's Alpha (α) ranges in value from 0 to 1; the higher the score, the more reliable is the scale. An α value of greater than (or equal to) 0.7 is generally accepted as indicative of a reliable scale (Borg and Gall 1989; Nunnaly 1978). For this study, the Cronbach's alpha values for the 13-item distributive negotiating tactics, the 13-item integrative negotiating tactics and the 20-item negotiation outcomes were 0.820, 0.860, and 0.855 respectively. The reliability of the scales was thus confirmed.

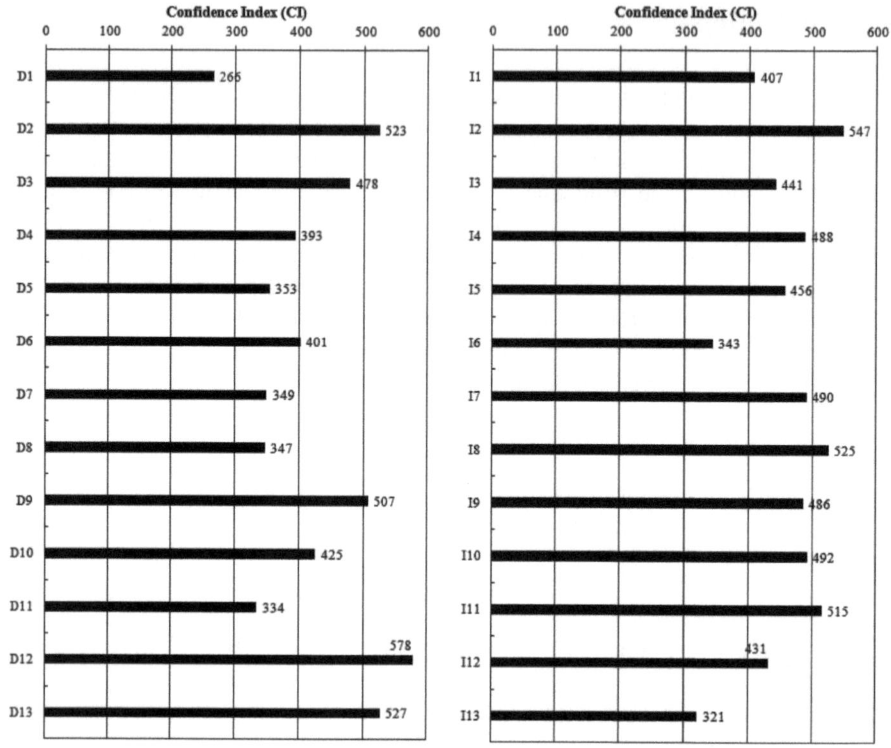

Fig. 15.2 The confidence index of distributive and integrative negotiating tactics

15.5.3 Confidence Indices

Confidence indices were calculated by the aforementioned approach for the negotiating tactics reported by the respondents (Fig. 15.2 refers). By averaging these indices, aggregate confidence indices, representing the indices of each negotiating tactic for the entire sample, could be obtained. Based on the aggregate indices, the confidence indices of using both distributive and integrative negotiating tactics ranged from 266 to 547. This result reveals that the majority of confidence indices stayed at a medium level. The confidence indices of two negotiating tactics, D12 (*I argued in support of my position*) and I2 (*I began with easy issues on common ground*), reached relatively high levels. Hence, construction negotiators appear to use these tactics frequently and confidently, probably because they often belong to and represent the interests of their own organisations (Loosemore 1999), which have minimum positions and objectives in each negotiation. In such situations, negotiators tend to use different arguments to support what is being sought in each situation and to justify the minima (Fells 1996). Furthermore, as Watkins (2003) suggested, negotiating easy issues first is a complexity reduction approach. This tactic can help in building momentum and confidence among negotiating parties at the beginning of negotiation. Interestingly, the confidence indices of two negotiating tactics, D1 (*I expressed my anger to my opponent*) and I13 (*I attempted to reveal and acknowledge personal feelings*) remained at relatively low levels. The expression of anger would have reduced leverage/trust and encouraged the other side to withdraw from the negotiation. In the construction industry, most negotiators primarily deal with negotiation in trustful situations to maintain ongoing relationships and their own reputations. However, construction negotiators are relatively less confident in revealing and acknowledging personal feelings.

15.5.4 Negotiation-Efficacy Constructs

As previously discussed, two types of negotiating tactics, distributive and integrative, were classified. To analyse the construct of the interrelationship among a large number of confidence indices calculated from the data set, principle component factor analyses (PCFA) were used to define a set of common underlying dimensions, known as factors, for a better understanding of the structure of the data. In this study, two PCFA were performed to form two constructs: distributive negotiation-efficacy (DNE) and integrative negotiation-efficacy (INE). DNE and INE refer to the aggregate 'strength' of negotiation-efficacy in executing distributive and integrative tactics respectively. These constructs can also reduce the number of variables into a smaller but more manageable representative of subsets for further analyses. According to the results of PCFA, the suitability of data was first assessed by examining the Kaiser–Meyer–Olkin (KMO) measure of sampling adequacy and the Bartlett test of Sphericity. The KMO values for both PCFA were 0.774 and 0.847, which indicated that the data were adequate as they met the threshold requirement of 0.50 (Cheung and Yeung 1998). The low significance of the Bartlett test of Sphericity also supports the adequacy of the dataset to perform PCFA. To shortlist factors, those with eigenvalues of greater than 1 were extracted. This method can explain an important amount of variability in the data (Schow et al. 2002). Furthermore, Hair et al. (2006) suggested that a factor loading value of 0.60 is considered to be a good demarcation for variable selection within a factor. Variables with factor loadings of less than 0.60 were thus discarded. The distributive and integrative negotiation-efficacy constructs are shown in Tables 15.3 and 15.4.

Four factors were extracted from the DNE construct. Six negotiating tactics were extracted for Factor 1. This factor is described as '*Making threats/issuing ultimatums*', addressing the most typical type of distributive negotiating tactics. The use of this tactic may cause the opponent to lose face and trigger a response of inflexibility (Young 1991). Factor 2 is composed of three negotiating tactics and is named '*Expressing anger and resentment*'. These types of tactics display negative emotions in the form of anger and frustration, and indicate that the negotiators regard the negotiation process as unfair (Hegtvedt and Killian 1999). Factor 3 is described as

Table 15.3 The distributive negotiation-efficacy (DNE) construct

	Factors			
	1	2	3	4
Factor 1: Making threats/issuing ultimatums				
I argued in support of my own position	**0.823**	0.053	0.149	0.210
I escalated my arguments, threats and assertions of my needs	**0.812**	0.232	0.063	−0.067
I attempted to increase time pressure by indicating the negotiation deadline	**0.711**	0.322	−0.260	0.154
I tried to control the pace of the negotiation	**0.673**	−0.244	0.034	0.429
I used my power to change the perceptions of my opponent	**0.635**	0.246	0.399	−0.177
I tried to hide my bottom line[a]	**0.588**	−0.160	0.459	0.220
I was silent when there was an unacceptable offer[a]	**0.550**	0.090	0.329	−0.290
Factor 2: Expressing anger and resentment				
I expressed my anger to my opponent	0.145	**0.803**	0.005	0.204
I attributed bad faith to the opponent when there was disagreement	−0.083	**0.659**	0.369	0.010
I used my authority to make decisions in my favor	0.464	**0.601**	−0.004	−0.239
Factor 3: Being non-reciprocal				
I maximised the information received and minimised information given	−0.010	0.056	**0.825**	−0.059
I tried to object to issues that were unfavorable to me[a]	0.306	0.255	**0.549**	0.140
Factor 4: Persuading others to give in				
I accepted the other parties' point of view but not their offer	0.007	0.133	0.025	**0.834**
% of Variance	33.718	12.532	10.298	8.496
Eigenvalue	4.383	1.629	1.339	1.104

[a] discarded item

'*Being non-reciprocal*' which is composed of one negotiating tactic. The aim of using this tactic is to secure the best benefit, without regard for the opponent's outcome. Factor 4 is named '*Persuading others to give in*'. The rationale for using this tactic is to give as little as possible and to force the opponent to give in.

Three factors were extracted for the INE construct. Factor 1 is described as '*Searching for joint gains*'. This is a typical integrative negotiator tactic whereby options are explored to increase joint gain without respect to the division of payoff (Lewicke et al. 1988; Heydenfeldt 2000). It is often regarded as co-operative tactic that can reinforce on-going relationships among negotiating parties. Factor 2 is named '*Emphasising common goals, objectives and interests*'. This is a basic tactic of integrative negotiation that aims at probing the needs of all parties, and creating an open flow of information for negotiation. Factor 3 is described as '*Reviewing/clarifying positions*'. If the negotiation process ends in deadlock, an integrative negotiating tactic such as giving the opponent a pause for thought to allow the opponent to rethink his or her position. It is therefore important to view deadlock as one of the ways to reframe the entire negotiation process.

Table 15.4 The integrative negotiation-efficacy (INE) construct

	Factors		
	1	2	3
Factor 1: Searching for joint gains			
I suggested a range of options or trade-offs across issues	**0.792**	0.250	−0.012
I provided others with more information	**0.775**	0.372	0.036
I brainstormed various options based on the interests of all parties	**0.752**	0.065	0.119
I tried to identify the core issue and clarify where each party stood	**0.751**	0.026	0.465
I gave counter-proposals to my opponent	**0.709**	0.208	0.242
I attempted to exchange concessions with my opponent[a]	**0.570**	0.291	0.015
I tried to understand the situation from my opponent's point of view[a]	**0.551**	0.525	−0.030
I appeared patient during the negotiation[a]	**0.481**	0.297	−0.337
Factor 2: Emphasising common goals, objectives and interests			
I began with easy issues on common ground	0.228	**0.715**	0.027
I used information exchange as a mechanism for establishing trust	0.189	**0.686**	0.146
I tried to get chances for caucus and break	0.181	**0.608**	0.075
Factor 3: Reviewing/clarifying positions			
I asked for a time-out whenever a deadlock occurred	0.242	−0.003	**0.730**
I attempted to reveal and acknowledge personal feelings	−0.065	0.358	**0.672**
% of Variance	39.580	10.790	8.468
Eigenvalue	5.145	1.403	1.101

[a] discarded item

15.5.5 Negotiation Outcomes

Similar to the aim of developing the DNE and INE constructs, PCFA was also performed to investigate the underlying dimensions of the negotiation outcomes from the use of distributive and integrative negotiating tactics. The suitability of data was also assessed by examining the Kaiser–Meyer–Olkin (KMO) measure of sampling adequacy and the Bartlett test of Sphericity. The KMO value was 0.812, and a low significance of the Bartlett test of Sphericity was also obtained. To shortlist factors, the same approaches used to develop the DNC and INC constructs were used. The factor matrix of negotiation outcomes is shown in Table 15.5.

Five factors were suggested by the use of PCFA. Factors 1 and 5, described as '*Conflict escalation*' and '*Negotiation stall*' respectively, are generally regarded as negative negotiation outcomes. These outcomes generally result from the competitiveness and adversity of negotiation, with disputes unlikely to be resolved as the conflict escalates. Factors 2, 3, and 4 are the positive negotiation outcomes. These outcomes involve looking for hidden agendas, achieving mutual beneficial solutions, and establishing a good relationship among negotiating parties.

Table 15.5 Factor matrix of negotiation outcomes

	Factors				
	1	2	3	4	5
Factor 1: Conflict escalation					
Stalemate, deadlock or impasse occurred	**0.849**	0.142	−0.119	0.149	−0.022
Organisational commitment and loyalty were affected	**0.813**	0.142	−0.109	0.005	0.155
A higher level of ongoing conflict was experienced	**0.749**	0.051	−0.058	−0.290	−0.044
Task conflict was turned into relationship conflict	**0.733**	0.079	0.185	−0.142	0.343
Solution development was unlikely to be achieved, resulting in wasted resources	**0.654**	0.122	−0.191	0.107	0.377
Information exchange was minimised and even false information was used	**0.629**	0.161	0.014	0.003	0.555
Factor 2: Effective and creative solutions					
Innovation and creativity in the organisation were stimulated	0.134	**0.891**	0.178	0.006	0.167
The time for solving problems was reduced	0.254	**0.883**	0.145	0.101	0.028
Organisational decision making was improved	0.123	**0.798**	0.258	0.120	0.161
Less future disagreements are likely to be made	0.062	**0.790**	0.073	0.244	0.087
Factor 3: Mutual beneficial solutions					
The solution satisfied the goals and needs of the both parties	0.029	0.022	**0.853**	0.163	−0.084
Information exchange was increased	−0.115	0.188	**0.739**	0.086	0.176
Mutual beneficial solution for both parties was achieved	−0.229	0.200	**0.690**	0.198	−0.080
Trust was developed between parties	0.061	0.393	**0.601**	0.296	−0.066
Factor 4: Sustainability of relationship					
A conflict-laden environment was eliminated	−0.059	0.089	0.130	**0.877**	0.029
The level of conflict was reduced	0.027	0.145	0.138	**0.838**	0.101
The relationship between parties was harmonious and the possibility of dealing with each other in the future improved	−0.128	0.309	402	**0.661**	−0.013
The quality of communication was enhanced[a]	−0.001	0.145	0.512	**0.599**	−0.152
Factor 5: Negotiation stall					
The opponents' needs could not be clearly defined	0.105	0.052	−0.081	0.065	**0.790**
Conflicts arose due to a lack of basic information	0.335	0.275	0.052	−0.040	**0.714**
% of Variance	28.651	22.812	7.755	6.699	5.747
Eigenvalue	5.730	4.562	1.551	1.340	1.149

[a] discarded item

15.5.6 Relating the 'Strength' of Negotiation-Efficacy to Negotiation Outcomes

The relationship between the 'strength' of negotiation-efficacy and negotiation outcomes was investigated by multiple regression analyses. The dependent variable was one of the five negotiation outcomes, and the independent variables were the factors extracted from the DNE and INE constructs. A total of five regression models were thus developed:

$$O_i = a_0 + a_1 I_{DNE1} + a_2 I_{DNE2} + a_3 I_{DNE3} + a_4 I_{DNE4} + a_5 I_{INE1} + a_6 I_{INE2} + a_7 I_{DNE3} \quad (15.1)$$

where O is the dependent variable (i.e. the negotiation outcome), and i = 1, 2, 3, 4, 5., and I_{DNE} (are I_{INE}) are the independent variables (i.e. the DNE and INE constructs).

In the multiple regression analyses, the above variables were the factor scales identified by the PCFAs of the negotiation-efficacy constructs and negotiation outcomes. These scales were the composite measures created for each observation on each factor extracted in the PCFA (Hair et al. 1995). Hence, four variables of DNE construct and three variables of INE construct were entered simultaneously into each regression equation to predict negotiation outcomes. The results of the five regression models are shown in Table 15.6. The F-statistics from the ANOVA indicate that each regression model is significant. The values of R^2 range from 0.489 to 0.789 and are significant among the five regression models. These values are comparable to those obtained in similar studies (Cheung et al. 2006; Gross and Guerrero 2000; Oetzel 1998; Sharland 2001).

The findings of this study indicate that the 'strength' of negotiation-efficacy appears to relate significantly to the achievement of certain negotiation outcomes, which can be revealed by observing the standardised regression coefficients (β) of each regression model. To facilitate interpretation, normalised regression coefficients (γ) were devised to quantify the relative contribution (Table 15.6): the higher the normalised coefficient, the greater is its contribution toward the prediction of the outcome. Among the five regression models, the following four major observations can be made.

(i) The 'strength' of negotiation-efficacy in executing the distributive tactic of making threats/issuing ultimatums (Model 1: β =+0.824; $\gamma = 0.767$; $p < 0.001$ and Model 3: β =+0.500; $\gamma = 0.383$; $p < 0.001$) is positively related to the negotiation outcome of conflict escalation (*Model 1*) or mutual beneficial solution (*Model 3*).

(ii) The 'strength' of negotiation-efficacy in executing the integrative tactic of searching for joint gains (β =+0.640; $\gamma = 0.516$; $p < 0.001$) is positively related to the negotiation outcome of effective and creative solutions (*Model 2*).

Table 15.6 Overall results of multiple regression analyses

Dependent variables	Independent variables	Standardised regression coefficients (β)	Normalised regression coefficients (γ)
Model 1: Conflict escalation (F-statistic: 46.357***; R^2: 0.789***)	DNE—Making threats/issuing ultimatums	0.824***	0.767
	DNE—Expressing anger and resentment	0.015	0.014
	DNE—Being non-reciprocal	0.103	0.096
	DNE—Persuading others to give in	−0.023	0.021
	INE—Searching for joint gains	0.042	0.039
	INE—Emphasising common goals, objectives and interests	−0.015	0.014
	INE—Reviewing/clarifying positions	0.052	0.048
Model 2: Effective and creative solutions (F-statistic: 31.008***; R^2: 0.714***)	DNE—Making threats/issuing ultimatums	0.100	0.081
	DNE—Expressing anger and resentment	0.095	0.077
	DNE—Being non-reciprocal	−0.002	0.002
	DNE—Persuading others to give in	0.065	0.052
	INE—Searching for joint gains	0.640***	0.516
	INE—Emphasising common goals, objectives and interests	0.202**	0.163
	INE—Reviewing/clarifying positions	−0.137*	0.110
Model 3: Mutual beneficial solution (F-statistic: 36.964***; R^2: 0.748***)	DNE—Making threats/issuing ultimatums	0.500***	0.383
	DNE—Expressing anger and resentment	0.243***	0.186
	DNE—Being non-reciprocal	−0.044	0.034
	DNE—Persuading others to give in	0.118*	0.090
	INE—Searching for joint gains	0.090	0.069
	INE—Emphasising common goals, objectives and interests	0.276***	0.211
	INE—Reviewing/clarifying positions	−0.036	0.028
Model 4: Sustainability of relationship (F-statistic: 11.878***; R^2: 0.489***)	DNE—Making threats/issuing ultimatums	0.116	0.096
	DNE—Expressing anger and resentment	0.388***	0.321
	DNE—Being non-reciprocal	0.167	0.138
	DNE—Persuading others to give in	−0.035	0.029
	INE—Searching for joint gains	−0.003	0.002
	INE—Emphasising common goals, objectives and interests	0.451***	0.373
	INE—Reviewing/clarifying positions	−0.048	0.040

(continued)

Table 15.6 (continued)

Dependent variables	Independent variables	Standardised regression coefficients (β)	Normalised regression coefficients (γ)
Model 5: Negotiation stall (F-statistic: 46.240***; R^2: 0.788***)	DNE—Making threats/issuing ultimatums	0.101	0.068
	DNE—Expressing anger and resentment	0.039	0.026
	DNE—Being non-reciprocal	0.058	0.039
	DNE—Persuading others to give in	0.631***	0.427
	INE—Searching for joint gains	0.020	0.014
	INE—Emphasising common goals, objectives and interests	−0.056	0.038
	INE—Reviewing/clarifying positions	0.573***	0.388

*** $p < 0.001$; ** $p < 0.01$; * $p < 0.05$

(iii) The 'strength' of negotiation-efficacy in executing the integrative tactic of emphasising common goals, objectives, and interests (β =+0.451; $\gamma = 0.373$; $p < 0.001$) is positively related to the negotiation outcome of sustaining relationships among negotiating parties (*Model 4*).

(iv) The 'strength' of negotiation-efficacy in executing the distributive tactic of persuading others to give in (β =+0.631; $\gamma = 0.427$; $p < 0.001$) is positively related to negotiation stall (*Model 5*).

The self-efficacy theory not only offers a promising avenue to better understand the behaviour of construction negotiators, but also helps researchers and practitioners to understand the important issue of building competence in adopting appropriate negotiating tactics through confidence. In general, the foregoing analyses affirm the proposition of Sullivan et al. (2006) that Bandura's (1977, 1986) self-efficacy theory could be used to explain the influence of self-efficacy on negotiation outcomes. Specifically, negotiators with confidence in executing integrative (or distributive) tactics are likely to sample those tactics, and consequently influence the negotiation outcomes. As Bandura (1982) suggested, the development of this confidence can be caused by past performance, vicarious experience, verbal persuasion and psychological arousal. Among these, past performance and vicarious experience should be the important contributors to the gains in negotiation-efficacy of construction negotiators because negotiation is seldom taught as part of a formal education and the related skills are usually gained through hard experience. If a negotiator has successfully used a particular negotiating tactic in the past, he or she will probably believe that this tactic can be used successfully again in the next negotiation. However, negotiators should deal with the issue of negotiation-efficacy with caution, because overconfidence in a particular set of negotiating tactics that may represent a bias that reduces concessionary behaviour and thus the ability to reach a desirable negotiation outcome.

Negotiators should understand and anticipate the reaction of the parties with whom they are negotiating to objectively manipulate their own level of confidence.

Finally, the authors would like to give the following two recommendations on how to gain negotiation-efficacy for construction negotiators:

(i) Successful experiences of using a particular set of tactics would boost negotiation-efficacy. Due to the fact that it is unusual to take any formal record for every negotiation case, construction negotiator is advised to write his or her log for every successful negotiation case. Each log shall contain the background information of the project dispute, negotiation process, the use of tactics and the outcomes. Most importantly, personal comments and reflection shall be included for future references.

(ii) Observing a peer's successful negotiation can also strengthen one's negotiation-efficacy. However, construction dispute negotiations are mostly conducted in private, direct observation may not always be possible. In this connection, the sharing of successful stories in construction dispute negotiations among peers may be a more pragmatic approach to gain negotiation-efficacy. Organising negotiation workshops or seminars may be a good option.

15.6 Chapter Summary

In construction dispute negotiation, it has long been an important issue to understand what drives the choice of tactics. Underpinned by Bandura's well-established self-efficacy theory, a concept of 'negotiation-efficacy' was engineered in this study. 'Negotiation-efficacy can be defined as 'the negotiator's confidence in his or her ability to successfully use tactics to conduct a negotiation to produce desired outcomes'. The study thus investigated the relationships between negotiation-efficacy and the achievement of negotiation outcomes. To achieve that aim, a survey was conducted to measure the frequency and confidence of using negotiating tactics, and the achievement of desired negotiation outcomes. With the collected data, confidence indices were created to reflect the strength of negotiation-efficacy for each negotiating tactic. Negotiation-efficacy constructs for distributive and integrative negotiating tactics were also developed. The relationships of negotiation-efficacy and the achievement of negotiation outcomes were finally examined by the use of multiple regression analyses. The findings show that the strength of negotiation-efficacy is significantly related to the achievement of certain negotiation outcomes. Generally, for those negotiators who have negotiation-efficacy in executing distributive (integrative) tactics, negative (positive) negotiation outcomes are likely.

The findings of this study make two important contributions to the construction dispute negotiation literature. First, they bridge the research gap in the behavioural study of construction negotiation. Negotiation-efficacy is an important variable that directs the choice of tactics, which is an uncharted area in construction dispute negotiation. Second, this study offers a means of quantifying negotiation-efficacy for construction dispute negotiators. It will then be possible to detect negotiator overconfidence, which has been proven to reduce concessionary behaviour and thus hamper chance of negotiation success. It can also serve as an instrument of reality testing to prevent negotiators from underestimating (or overestimating) the prospect of having a negotiated settlement.

Acknowledgments Special thanks to Miss Lai Ying Siu for collecting data for the study. The content of this chapter has been published in Volume 138(3) of the Journal of Construction Engineering and Management and is used with the permission from ASCE.

References

Bandura, A. (1977). Self-efficacy: Toward a unifying theory of behavioral change. *Psychological Review, 84*, 191–215.

Bandura, A. (1982). Self-efficacy mechanism in human agency. *American Psychologist, 37*(2), 122–147.

Bandura, A. (1986). *Social foundations of thought and action: A social cognitive theory*. Englewood Cliffs: Prentice-Hall.

Bandura, A. (1997). *Self-efficacy: The exercise of control*. New York: Freeman.

Bandura, A., & Cervone, D. (1983). Self-evaluative and self-efficacy mechanisms governing the motivational effects of goal system. *Journal of Personality and Social Psychology, 45*, 1017–1028.

Bandura, A., & Cervone, D. (1986). Differential engagement of self-reactive influences in cognitive motivations. *Organizational Behavior and Human Decision Processes, 38*, 92–113.

Barry, B., & Friedman, R. A. (1998). Bargainer characteristics in distributive and integrative negotiation. *Journal of Personality and Social Psychology, 74*(2), 345–359.

Bazerman, M. H. (1994). *Judgment in management decision making*. New York: Wiley.

Bordone, R., & Todd, G. S. (2005). *Have you negotiated how you'll negotiate?*. Watertown: Harvard Business Publishing Newsletters.

Borg, W. R., & Gall, M. D. (1989). *Education research* (5th ed.). New York: Longman.

Bouffard-Bouchard, T. (2001). Influence of self-efficacy on performance in a cognitive task. *Journal of Social Psychology, 130*(3), 353–363.

Brown, S. A., Carrello, P. D., Vik, P. W., & Porter, R. J. (1998). Change in alcohol effect and self-efficacy expectancies during addiction treatment. *Substance Abuse, 19*(4), 155–167.

Cervone, D., & Peake, P. K. (1986). Anchoring, efficacy, and action: The influence of judgmental heuristics on self-efficacy judgments and behaviour. *Journal of Personality and Social Psychology, 50*, 492–501.

Cheung, S. O., & Yeung, Y. W. (1998). The effectiveness of the dispute resolution advisor system: A critical appraisal. *International Journal of Project Management, 16*(6), 367–374.

Cheung, S. O., Yiu, T. W., & Yeung, S. F. (2006). A study of negotiating styles and negotiation outcomes in construction dispute resolution. *Journal of Construction Engineering and Management, 132*(8), 805–814.

Cheung, S. O., Chow, P. T., & Yiu, T. W. (2009). Contingent use of negotiators' tactics in construction dispute negotiation. *Journal of Construction Engineering and Management, 135*(6), 466–476.

Churchman, D. (1993). *Negotiation Tactics*. Lanham: University Press of America.

De Dreu, C. K. W., & Van Kleef, G. A. (2004). The influence of power on the information search, impression formation, and demands in negotiation. *Journal of Experimental Social Psychology, 40*, 303–319.

Egaly, A. H., & Crowlely, M. (1986). Gender and helping behavior: A meta-analytic review of the social psychological literature. *Psychological Bulletin, 100*(3), 283–308.

Einhorn, H. J., & Hogarth, R. M. (1978). Confidence in judgment: Persistence of the illusion of validity. *Psychology Review, 85*(11), 395–416.

Fells, R. (1996). Preparation for negotiation. *Personnel Review, 25*(2), 50–60.

Friedman, R., Curral, S. C., & Tsai, J. C. (2000). What goes around comes around: The impact of personal conflict style on work conflict and stress. *International Journal of Conflict Management, 11*(1), 32–55.

Funk, R., Ives, M., & Dennis, M. (2007). Reliability: Calculating Cronbach's alpha. http://www.chestnut.org/LI/downloads/training_memos/Alpha.pdf (Jan. 23, 2009).

Giordano, G., Stoner, S., Brouer, R., & George, J. (2007). The influences of deception and computer-mediation on dyadic negotiations. *Journal of Computer-Mediated Communication, 12*(2), 362–383.

Gist, M. E., & Mitchell, T. R. (1992). Self-efficacy: A theoretical analysis of its determinants and malleability. *Academy of Management Review, 17*(2), 183–211.

Gross, M. A., & Guerrero, L. K. (2000). Managing conflict appropriately and effectively: An application of the competence model to Rahim's organizational conflict styles. *International Journal of Conflict Management, 11*(3), 200–226.

Gulliver, P. H. (1979). *Disputes and negotiation: A cross-culture perspective*. San Diego: Academic Press Inc.

Gutkind, D., Ventura, J., Barr, C., Shaner, A., Green, M., & Mintz, J. (2001). Factors affecting reliability and confidence of DSM-III-R psychosis-related diagnosis. *Psychiatry Research, 101*(3), 269–275.

Hair, J. F., Anderson, R. E., Tatham, R. L., & Black, W. C. (1995). *Multivariate data analysis with readings* (4th ed.). Englewood Cliffs: Patience Hall.

Hair, J. F., Anderson, R. E., Tatham, R. L., & Black, W. C. (2006). *Multivariate data analysis* (6th ed.). New Jersy: Patience Hall.

Hegtvedt, K. A., & Killian, C. (1999). Fairness and emotions: Reactions to the process and outcomes of negotiations. *Social Forces, 78*, 269–302.

Heydenfeldt, J. A. G. (2000). The influence of individualism/collectivism on Mexican and US business negotiation. *International Journal of Intercultural Relations, 24*(3), 383–407.

Lewicke, R., Weiss, S., & Lewin, D. (1988). *Models of conflict, negotiation, and third party intervention: a review and synthesis*, Ohio State University College of Business, Working paper series pp. 88–33.

Lichtenstein, S., Fischhoff, B., & Phillips, L. D. (1982). Calibration of probabilities: Stage of the art to 1980. In D. Kahneman, P. Slovic & A. Tversky (Eds.), *Judgment under uncertainty: Heuristics and biases* (pp. 306–334). New York: Cambridge University Press.

Loosemore, M. (1999). Bargaining tactics in construction disputes. *Construction Management and Economics, 17*, 177–188.

Loucks-Atkinson, A., & Mannell, R. C. (2007). Role of self-efficacy in the constraints negotiation process: The case of individuals with Fibromyalgia Syndrome. *Leisure Sciences, 29*, 19–36.

Mnookin, R. H., Peppet, S. R., & Tulumello, A. S. (2000). *Beyond winning: Negotiation to create value in deals and disputes*. Cambridge: Belknap Press.

Neale, M. A., & Bazerman, M. H. (1985). The effects of framing and negotiator overconfidence on bargaining behaviours and outcomes. *The Academy of Management Journal, 28*(1), 34–49.

Nunnaly, J. (1978). *Psychometric theory*. New York: McGraw-Hill.

O'Connor, K. M., & Arnold, J. A. (2001). Distributive spirals: Negotiation impasses and moderating effects of disputant self-efficacy. *Organizational Behaviour and Human Decision Processes, 84*, 148–176.

Oetzel, J. G. (1998). The effects of self-control and ethnicity on self-reported conflict styles. *Communication Reports, 11*(2), 133–143.

Olekalns, M., Smith, P. L., & Walsh, T. (1996). The process of negotiating: Strategy and timing as predictors of outcomes. *Organizational Behaviour and Human Decision Processes, 68*, 68–77.

Pruitt, D. (1991). Strategic Choice in Negotiation in Negotiation Theory and Practice. In J. William Breslin & Jeffery Z. Rubin (Eds.), *The Program on Negotiation*. Cambridge: Harvard Law School.

Pruitt, D. G. (1981). *Negotiation behaviour*. New York: Academic Press.

Pruitt, D. G., & Lewis, S. A. (1975). Development of integrative solutions in bilateral negotiations. *Journal of Personality and Social Psychology, 31*, 621–630.

Rahim, M. A. (1983). A measure of styles of handling interpersonal conflict. *Academy Management Journal, 26*, 367–368.

Rahim, M. A. (2001). *Managing conflict in organizations* (3rd ed.). Westport: Quorum Books.

Rahim, M. A., Manger, N. R., & Shapiro, D. L. (2000). Do justice perceptions influence styles of handling conflict with supervisors? What justice perceptions, precisely? *International Journal of Conflict Management, 11*(1), 9–31.

Rao, A., & Schmidt, S. M. (1998). A behavioural perspective on negotiating international alliances. *Journal of International Business Studies, 29*(4), 665–689.

Ren, Z., Anumba, C. J., & Ugwu, O. O. (2002). Negotiation in a multi-agent system for construction claims negotiation. *Applied Artificial Intelligence, 16*, 359–394.

Schawarz, R., & Peutsch, C. (2001). *Negotiation skills development*. Invited paper at the IEEE Symposium Extra Skills for Young Engineers, Maribor, Slovenia, October 19, ISBN 86-435-0440-8, pp. 15–18.

Schow, R. L., Seikei, J. A., Chermak, G., Berent, M., & Domitz-Vieira, D. M. (2002). Support for a multiple-factor model of auditory processing. *American Journal of Audiology, 11*, 10–12.

Sharland, A. (2001). The negotiation process as a predictor of relationship outcomes in international buyer-supplier arrangements. *Industrial Marketing Management, 30*(7), 551–559.

Sheppard, B. H., Blumenfeld-Jones, K., & Roth, J. (1989). Informal third-partyship: Studies of everyday conflict intervention. In K. Kressel & D. G. Pruitt (Eds.), *Mediation research* (pp. 166–189). San Francisco: Jossey-Bass.

Smith, M. L. (1992). Planning your negotiation. *Journal of Management in Engineering, 8*(3), 254–260.

Steen, R. H., & MacPherson, R. J. (2000). Resolving construction disputes out of court through ADR. *Journal of Property Management, 65*(5), 58.

Stevens, C. K., & Gist, M. E. (1997). Effects of self-efficacy and goal-orientation training on negotiation skills maintenance: what are the mechanisms? *Journal of Personality and Psychology, 50*, 955–978.

Sullivan, B. A., O'Connor, M. O., & Burris, E. R. (2006). Negotiator confidence: The impact of self-efficacy on tactics and outcomes. *Journal of Experimental Social Psychology, 42*, 567–581.

Thompson, L. (1990). Negotiation behavior and outcomes: Empirical evidence and theoretical issues. *Psychological Bulletin, 108*(3), 515–532.

Volkema, R. J., & Fleury, M. T. (2002). Alternative negotiating conditions and the choice of negotiation tactics: A cross-cultural comparison. *Journal of Business Ethics, 36*(4), 381–398.

Walton, R. E., & McKersie, R. B. (1993). *A behavioral theory of labor negotiations: An analysis of a social interaction system*. Ithaca: ILR Press.

Walton, R. E., & McKersie, R. B. (1965). *A behavioural theory of labour negotiations*. New York: McGraw-Hill.

Watkins, M. (2003). Strategic simplification: Toward a theory of modular design in negotiation. *International Negotiation, 8*(1), 149–167.

Weingart, L. R., Hyder, E. B., & Prietula, M. J. (1996). Knowledge matters: The effect of tactical descriptions on negotiation behaviour and outcome. *Journal of Personality and Social Psychology, 70*, 1205–1217.

Wilson, S. R., & Putnam, L. L. (1993). Interaction goals in negotiation. In J. A. Anderson (Ed.), *Communication yearbook* (vol. 13, pp. 374–406). Newbury Park: Sage.

Young, H. P. (1991). *Negotiation analysis*. Ann Arbor: University of Michigan Press.

Part IV
Mediation: A Form of Assisted Negotiation—Best Alternative to a Negotiated Agreement (BATNA)

Chapter 16
The Effective Use of ADR Processes in Construction

Sai On Cheung

Abstract The formality of litigation and arbitration, with its concomitant escalation in costs, delays and adversarial relationship, have encouraged the rapid growth of alternative dispute resolution (ADR) processes, namely conciliation, mediation, adjudication and other hybrid processes (Brown and Marriott 1999; Fenn and Gameson 1992; Kaplan et al. 1991). These processes have been widely used and well received. For example, mediation is an integral part of dispute settlement clause in all conditions of contract for Hong Kong Government construction projects. Moreover, use of multiple-tier of ADR renders it impossible to obtain speedy and economic resolution. Overtly complicated ADR based resolution procedures destroy the original intents of having flexible and direct dispute resolution. In this study, a hierarchical model is developed to organise the different attributes of an ADR process. This arrangement fits neatly with the Analytical Hierarchy Process (AHP) methodology. AHP can be used to prioritise the attributes. The top ten ranked attributes are identified as critical attributes. These include, among others, preservation of relationship, enforceability, neutrality and consensus. This study also reports suggestions by experts on the means to enshrine these attributes. It is recommended that by focusing on these critical attributes, the dispute resolution process can be kept simple and effective.

16.1 Introduction

Reviews of the construction industry (CIRC 2001; Egan 1998; Latham 1994; Merna and Bower 1997) have pointed to some fundamental causes of project failure, one of which is the proliferation of disputes. If disputes are not resolved

S. O. Cheung (✉)
Construction Dispute Resolution Research Unit, Department of Civil and Architectural Engineering, City University of Hong Kong, Hong Kong, People's Republic of China
e-mail: saion.cheung@cityu.edu.hk

S. O. Cheung (ed.), *Construction Dispute Research*,
DOI: 10.1007/978-3-319-04429-3_16,

promptly, they can cause further project delays, lead to claims, may require litigation proceedings for resolution and, ultimately, destroy business relationships. Kumaraswamy (1997) has summarised twenty common causes of construction disputes, including the speed of construction, cost and quality control, technological advances, stringent building regulations, and economic difficulties. All are common features of the construction industry in Hong Kong. These, together with the increasingly complex construction activities and intense competition among contractors, have aggravated the problem. Thus, it is not surprising that the number of construction disputes has increased dramatically. This is consistent with the report by the Hong Kong International Arbitration Centre (HKIAC), which revealed that the number of disputes referred to the center has tripled in the last decade. In 2012, a total of 71 cases of construction dispute arbitration were handled by the HKIAC (2012). Hence, the skill of dispute resolution should be part of the tool kit for practitioners, especially for those in managerial capacity or senior position. This study aims to unveil the critical attributes of alternative dispute resolution (ADR).

16.2 The Study

The study has four stages:

1. Identification of ADR process attributes:
2. Development of a hierarchical model of ADR process attributes:
3. Identification of critical attributes through prioritisation: and
4. Collecting suggestions on means to enshrine the critical attributes.

16.2.1 Stage One: Identification of ADR Process Attributes

The research on dispute resolution has attracted the interest of many researchers and practitioners. Goldberg et al. (1992) offered a comprehensive list of attributes of dispute resolution, including voluntary, third party, degree of formality, nature of proceeding, outcome, and privacy. York (1996) is more concerned with the practical issues. The attributes that he identified are time, cost, preservation of relationship, binding decision, degree of control by parties, flexibility in procedure and confidentiality. In contrast, David (1988) focuses on social and human issues such as impartiality, consensus and continuing business relationship. Together with the works of Hibberd and Newman (1999), Cheung (1999), and Brown and Marriott (1999), a summary of the critical attributes is derived and summarised in Table 16.1.

Table 16.1 Summary of attributes of ADR processes

Attributes	a	b	c	d	e	f
1. Cost		√	√	√	√	√
2. Confidentiality		√	√	√	√	√
3. Consensus				√		√
4. Control by parties		√	√	√	√	√
5. Creative agreement		√	√		√	
6. Enforceability	√		√	√		√
7. Fairness		√	√	√		
8. Flexibility		√	√	√	√	
9. Formality	√			√		
10. Knowledge in construction		√				
11. Liabilities to opponent's cost					√	
12. Neutrality	√			√	√	
13. Power to compel consolidation				√		
14. Preservation of relationship		√	√	√	√	√
15. Privacy	√	√	√	√		
16. Speed		√	√	√	√	√
17. Range of issue				√	√	
18. Width of remedy		√	√	√	√	√
19. Willingness	√			√		

Keys
a Goldberg et al. (1992)
b Hibberd & Newman (1999)
c Cheung (1999)
d Brown and Marriott (1999)
e York (1996)
f David (1988)

16.2.2 Stage Two: Development of a Hierarchical Model of ADR Process Attributes

In Stage One, a total of nineteen attributes were identified from previous research works, ranging from intangibles such as consensus, control by parties, preservation of relationship and voluntariness, to tangibles such as cost, time, and speed. Due to the relatively large number of attributes involved, it would be difficult to compare the relative standings of these attributes with the same level of focus. This nature of the problem fits nicely with the functions of the Analytical Hierarchy Process (AHP). The successful use of AHP to assess priorities within a given set of attributes has been reported in the selection of design/build proposals (Alhazmi and McCaffer 2000), the prioritisation of maintenance schedule (Shen et al. 1998), the selection of contractors (Fong and Choi 2000) and the selection of procurement

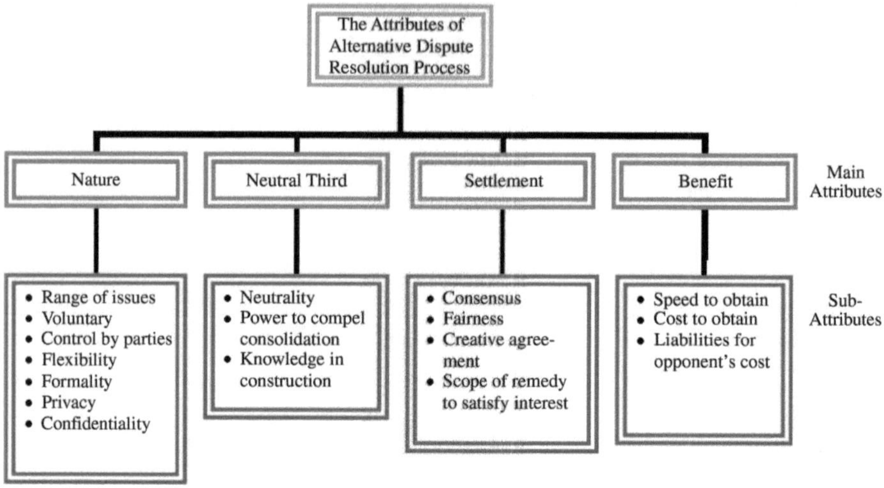

Fig. 16.1 Hierarchical structure of ADR attributes

Fig. 16.2 Phases of a resolution process

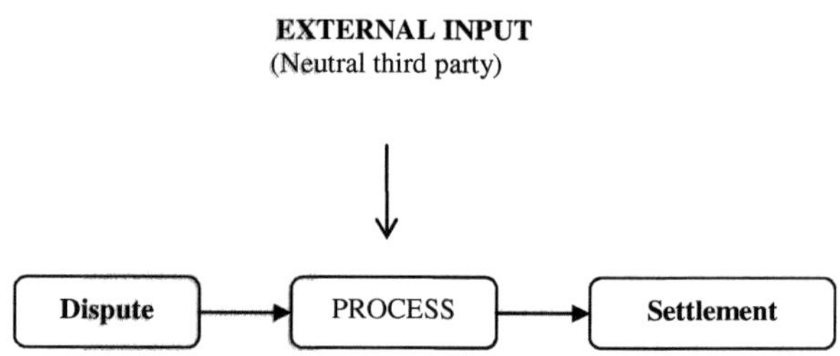

approaches (Cheung et al. 2001). The AHP employs pair-wise comparison between attributes by the decision maker and is used in this study to prioritise the set of alternative dispute resolution process attributes obtained in Stage One. The hierarchical model of dispute resolution process attributes is shown in Fig. 16.1.

The nineteen attributes are arranged under four main headings: *Nature*; *Neutral third party*; *Settlement*, and *Benefit*. The rationale for such categorisation is based on Walker's (1996) system view of a process. A typical alternative dispute resolution process consists of input (dispute), process (assisted negotiation) and output (settlement). Figure 16.2 shows a typical dispute resolution process with the input of a neutral third party.

The process begins once the parties agree to adopt a resolution strategy to resolve their dispute. It is an inherent nature of dispute resolution that human factors play an important role in the process. In ADR processes in particular, parties have control over the content and procedure of the process, which are not possible in litigation and arbitration. Therefore, attributes including range of issues; voluntariness; control by parties; flexibility; informality; privacy; and confidentiality naturally fall under the *Nature* category. The involvement of a neutral third party to assist the parties in reaching a settlement is another main attribute of dispute resolution. In mediation, the impartial third party takes a facilitative role to ensure that the process is fair and that the mediated settlement is satisfactory to both parties. The effectiveness of the process depends heavily on the competence and experience of the facilitator (Brown and Marriott 1999; Goldberg et al. 1992). Hence, attributes such as neutrality; power to compel consolidation; and knowledge in construction are grouped under the *Neutral third party* category. In reality, a mutually agreed settlement is what the parties would want to achieve at the end of the process. A settlement is the result of the collaborative efforts of the parties and the 3rd party neutral. Therefore, consensus, fairness, creative agreement, scope of remedy to satisfy interest, and enforceability all fall into the *Settlement* category. As for the remaining factors, namely speed, cost, liabilities for opponent's cost, and preservation of business relationship, they are arranged under the *Benefit* category.

16.2.3 Stage Three: Identification of Critical Attributes Through Prioritisation

The ExpertChoice software (ExpertChoice 1998) featuring the Analytical Hierarchy Process (AHP) methodology is employed to prioritise the different attributes identified so far. The top ten ranked attributes are then classified as the critical attributes of the alternative dispute resolution process. The AHP employs a pair-wise comparison between attributes, thus enabling the checking of consistency in scale assignments. The section below explains the working procedure of the prioritisation, followed by an illustration.

Figure 16.1 presents the hierarchical arrangement of the 19 attributes. Level 1 (top level) is entitled 'The Attributes of ADR Processes'. The main attributes at Level 2 are: *Nature*, *Neutral third party*, *Settlement*, and *Benefit*. Level 3 consists of sub-attributes to each of the main attributes at Level 2. These sub-attributes are the nineteen attributes as listed in Fig. 16.1. Such arrangement allows pair-wise comparison to be performed between sub-attributes of the same main attribute group at first. Using Fig. 16.1 as an example, under the *Neutral third party* group, the expert is required to make judgments on the relative importance weightings of the sub-attributes. Hence, *neutrality* is compared with *power to compel consolidation,* and then with *knowledge in construction*, followed by a comparison between *power to compel consolidation* and *knowledge in construction*. The pair-wise comparison matrix for *Neutral third party* is shown in Fig. 16.3 below.

Pair-wise comparison has the advantage of focusing exclusively on two sub-attributes at a time. Another advantage of using AHP is its ability to check the consistency of scale assessments. An Inconsistency Index (II) is calculated automatically by ExpertChoice for each matrix by the software. A scale assessment can be accepted if its Inconsistency Index is 0.1 or less (Saaty 1980). The measure of inconsistency is useful in identifying possible errors in judgments, as well as the

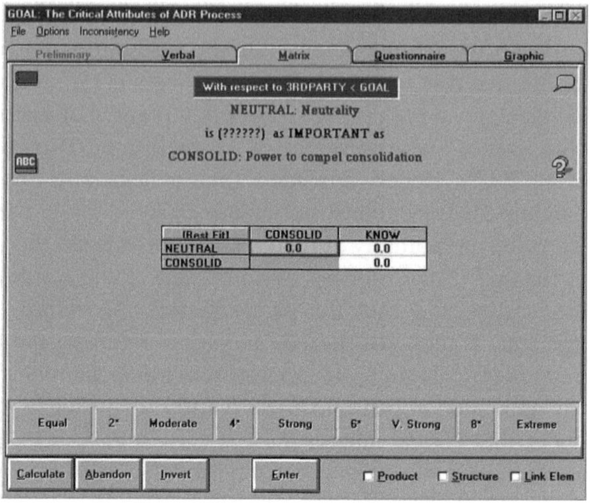

Fig. 16.3 Pair-wise comparison of sub-attributes

Table 16.2 9-point pair-wise comparison scale

Numerical scale	Verbal meaning
1	Equal importance of both elements
3	Moderate importance of one element over the other
5	Strong importance of one element over the other
7	Very strong importance of one element over the other
9	Extreme importance of one element over the other
2, 4, 6, 8	Intermediate values between the above adjacent values

extent of inconsistencies in the judgment themselves. This distinctive feature of AHP makes it an appealing tool when compared with other approaches dealing with subjective ranking (Chua et al. 1999). This approach to solicit an expert's judgment has made it a reliable tool to determine the priorities of a set of factors (Chua et al. 1999). The pair-wise comparisons are guided by a nine-point scale as shown in Table 16.2. The experts formulate their assessments based on this scale (Saaty 1980). There are therefore five matrices to be completed by each expert—four for the four main attribute groups, and one for the comparison of the four main attributes themselves.

The following serves to illustrate how the experts would perform the prioritisation of the nineteen attributes. In the *Neutral third party* group, if the sub-attribute *NEUTRAL* is considered to be moderately important compared with the sub-attribute *CONSOLID*, a "3" is inserted into the matrix table. If the sub-attribute *NEUTRAL* is considered to be extremely important compared with the sub-attribute *KNOW*, a "7" would be inserted. If the sub-attribute *CONSOLID* is considered to be equally important compared with the sub-attribute *KNOW*, a "1"

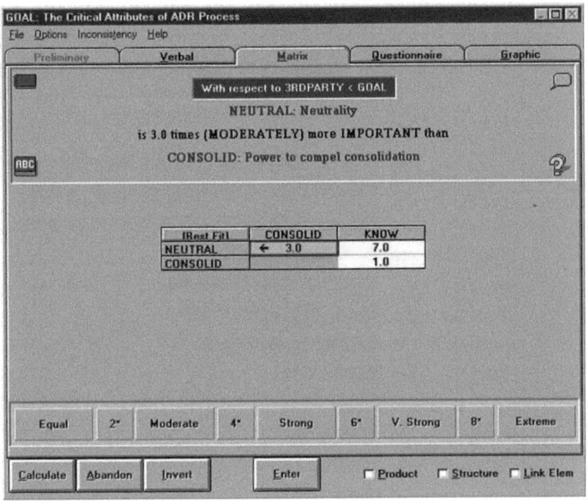

Fig. 16.4 Pair-wise comparison matrix

is inserted. The shaded portion of the comparison matrix need not be completed because these cells should be the reciprocals of the corresponding cells in the non-shaded portion. Figure 16.4 shows the pair-wise comparison matrix for *Neutral* with scale assignments inserted.

The same scale assignment applies to the other three main attribute groups as well as the matrix involving the four main attributes. The ExpertChoice (1998) provides a summary analysis of the results automatically. Figure 16.5 shows the distributive summary of the results. It clearly shows that the relative standings of the nineteen attributes in descending order upon the completion of the five matrices.

The mathematics underlying the use of the AHP techniques to generate the relative importance weightings for the critical attributes are based on linear algebra and graph theory. It is beyond the scope of this study to discuss the mathematical theory in depth. Details of the mathematical treatment and proof can be found in the works of Saaty (1980, 1988), Saaty and Vargas (1991). Having explained the procedures involved in the prioritisation of the nineteen attributes and the rationale for the hierarchical structure, the following explains the procedures involved in the collection of data, and the results so obtained.

16.3 Pilot Study and Main Survey

A pilot study was conducted with a panel of experienced experts in alternative dispute resolution. The pilot study seeks to test the running of the prioritisation exercise and to identify system deficiencies, so as to ensure the efficient running of the formal exercise with the experts.

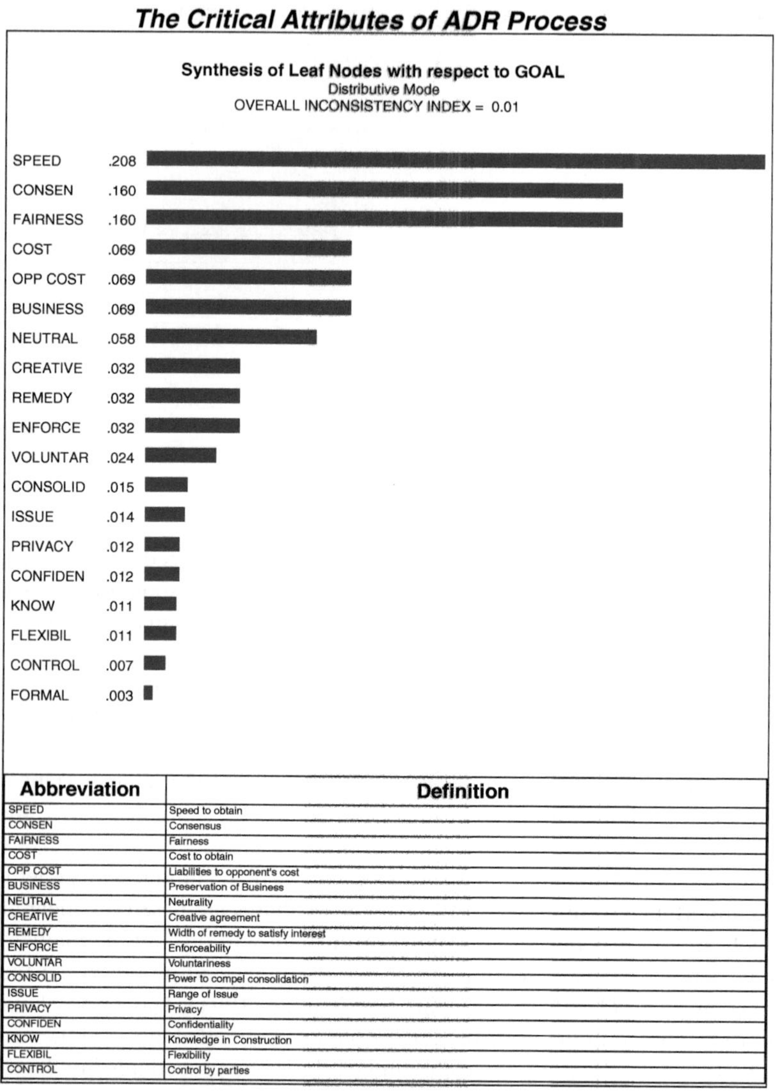

Abbreviation	Definition
SPEED	Speed to obtain
CONSEN	Consensus
FAIRNESS	Fairness
COST	Cost to obtain
OPP COST	Liabilities to opponent's cost
BUSINESS	Preservation of Business
NEUTRAL	Neutrality
CREATIVE	Creative agreement
REMEDY	Width of remedy to satisfy interest
ENFORCE	Enforceability
VOLUNTAR	Voluntariness
CONSOLID	Power to compel consolidation
ISSUE	Range of Issue
PRIVACY	Privacy
CONFIDEN	Confidentiality
KNOW	Knowledge in Construction
FLEXIBIL	Flexibility
CONTROL	Control by parties

Fig. 16.5 Distributive summary of the results

The pilot study brought out several improvements. One of them was to put the largest matrix towards the end of the exercise. By going through the smaller matrices first, the experts can acquire more experience with the scale assignment, before dealing with the more complicated ones. This improvement was proved valuable, as it prevents frustration from building up at the early stage as a result of any failure to achieve an II of 0.1 or below. Also, as the experts get more familiar with the scale assignment, naturally it becomes less difficult to achieve the required IR even for the larger matrices.

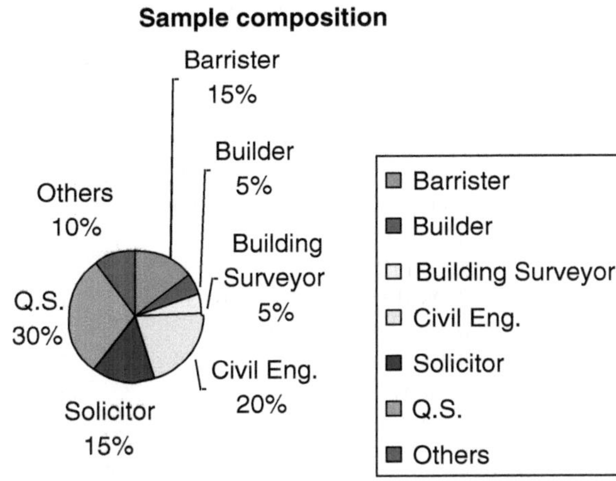

Fig. 16.6 Composition of experts by profession

The prioritisation of attributes was then performed face to face with each of the participating experts. The experts were selected according to the following criteria:

- Practitioners who have extensive working knowledge of the construction industry in Hong Kong; and
- Practitioners who have profound knowledge of and experience with the various ADR processes; must be a member of the HKIAC; having at least 2 years ADR experience;

A sample of forty-nine accredited mediators and arbitrators were approached. The panel of experts consisted of arbitrators and mediators from the HKIAC who are also practicing barristers, solicitors, quantity surveyors, civil engineers and construction related professionals. A total of twenty experts agreed to participate in the research. This translates to a response rate of 40 %. The aggregated sample composition by professions is given in Fig. 16.6.

All respondents are in senior management positions of leading private construction companies or consultant firms in Hong Kong. Some are serving in governmental departments, and some are barristers in chambers or solicitors in law firms. They are well-recognised experts in construction dispute resolution in Hong Kong. 70 % of the respondents have more than 10 years of experience with dispute resolution; 15 % of them have between 5 and 10 years of experience; and the remaining 15 % between 1 and 5 years. Although the size of the sample is relatively small, the accumulated knowledge of the respondents in construction dispute resolution is beyond doubt. In fact, some of the experts are founders and major proponents of the ADR process in Hong Kong. Therefore, their views should be reliable and are reflective of the current situations of the industry. To ensure good, truthful replies and good quality data, a brief introduction was made to each respondent to explain the objectives of the study and the methodology adopted. It was made clear at the beginning of the interviews that the study is focused on the group of alternative dispute resolution processes. Litigation and

Table 16.3 RIW of SA assessed by participant no. 1

Order	SA	RIW	Order	SA	RIW
Under *NATURE* Group			Under *SETTLEMENT* Group		
1.	VOLUNTAR	0.398	1.	CONSEN	0.330
2.	FLEXIBIL	0.159	2.	REMEDY	0.222
3.	ISSUE	0.149	3.	ENFORCE	0.222
4.	CONTROL	0.100	4.	FAIRNESS	0.174
5.	PRIVACY	0.080	5.	CREATIVE	0.052
6.	CONFIDEN	0.080	Under *BENEFITS* Group		
7.	FORMAL	0.035	1.	BUSINESS	0.696
Under *NEUTRAL 3rd PARTY* Group			2.	COST	0.140
1.	NEUTRAL	0.735	3.	SPEED	0.082
2.	KNOW	0.207	4.	OPP COST	0.082
3.	CONSOLID	0.058	–	–	–

Keys Sub-attributes (*SA*), Relative Importance Weightings (*RIW*)

Table 16.4 Relative weightings of main attributes

Order	Attributes	Relative importance weightings
1.	*SETTLE*	0.354
2.	*BENEFITS*	0.354
3.	*NEUTRAL 3rd PARTY*	0.161
4.	*NATURE*	0.131

arbitration are regulated by existing laws and therefore fall outside the scope of this study. The twenty practitioners were asked to prioritise the nineteen attributes listed in Table 16.3 using the Expert Choice software. The top ten ranked attributes were identified as the critical attributes.

16.4 Stage Four: Collecting Suggestions on Means to Enshrine the Critical Attributes

Suggestions on the means to enshrine the top ten ranked attributes were collected from each of the experts. These suggestions are summarised and discussed in the later sections of this chapter.

16.4.1 Result of Prioritisation

The following presents the results from one of the experts. Table 16.3 gives the relative weightings of the sub-attributes under each of the main attributes. Table 16.4 presents the relative weightings between the four main attributes themselves. Table 16.5 gives the overall weightings of the nineteen attributes.

Table 16.5 Overall relative standings of attributes

Order	Attributes	RIW	Order	Attributes	RIW
1.	BUSINESS	0.246	10.	SPEED	0.029
2.	NEUTRAL	0.118	11.	OPP. COST	0.029
3.	CONSEN	0.117	12.	FLEXIBIL	0.021
4.	REMEDY	0.079	13.	ISSUE	0.020
5.	ENFORCE	0.079	14.	CREATIVE	0.019
6.	FAIRNESS	0.061	15.	CONTROL	0.013
7.	VOLUNTAR	0.052	16.	PRIVACY	0.010
8.	COST	0.050	17.	CONFIDEN	0.010
9.	KNOW	0.033	18.	CONSOLID	0.009
–	–	–	19.	FORMAL	0.005

Table 16.6 Relative standings of the nineteen attributes

Order	Attributes	Main attributes	RIW
1.	Preservation of business relationship	Benefits	0.119
2.	Enforceability	Settlement	0.101
3.	Neutrality	Neutral	0.097
4.	Consensus	Settlement	0.091
5.	Cost to obtain	Benefits	0.083
6.	Speed to obtain	Benefits	0.071
7.	Fairness	Nature	0.059
8.	Scope of remedy to satisfy interest	Nature	0.050
9.	Creative agreement	Nature	0.041
10.	Confidentiality	Nature	0.039
11.	Voluntariness	Nature	0.036
12.	Knowledge in construction	Neutral	0.035
13.	Privacy	Nature	0.034
14.	Liabilities to opponent's cost	Benefits	0.033
15.	Power to compel consolidation	Neutral	0.031
16.	Flexibility	Nature	0.027
17.	Control by parties	Nature	0.025
18.	Range of issue	Nature	0.013
19.	Formality	Nature	0.009

A total of twenty prioritisation exercises were conducted. The relative standings of the nineteen attributes, based on the average of the relative weightings obtained in the twenty exercises, are presented in Table 16.6.

16.4.2 Suggestions on Means to Enshrine the Critical Attributes

Upon completion of the prioritisation exercise, each expert was asked to suggest means to enshrine the top ten attributes ranked by him/her. Some experts suggested that some means are common to more than one attributes. These include: neutrality and fairness, scope of remedy to satisfy interest and creative agreement. Taking into account of these comments, the final ten most important critical attributes were: voluntariness, confidentiality, neutrality/fairness, knowledge in construction, consensus agreement, creative agreement, enforceability, speed, cost, and preservation of relationship. These are found to be consistent with the data from similar studies in the UK (Brooker and Lavers 2000a). In that study, it was found that where on-going relationship, privacy, speed or economy of resolution were desired, mediation and mini-trial were considered to be suitable strategies to be adopted. Table 16.7 presents a summary of the means to enshrine the critical attributes as suggested by the panel of experts. These are discussed seriatim in ascending order of criticality, as ranked in this study.

16.4.2.1 Voluntariness

In a purely consensual ADR process, nothing is binding on the parties until the parties sign an enforceable settlement agreement. In other words, the parties can walk out at any time during the process without interfering with their legal rights. Therefore, the use of ADR process will not be successful unless there is a basic willingness to take part and attempt a settlement (Bevan 1992; Cheung 1999; Pengilley 1990).

To foster voluntariness, the disputants need to be educated on the benefits of the process, as compared with arbitration and litigation. The 3rd party neutral should make the parties fully aware of the consequences and costs involved if the dispute is to be settled by arbitration or litigation. The best time to advise the parties about costs is before the process begins, rather than during the often emotionally-charged process. During a heated argument, parties tend to forget about the real issues, and focus instead on personal issues. It is also part of the neutrals' responsibilities to educate their clients on the perceived benefits of ADR, such as being less expensive, confidential, voluntary, capable of more remedies, and maintenance of relationship etc.

16.4.2.2 Confidentiality

Confidentiality is one of the essential terms governing the conduct of the parties in a purely consensual ADR process (York 1996). It is an implied and inherent feature of ADR processes that parties to a dispute are not allowed to disclose any information or materials to the public unless by mutual consent of the parties. This is normally achieved by establishing 'house rules' in the form of a written agreement between the parties to that effect. House rules should be laid out at the very beginning, requiring communications between the parties and the 3rd party neutral to be made in confidence. Rules can be stipulated into an agreement that

Table 16.7 Summary of means to enshrine the critical attributes

Critical attributes/means suggested by experts	Critical attributes/means suggested by experts
Voluntariness • Parties' willingness to settle • Use ADR process instead of litigation • Neutrals make parties fully aware of the consequences if the dispute was to be resolved by traditional court process • Neutrals explain to the parties the benefits of using ADR process	Creative agreement • Neutrals should explore alternative settlement options • Careful selection of the neutrals, those with good all-round experience in the various dispute resolution strategies, and the ability to critically analyse disputes and come up with creative answer that can satisfy the parties
Confidentiality • Written agreement between parties • Keep the process and related materials strictly confidential • Allow only the parties involved in a dispute to participate the process	Enforceability • Selecting neutrals with the competence in drawing up agreement/award • Signed agreement between parties • Parties are committed to settle
Neutrality/Fairness • Improve competence and judicial quality of neutrals; continuing profession development • Maintain a pool of experienced mediators and arbitrators • Neutrals to make mandate statement declaring any conflict of interest • Careful selection of neutrals by parties, study neutral's track records	Speed • Neutrals keep close eye on the process • Parties' willingness to co-operate • Documents-only process. Avoid unnecessary procedures • Use ADR as oppose to arbitration and litigation • Make sure parties are thoroughly briefed about all the facts • Focus on key issues • Good time management. Don't waste time on things that do not matter
Consensus agreement • Neutrals help to consolidate differences • Parties' willingness to adopt dispute resolution • Neutrals advise the most appropriate contract provision • Neutrals make sure the parties understand the critical issues of dispute • Neutrals ensure the process is conducted in a non-adversarial manner	Cost • If the nature of dispute allows, avoid the involvement of lawyer • Limiting discoveries • Good time management. Set time limit in hearing • Neutrals explain to parties the importance of time, and the implications if the dispute drags on without resolution • Documents-only process • Proactive neutrals. Neutrals are given authority/teeth to control the process • Less legal input and more parties involvement • Focus on major issues and not to be caught up by minor details
Knowledge in construction • Continuing profession development • Promotion of ADR workshops	Preservation of business relationship • Parties' willingness to adopt dispute resolution scheme • Parties are willing to accommodate differences • Emphasis on consensus through ADR, avoid going to arbitration and litigation • Parties show respect to each other • Avoid heated confrontation. Emphasise win/win solution • Don't take issues personal • Neutrals encourage open and honest discussion

should be signed by the parties before the commencement of the process. The neutrals also have a duty to remind the parties to follow the house rules strictly during the process.

16.4.2.3 Neutrality/Fairness

Neutrality and fairness depend heavily on the competence, training and integrity of the 3rd party neutrals. During the resolution process, a 3rd party neutral owes a duty of care to his/her clients to remain impartial. He/she facilitates the parties to reach a settlement. In addition, he/she must make conscientious effort to avoid personal biases creeping in.

The possible means to achieve a fair process include:

(a) Maintenance of a panel of experienced 3rd party neutrals: Since the skills, knowledge and experience of a 3rd party can be the determinant of the success or failure of a resolution process, the choice of the 3rd party neutral is therefore of paramount importance. For example, the HKIAC has developed a code of conduct to monitor the standard of professional mediators. The Centre also maintains two panels of mediators and a roster of arbitrators. Only those who have successfully fulfilled the qualifications or requirements set out by the Centre can apply to be an accredited member.
(b) Choice of the 3rd party neutral: A 3rd party neutral is often appointed through mutual agreement by the parties. The selection criteria may include past track record, experience, knowledge, and professional/academic background. The parties should only appoint a person whom they can trust and feel comfortable with. In this regard, a mandate statement by the appointee to declare his interests is recommended.

16.4.2.4 Knowledge in Construction

The expertise of a 3rd party neutral can be a major advantage of the ADR process. If the dispute involves issues of a very technical nature, it is desirable to have a 3rd party neural who have at least some related background knowledge. Suggested means to assist the 3rd party neutral include the use of continuing professional programme to keep facilitators updated with both skills and technical matters. For example, the Hong Kong Mediation Centre, in collaboration with the Law Society of Hong Kong, runs introductory mediation training courses for the training of general and commercial mediators (Wall 2000).

16.4.2.5 Consensus Agreement

Without the parties' commitment to the process, it is often difficult, if not impossible, to reach an agreement. A 3rd party neutral should ensure that the process itself is conducted in a non-adversarial manner. He/she should make sure that the parties are aware of each other's needs. He/She should act as a facilitator rather than an adjudicator, leaving all major decisions to the parties. It is part of his/her responsibility to help identify common grounds, so that the parties can begin negotiation more easily, and to advise on the relevant procedures.

16.4.2.6 Creative Agreement

Depending on the nature and the requirements of parties, the 3rd party neutral should try to come up with a solution that can satisfy both parties' needs. ADR process can offer a greater range of settlement options than in litigation or arbitration. Settlements may include face-saving concessions, in which case human factors such as business relationships will be considered. Lateral thinking is vital given the number and variety of factors that must be taken into consideration before a settlement can be reached. Possible means to achieve a creative agreement are:

(a) *Reality testing:* The 3rd party neutrals can help by writing down all the possible solutions and testing them against one another. Some solutions may be more preferable in certain situations and it is very much up to the parties to decide after weighing the pros and cons.
(b) *Training of neutrals:* The more imaginative the facilitator is, the more likely that he will put forward creative settlement suggestions.

16.4.2.7 Enforceability

In practice, a facilitated settlement cannot be enforced unless an agreement has been concluded. The settlement agreement should always reflect the true intention of both parties. How capable the facilitator is in producing a draft to that effect will therefore be crucial. As each dispute is unique, the facilitator should be prepared to come up with solutions that the parties would be willing to agree upon.

16.4.2.8 Speed

In Hong Kong in particular, people favour speedy settlement because they are often under time constraints and have busy schedules. It is difficult to tell how long a settlement typically takes as each dispute is unique. The duration of an ADR process can be measured in days or weeks, rather than months or years, as can be the case in litigations or arbitrations. To a large extent, it depends on the complexity and nature of the dispute, together with the number of parties involved. Other external factors, such as political, financial and human factors, may also cause delays in resolution. Hence, it lies heavily on the skills of the 3rd party neutrals to control the pace of proceeding. It was suggested that a speedy resolution could be achieved through:

(a) Documents-only-process: A documents-only process can greatly reduce the overall time to reach a decision. Time is saved from not having to involve discovery and expert witnesses, which could take up months before the commencement of process. However, in cases where technical issues predominate, discovery or expert witnesses are inevitable.
(b) Time management: Unnecessary procedures should be avoided. The 3rd party neutral should make sure that the parties are thoroughly briefed about all the facts and procedures before actually dealing with the issue. He/She should keep a close eye on the process, set time limits if necessary, and constantly remind the parties to focus on the most important issues and not being carried away by the minor ones.
(c) Early settlement: As soon as disputes arise, do not wait until they turn into full-blown problems. The parties should seek third-party assistance when they find that it is not possible to resolve the dispute between them.

16.4.2.9 Cost

Cost and time are twin attributes. A speedy resolution means lower costs. Sources of costs involved in reaching a settlement include: expense related to venue, the hiring of the 3rd party neutral, documentation, and settlement costs. Some of the suggested means to curb costs are:

(a) Use of Partnering: The central idea of using Partnering is to reduce the adversarial relationship between two parties and to encourage the parties to work in a cooperative manner. By being open and honest to each other, it is anticipated that the chance of conflict will be greatly reduced, resulting in fewer disputes ultimately. Partnering also helps the parties to establish long-term working relationship.
(b) Competent 3rd party neutral: The 3rd party neutral can help to reduce overall cost by ensuring that parties are working towards the same common goal.
(c) Less legal input and greater involvement of the disputants: Where the nature of dispute allows, especially in cases where no point of law is involved, the disputants should have a substantial involvement in the resolution process instead of leaving it to the lawyers.
(d) Cost-benefit analysis: In assessing the suitability of a case for ADR, a cost-benefit analysis must be undertaken. Cost-benefit analysis enables the parties to have better understanding of the key critical issues and the likely expense should the dispute continues. "Reality testing" can also be performed (Brooker and Lavers 2000b).

16.4.2.10 Preservation of Relationship

Continuing relationship with business partners is one of the key objectives of using ADR. An enduring relationship is always founded on common interests, mutual trust and respect. It requires the efforts and commitments from both parties to maintain the relationship. The means suggested to preserve a relationship are:

(a) Avoidance of arbitration and litigation: Many cases have demonstrated that relationship fractures when a dispute is resolved by means of arbitration and litigation.
(b) Avoidance of confrontation: There is no point in continuing a discussion if the parties are not willing to compromise, or when they are emotionally charged. In order to achieve a win/win situation, both parties must learn to focus on the real issue and not be caught by emotional desires.
(c) Be reasonable: It is easier to remain reasonable with the assistance of a 3rd party neutral, who should always try to prevent confrontations during the resolution process.

In sum, it can be seen that ADR processes need not be complex. The success of such a process depends very much on the attitude of the disputants. If they have no desire to settle, or are unwilling to make compromises at all, no matter how detailed the process is planned, there will be little hope for success. The 3rd party neutral can be extremely instrumental. He/she must have the confidence of the parties. With his/her skills and knowledge, amicable outcomes can be achieved out of the resolution process.

16.5 Chapter Summary

Resolving construction dispute is no easy task, especially when the available resources are limited or when the dispute is complex. The use of ADR processes in construction is an attempt to overcome the shortcomings of litigation and arbitration. However, overtly complex resolution procedures, which involve the sequential use of a range of ADR techniques and arbitration, destroy the intended positive effects, especially in terms of time and cost savings. By focusing on the critical attributes, ADR processes can be kept simple and effective. To achieve this objective, a hierarchical model is used to structure the nineteen attributes identified in literature reviews. With the use of AHP methodology and the profound knowledge of a panel of experts in the field of construction dispute resolution, the nineteen attributes are prioritised. The top ten ranked attributes are: voluntariness, enforceability, creative agreement, knowledge in construction, consensus agreement, confidentiality, neutrality/fairness, speed, cost, and preservation of relationship. The panel of experts also suggested means to enshrine these critical attributes. It is suggested that the 3rd party neutral can be extremely instrumental to facilitate a settlement. Nonetheless, a settlement can hardly be reached if the disputing parties are not committed to the process.

Acknowledgments The authors would like to express their appreciation to the accredited mediators and arbitrators of the HKIAC for their information and participation in the research. The content of this chapter has been published in Volume 128(5) of the Journal of Construction Engineering and Management and is used with the permission from ASCE.

References

Alhazmi, T., & McCaffer, R. (2000). Project procurement system selection model. *Journal of Construction Engineering and Management ASCE, 126*(3), 176–184.

Bevan, A. H. (1992). *Alternative dispute resolution*. London: Sweet & Maxwell.

Brooker, P., & Lavers, A. (2000a). Appropriate ADR—identifying features of construction disputes which affect their suitability for submission to ADR. *ICLR*, vol. 17, part 2.

Brooker, P., & Lavers, A. (2000b). Appropriate ADR and the CPR rules. *International Council for Research and Innovation in Building and Construction, Working Commission 103 (W103) Construction Conflict: Avoidance and Resolution*, Manchester.

Brown, H., & Marriott, A. (1999). *ADR principles and practice* (2nd ed.). London: Sweet & Maxwell.

Cheung, S. O. (1999). Critical factors affecting the use of alternative dispute resolution processes in construction. *International Journal of Project Management, 17*(3), 189–194.

Cheung, S. O., Lam, T. I., Leung, M. Y., & Wan, Y. W. (2001). An analytical hierarchy process bases procurement selection method. *Construction Management and Economics, 19*(4), 427–437.

Chua, D. K. H., Kog, Y. C., & Loh, P. K. (1999). Critical success factors for different project objectives. *Journal of Construction Engineering and Management, 125*(3), 142–150.

Construction Industry Review Committee (CIRC). (2001). Construct for excellence, construction industry review committee, The government of the Hong Kong Special Administrative Region.

David, J. (1988). Dispute resolution for lawyers-overview of range of dispute resolution processes. The University of Sydney, Faculty of Law, Continuing Legal Education, Sydney.

Egan, J. (1998). Rethinking construction, Department of the Environment, Transport and the Regions, HMSO, London.

ExpertChoice. (1998). *Team expert choice: Advanced group decision support software*. USA: ExpertChoice Inc.

Fong, S. W., & Choi, K. Y. (2000). Final contractor selection using the analytical hierarchy process. *Construction Management and Economics, 18*(5), 547–557.

Fenn, P., & Gameson, R. (1992). Construction conflict management. *Proceedings of the First International Construction Management Conference*. The University of Manchester. Institute of Science and Technology 25–27 September 1992, E & FN Spon., London.

Goldberg, S. B., Sander, F. E. A., & Roger, N. H. (1992). *Dispute resolution-negotiation, mediation, and other processes*. Canada: Little, Brown and Company.

Hibberd, P., & Newman, P. (1999). *ADR and adjudication in construction disputes*. London: Blackwell Science Ltd.

Hong Kong International Arbitration Centre (HKIAC). (2012). Hong Kong International Arbitration Centre Homepage: Services and Facilities. http://www.hkiac.org.

Kaplan, N., Spruce, J., & Cheng, T. Y. W. (1991). *Hong Kong arbitration cases and materials*. Hong Kong: Butterworths.

Kumaraswamy, M. M. (1997). Common categories and causes of construction claims. *Construction Law Journal, 13*(1), 21–34.

Latham, M. (1994). *Constructing the team: final report by Sir Michael Latham*. Joint Review of Procurement and Contractual Arrangements in the United Kingdom Construction Industry, HMSO, London.

Merna, A., & Bower, D. (1997). *Dispute resolution in construction & infrastructure projects*. Hong Kong: Asia Law & Practice Publishing Ltd.

Pengilley, W. (1990). Alternative dispute resolution: the philosophy and the need. *Australian Dispute Resolution Journal, 1*(2), 81–95.

Saaty, T. L. (1980). *The analytical hierarchy process*. New York: McGraw.

Saaty, T. L. (1988). *The analytic hierarchy process: planning, priority setting, resources allocation*. London: McGraw-Hill.

Saaty, T. L., & Vargas, L. G. (1991). *Prediction, Projection and Forecasting*. Boston: Kluwer Academic.

Shen, Q., Lo, K. K., & Wang, Q. (1998). Priority setting in maintenance: a modified multi-attribute approach using analytical hierarchy process. *Construction Management and Economics, 16*(6), 693–702.

Walker, A. (1996). *Project management in construction*. London: Blackwell Science.

Wall, C. (2000). What's new in member countries: Hong Kong-monograph of practice and procedure in Hong Kong. *International Council for Research and Innovation in Building and Construction, Working Commission 103 (W103) Construction Conflict: Avoidance and Resolution*, Manchester.

York, S. (1996). *Practical ADR*. London: Person Professional Limited.

[illegible]

Chapter 17
Contractual Use of Alternative Dispute Resolution

Sai On Cheung

Abstract Courtroom is conventionally recognised as the place for justice. Subjecting a dispute to formal processes like litigation and arbitration is thus considered as the most natural and logical by many people. However, it is virtually impossible for disputing parties who had ruined their relationships in adversarial proceedings. Evidently a better form of dispute resolution that directs problem solving shall be employed. Alternative dispute resolution (ADR) techniques have been viewed as effective means to speedily and economically resolve construction dispute. This chapter firstly reviews some of the ADR initiatives in Hong Kong. The approaches taken in several common law jurisdictions in the use of alternative dispute resolution to deal with construction disputes are compared. In addition, the voluntary mediation procedures introduced under the Civil Justice Reform in Hong Kong is outlined. Adverse cost order is used to discourage 'refusal to mediate' and 'failing to attempt to mediate'. Nonetheless, the cost sanction may make the voluntary use of mediation less voluntary.

17.1 Introduction

Courtroom is conventionally recognised as a place for justice. Subjecting a dispute to a court order is thus a natural and acceptable option to many people. Notwithstanding this legal perspective of dispute resolution, Bryan and Philips (2007) of the International Institute for Conflict Prevention and Resolution aptly reminded the importance of bringing 'business sense' back to dispute resolution. This is

S. O. Cheung (✉)
Construction Dispute Resolution Research Unit, Department of Civil and Architectural Engineering, City University of Hong Kong, Hong Kong, People's Republic of China
e-mail: saion.cheung@cityu.edu.hk

S. O. Cheung (ed.), *Construction Dispute Research*,
DOI: 10.1007/978-3-319-04429-3_17,

advocated in the light of the dissatisfaction of the exorbitant cost involved in litigation and arbitration as well as the draconian relationships between the disputing parties. In fact, it is virtually impossible for disputing parties who had ruined their relationships through arbitration or litigation to have further business. Evidently a better form of dispute resolution that directs problem solving shall be employed. Alternative dispute resolution (ADR) techniques have been viewed as effective means to speedily and economically resolve construction disputes. In this regard, some jurisdictions have opted to use mandatory adjudication to deal with construction, in particular payment-related disputes. Moreover, Hong Kong has opted for voluntary mediation. In his 2007–2008 Policy Address, *A New Direction for Hong Kong*, the Chief Executive of the Hong Kong Special Administrative Region (HKSAR) pledged to develop arbitration and mediation services in Hong Kong. With this policy decision, mediation will become the mainstream ADR technique to resolve civil disputes in Hong Kong. This chapter firstly reviews some of the ADR initiatives in Hong Kong. A comparison of the approaches taken in a number of common law jurisdictions to deal with construction disputes is also presented. Furthermore, voluntary mediation has also been introduced in the civil procedures rules of the High Court as part of the Civil Justice Reform launched in 2009. Adverse cost order is used to discourage 'refusal to mediate' and 'failing to attempt to mediate'.

17.2 Dispute Resolution/Settlement Provisions in Hong Kong Construction Contracts

The dispute resolution/settlement provisions of the following standard forms of construction contract are examined to identify any trend in the choice of dispute resolution method. Flow charts are used to illustrate the working process.

- The Government of Hong Kong, General Conditions of Contract for the Airport Core Programme Civil Engineering Works, 1992 edition.
- The Government of Hong Kong, General Conditions of Contract for Building Works, 1999 edition.
- The Agreement and Schedule of Conditions of Building Contract for use in HKSAR, 2005 edition.
- The Dispute Resolution Advisor System.

In addition, the Dispute Resolution Advisor system promoted by the Architectural Services Department of the Hong Kong Special Administration Region will also be discussed.

17.2.1 The Hong Kong Government General Conditions of Contract for Airport Core Programme (HKACP92)

The Hong Kong Government General Conditions of Contract for Airport Core Programme (HKACP92) was published in 1992 for use in the ten projects under the Airport Core Programme (ACP). The ACP is part of the Port and Airport Development Strategy (PADS). In the late 1980s, the Hong Kong Government initiated the formulation of the Metroplan to restructure the city so as to bring about a better organised, more efficient and more desirable place in which to live and work. PADS is part of this Metroplan, and is seen as the major infrastructure investment of Hong Kong. ACP consists mainly of ten infrastructure projects including the Airport at Chap Lap Kok and the associated site formation, railway and roadwork. Smooth running of these construction projects was of prime concern and disruption should be kept to the minimum, so that the airport can be opened as scheduled. The rationale behind this dispute resolution provision therefore is to encourage early resolution of any disputes. One aspect of such is the strict procedural requirement. The dispute settlement procedure is presented as a flow chart (Fig. 17.1). In principle, a three-tier dispute resolution procedure is implemented. Similar to the other Conditions of Contract discussed before, a dispute arises if either party disagrees with the decision of the supervising officer. Clause 92(5) requires the dispute to be referred to mediation. And under Clause 92(11)(a), it shall be a *condition precedent* to the commencement of any reference of a dispute to adjudication, arbitration or an action at law that the issues arising in the dispute shall have been subject of a reference under the mediation procedure. If the dispute cannot be resolved by the mediation, the dispute can then be referred to adjudication. As seen in Fig. 17.1, adjudication can be bypassed if there is no request for adjudication. This has the similar effect of the voluntary use of mediation under the HKG99 Form. Arbitration can also be commenced without going through adjudication if the dispute is not related to payment or extension of time or the certificate of completion (or the last one if more than one) has been issued. Under these circumstances, reference to arbitration can be commenced without going through the adjudication process. Under Clause 92(8)(9), the decision of the adjudicator is final and binding unless and until the dispute has been settled or an award made in a subsequent arbitration. Again arbitration can only be opened after substantial completion.

The rules for arbitration, adjudication and mediation are included in part III of the General Conditions. While mediation appears to be compulsory, however, under Clause 92(11)(b), a dispute shall be deemed to have been the subject of a reference under mediation if a period of 42 days had elapsed after the service of a request for mediation in respect of such dispute being served strictly in accordance with Clause 92(6). This seriously erodes the compulsory intent of the use of mediation.

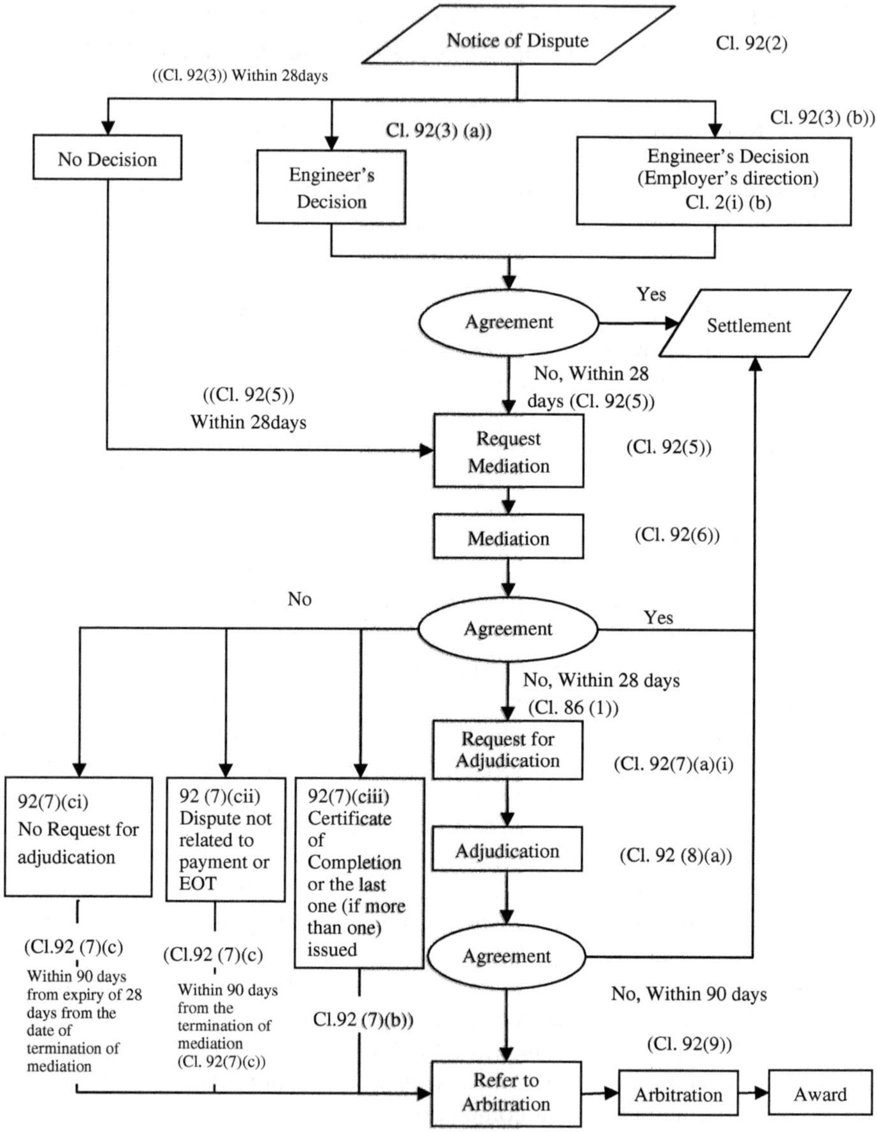

1. ANY and ALL disputes shall be settled in accordance with Clause 92. Cl. 92(1)
2. It shall be a condition precedent to the commencement of any reference of a Dispute to adjudication, arbitration or of an action at law that the issues arising in the Dispute shall have been subject of a reference under the Mediation Procedure. Cl. 92(11)(a)
3. A dispute shall deemed to have been the subject of a reference under the Mediation Procedure if a period of 42 days has elapsed after the service of a Request for Mediation in respect of such Dispute being served strictly in accordance with Cl. 92(5) or, as the case may be, if consent to late service thereof was given in accordance with the proviso to Cl. 92(6). Cl. 92(11)(b)

Fig. 17.1 The dispute resolution procedure under the HKACP92 contract

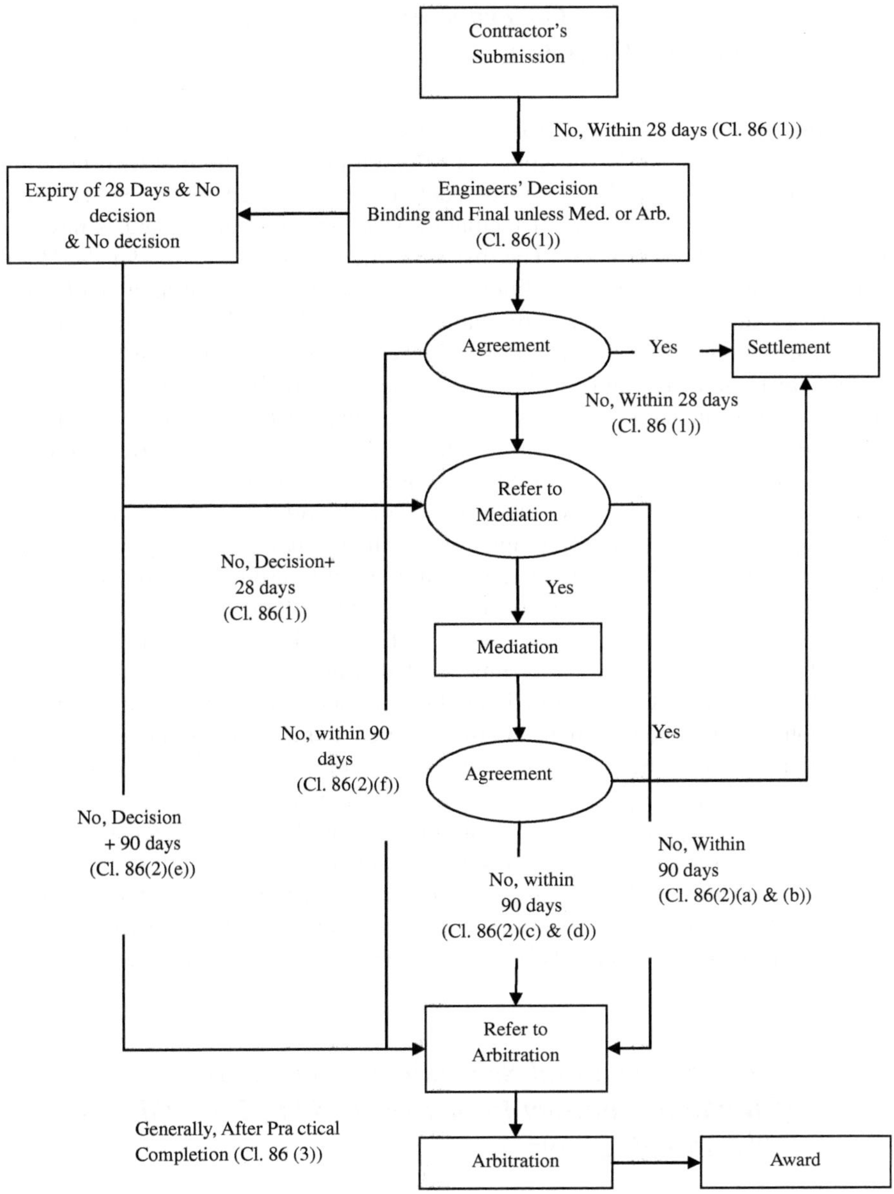

Fig. 17.2 The dispute resolution procedure under the HKG99 contract

17.2.2 The Hong Kong Government General Conditions of Contract (HKG99)

The Hong Kong Government was the pioneer in the use of alternative dispute resolution techniques in Hong Kong construction projects. The first incorporation of mediation into a standard form of contract was initiated by the Hong Kong Government in 1989. The use of mediation was further extended and is now for use in all government projects. The dispute settlement procedure under the Hong Kong Government General Conditions of Contract for both building works and civil works are the same and presented in Fig. 17.2. Mediation is availability before a dispute is referred to arbitration. Moreover, as can be noted in Fig. 17.2, it is possible to skip mediation. This may be attained if the parties refuse to use mediation. The mediation process may also be bypassed if the architect fails to make a decision within the time limit specified in the contract. The presentation in Fig. 17.2 allows easy recognition of the routes available pertinent to a particular point in time during the project duration. Typically, arbitration proceedings shall not be opened until practical completion or alleged practical completion of the works. Exceptions to this general provision proceed to arbitration immediately or the dispute is related to a question over the power of the architect to sanction remedies under forfeiture of the contract. The mediation rule for use in this contract is the Hong Kong Government Mediation rule. If arbitration is to be conducted, the Hong Kong International Arbitration Centre rule shall apply. Again arbitration will only be commenced after practical completion or alleged practical completion except where the consent of the employer and the contractor is obtained to proceed despite practical completion is not attained. The inclusion of mediation allows the introduction of a person neutral to the project to assist in resolving the dispute. One of the key success factors of mediation is the impartiality of the mediator. The suggestions, advice and/or opinions can then be more acceptable to the disputants. Although mediation is available to the parties, the use of which is totally voluntary. This is called the contractual use of mediation.

17.2.3 The Agreement and Schedule of Conditions of Building Contract for use in HKSAR, 2005 Edition (2005 Building Form)

This contract is primarily used for private building projects in Hong Kong. The contract is published jointly by the Hong Kong Institute of Architects (HKIA), the Hong Kong Institute of Construction Managers (HKICM) and the Hong Kong Institute of Surveyors (HKIS). The predecessor of this contract is the HKIA form that was modeled on the British Joint Contract Tribunal 1963 edition. The Hong Kong first ever construction industry review (CIRC 2001) called for a new conditions of contract that embraces equitable risk allocation so that claims and

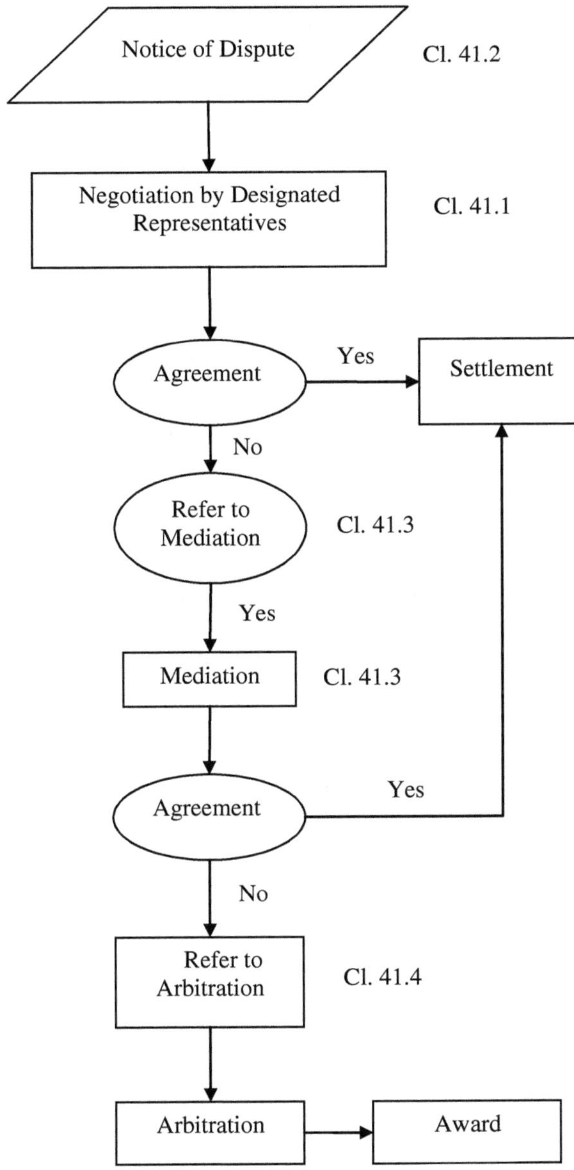

Fig. 17.3 Dispute resolution procedure under the 2005 building form

disputes can be reduced. This contract can be taken as the response to the CIRC's recommendations although its drafting had commenced long before the said recommendation. The procedure introduces the use of designated representatives who are not involved in the day-to-day administration of the contract to settle disputes that arise during the carrying out of the works. This is an admirable step in taking the settlement negotiation away from the parties directly involved in the works. Under Clause 41.2, if a dispute arises under or in connection with the contract, the

Architect shall, at the request of either party, immediately refer the dispute to the Designated Representatives who shall meet within 7 days of receipt of the Architect's notice.

If the dispute is not resolved by the Designated Representatives within 28 days, either party may refer the dispute to mediation. If the dispute is not settled by mediation within 28 days of the commencement of the mediation, either party may refer the dispute to arbitration that shall generally not commence until after Substantial Completion (Fig. 17.3).

17.3 The Dispute Resolution Advisor System

The Architectural Services Department (ASD) introduced a novel form of dispute resolution process, called the dispute resolution advisor (DRAd) system. The DRAd system was first used in the Queen Mary Hospital Extension and Renovation Project on a trial basis in 1991. The system was also used in two other hospital projects commencing in 1992 and 1993 respectively. As three projects were completed free from outstanding disputes and claims after completion. The ASD was satisfied with the system. DRAd system is a hybrid system combining elements from many positive attributes of ADR techniques (Wall 1993). This view has further been discussed in detail (Tsin 1997) claiming that the DRAd system embraces many of the features of both the preventive techniques like partnering and intervention techniques like facilitative mediation, expert determination/ appraisal and mini-trial.

The principal objectives in introducing the DRAd system include (ASD 1996):

(a) Encourage co-operation and joint problem-solving so as to prevent disputes from arising;
(b) Maximise the chances that any disputes that do arise will be resolved at site level; and
(c) Resolve any dispute that is not settled at site level as expeditiously and as cost-effectively as possible so that no dispute survives the completion of the contract.

In order to achieve the aforementioned objectives, the DRAd system has the following features:

(i) Time limits;
(ii) Involvement of nominated and specialist subcontractors;
(iii) Good faith requirement in negotiation;
(iv) The Dispute Resolution Advisor;
(v) Short form Arbitration.

The working procedures of the DRAd system are outlined in Fig. 17.4.

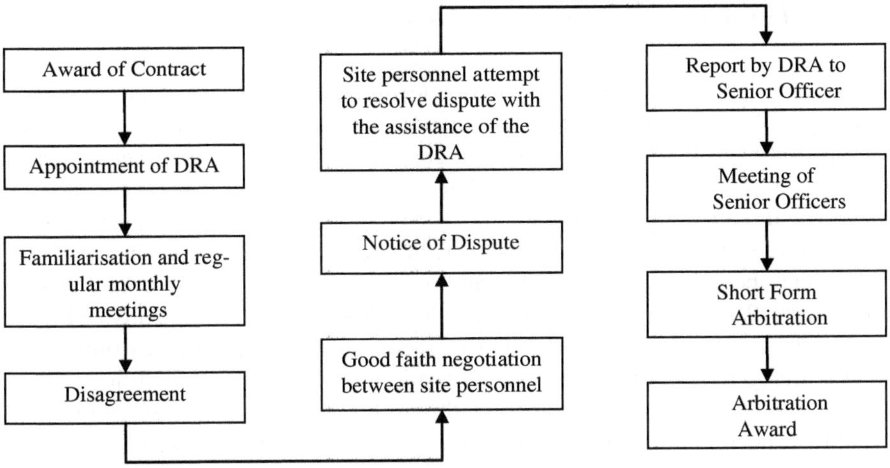

Fig. 17.4 Operation procedures of the dispute resolution advisor system

17.3.1 Time Limits

In order to avoid unattended problems escalate into disputes, the DRAd system requires timely responses from both the contractor and the contract administrator. In general, a 28 days' time limit is imposed in situation where notice has to be served or decisions have to be made. This time limit applies equally to both the contract administrator and the contractor. In the situations where further information is requested, the response time limit is 7 days.

17.3.2 Involvement of Nominated and Specialist Subcontractors

In Hong Kong, the use of nominated subcontractors to carry out specialist works is common. Experience reveals that many claims involve these specialist sub-contractors. It is believed that dispute resolutions should involve all who have an interest in the claim/dispute. The DRAd system obliges the nominated/specialist subcontractors to be represented in all forums where their interest is at issue.

17.3.3 Good Faith Requirement in Negotiation

The DRAd system requires the disputing parties to negotiate in good faith, a concept parallel to that adopted in partnering.

17.3.4 The Dispute Resolution Advisor

Unlike other alternative dispute resolution techniques where a neutral third party is to be agreed by the parties after a dispute has arisen. The DRAd system requires the appointment of a dispute resolution advisor (DRA) at the commencement of the project. The involvement of the DRA then is not only confined to holding meetings if called upon. Instead, on a monthly basis, the Employer and the Contractor, either separately or together, attempt to resolve problems that arise before they become formal disputes and to anticipate problems that may arise in the future. The contract also obliges the DRA to meet frequently with the Employer and the Contractor if either of them makes a request in writing.

Any disagreement over decision, instruction, order, direction, certificate of the Architect or valuation by the Surveyor should first attempt to be resolved through good faith negotiation. If negotiation fails, the aggrieved party may file a "Notice of Dispute" to the DRA who should promptly meet with the site level representatives of the relevant parties. The DRA has the flexibility of the choice of a dispute-resolution approach to help settle the dispute.

If the dispute cannot be resolved within 14 days of the service of "Notice of Dispute", then the DRA should submit a written report to senior officers of the disputing parties. The report also includes the DRA's non-binding recommendations or evaluation of the merits of the dispute. This report should not be admissible in any subsequent arbitration and litigation.

Upon receipt of the report from the DRA, the senior officers should meet to attempt to resolve the dispute. Furthermore, the DRA may recommend another form of dispute resolution although the employer and the contractor are not obliged to accept.

17.3.5 Short Form Arbitration

If the dispute is not settled within 14 days of the date of transmittal of the report to the senior officers, the employer and the contractor should participate in short form arbitration, a specially designed form of arbitration for use with the DRAd system.

Unlike most other arbitration provisions that require arbitration to be opened only after practical completion, short form arbitration is to be held during the currency of the contract. For single-issue dispute, a time limit of one day of hearing is imposed. For a dispute involving more than two parties or more than one distinct claim or issue, the Employer, the Contractor and the DRA shall agree upon the maximum length of time for the arbitration hearing which should be as short as possible. Failing agreement, the DRA should determine the duration of the hearing. The decision of the arbitration should be final and binding on the Employer and the Contractor.

Out of the five special features built in the DRAd system, it was found that the rigid time frame for action being critical to the success of such a system (Cheung and Yeung 1998).

17.4 Use of ADR in Construction

Disputes have been identified as epidemic in construction. Numerous attempts have been instigated to curb dispute occurrence. Notable examples include the use of partnering and equitable risk allocation. Notwithstanding, the nature of construction contracting appears to be conflict laden and dispute prone. Disputes have to be resolved and alternative dispute resolution (ADR) techniques have been introduced with the aims to alleviating the time and cost burden of the formal resolution method of litigation and arbitration. Owing to having similar legal system and industry structure, the use of ADR to resolve construction dispute in the U.K., Hong Kong, Singapore, Australia and New Zealand is compared. Based on a literature review, handling construction disputes in the five common law jurisdictions broadly falls into two approaches as presented in Fig. 17.5. Under the Type A approach, a dispute will be resolved firstly by ADR. If this fails, the dispute will then be referred to arbitration upon practical completion of the project. This approach has been embodied in a number of standard forms of contract, thus can be termed as contractual use of ADR. Take Hong Kong as an example, mediation has become an integral part of the dispute resolution clause in the major Government General Conditions of Contracts for viz.: Airport Core Program 1992 (The Government of Hong Kong 1992), Civil Engineering Works 1999 (Government of Hong Kong 1999a), Building Works 1999 (The Government of Hong Kong 1999b) and Design and Build Contracts 1999 (The Government of Hong Kong 1999c) and the latest version of the private forms of building contract published by the Joint Contract Working Committee (HKIA et al. 2005, 2006). Similarly, in the UK, mediation and adjudication have been introduced as an optional dispute resolution approach as stipulated in the Joint Contracts Tribunal (JCT) Standard Building Contract 2005 (JCT 2005a) and Standard Design and Build Contract 2005 (JCT 2005b).

It is noteworthy that, under the contractual framework, the use of ADR techniques before referring a dispute to arbitration is voluntary. If either of the contracting parties refuses, the use of the ADR techniques can be bypassed (Cheung and Yeung 1998). Furthermore, the prescribed voluntary ADR procedures typically involve appointing an independent neutral to give expert opinion. Nevertheless, in contrast to arbitration and litigation, the expert's recommendations are typically not binding on the parties (Jones 2006). Contrary to the Type A approach, a dispute is firstly be referred to statutory adjudication with the Type B approach. The arrangement has been considered effective to tackle two major deficiencies of the conventional contractual dispute resolution regime in construction: (1) the parties' right to bypass ADR before proceeding to arbitration and; (2) the enforcement of the non-binding experts' determination (Jones 2006). Through legislation, contracting parties now have the right to refer a dispute to adjudication. Furthermore, the decision of the adjudicator is binding unless and until the dispute has been settled by agreement, litigation or arbitration (Gaitskell 2007). In Hong Kong, the Type A approach has been used (Chau 2007; Leung 2007). In New

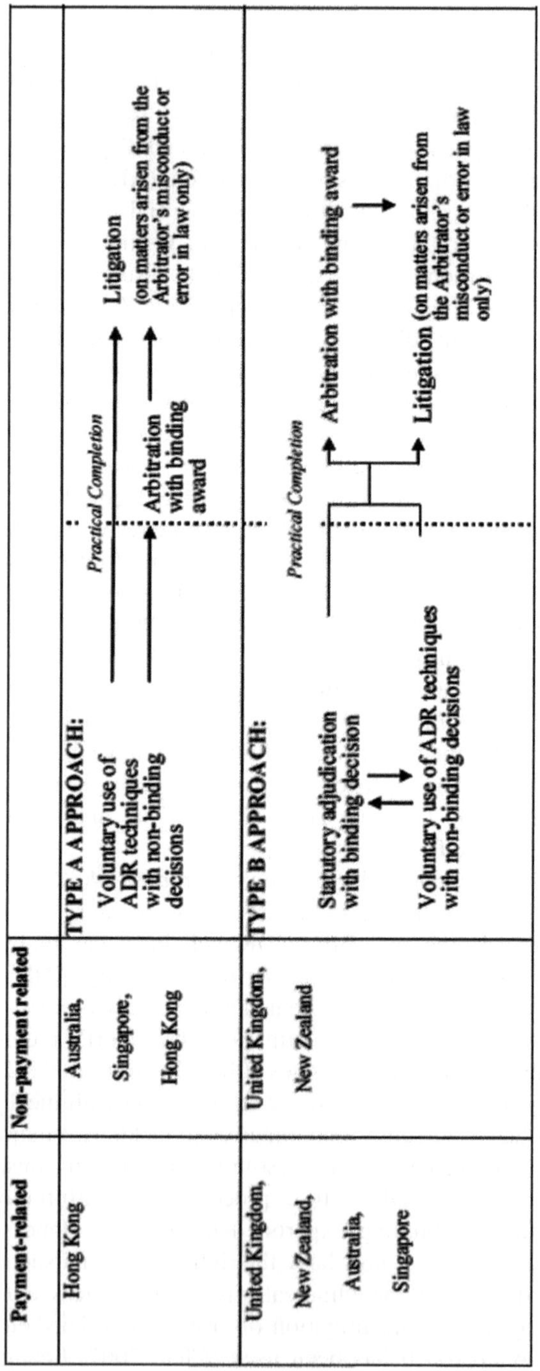

Fig. 17.5 Approaches in the use of ADR to resolve construction dispute

Zealand and the UK, in contrast, Type B approach is preferred (Gaitskell 2007). In Australia and Singapore, Type A approach applies to non-payment related disputes like the disputes about claims arisen from the extension of time, delay and disruption, personal liability, while Type B approach applies in handling payment related disputes (including progress, one-off and final payment) (Jones 2006). Furthermore, Hong Kong has taken a somewhat quite different approach from the other four jurisdictions; mediation is preferred over adjudication for all types of construction dispute.

17.5 Mediation Within the Civil Justice Reform in Hong Kong

In 2000, the Chief Justice appointed a working party to review the civil rules and procedures of the High Court and to recommend changes thereto with a view to ensuring and improving access to justice at reasonable cost and speed. ADR is considered a potentially useful process in appropriate cases as an alternative or adjunct to civil proceedings. The working party was asked to look into whether ADR should be introduced. The option of mandatory or voluntary use of ADR was also investigated. Mediation is not in law compulsory, but is at the heart of today's civil justice system, and any unjustified failure to give proper attention to the opportunities afforded by mediation, and in particular in any case where mediation affords a realistic prospect of resolution of dispute, there must be anticipated as a real possibility that adverse consequences may be attracted. The Working Party published an Interim Report and Consultative Paper in November 2001. The Interim Report included six proposals for how the Court might approach ADR. These were: (a) mandatory mediation by statutory rule for particular types of cases; (b) mediation as a condition for proceeding with the action; (c) mandatory mediation by election of one party; (d) mediation as a condition of legal aid; (e) unreasonable refusal of mediation reflected in costs; (f) encourage purely voluntary mediation.

The Final Report was published in March 2004 and recommended that courts should provide litigants with better information and support with a view to encouraging greater use of voluntary mediation. Proposal (d) suggests that the Director of Legal Aid can limit legal aid to ADR in appropriate cases. This will in effect make an attempt at ADR a condition of any further legal aid. The Final Report recommended that the Legal Aid Department should have power in suitable cases to limit its initial funding of persons who are qualified for legal aid to the funding of mediation while retaining its power to fund court proceedings where mediation is inappropriate or where mediation has failed.

Proposal (e) suggests using cost sanction to guard against unreasonable refusals of mediation. This has attracted a lively discussion since defining what 'an unreasonable refusal' is inherently difficult. After due consultation, the Working Party suggested the court should have power, after taking into account all relevant circumstances and adopting appropriate rules and proceedings, to make adverse costs orders in cases where mediation has been unreasonably refused after a party has served a notice requesting mediation to the other party or parties; or after mediation has been recommended by the court on the application of a party or of its own motion.

However, proposals (a), (b) and (c) of the Interim Report received strong objections. The main drawback of proposal (a) was the suggestion that cases unsuitable for mediation would inevitably be caught by the inflexibility of the rule. Proposal (b) was likely to raise doubts over the Court's inherent duty of conducting litigation if it is required to suggest mediation. Proposal (c) was considered to be a recipe for abuse by parties wishing to delay proceedings and likely to worsen the relationship between the parties. Therefore these proposals were rejected by the Working Party in the Final Report. The Final Report has subsequently been endorsed and came into force on 2 April 2009. It appears that the Hong Kong Judiciary is determined to promote voluntary use of mediation to resolve disputes in Hong Kong.

As far as construction disputes are concerned, on 4 July 2006, the Judiciary issued Practice Direction 6.3 titled "*Construction and Arbitration List- Pilot Scheme for voluntary Mediation*". The pilot scheme was to run from 1 September 2006 till 31 August 2008. The purpose of the pilot scheme is to encourage parties in construction cases on the Construction and Arbitration List to consider using mediation as a possible cost-effective means of resolving disputes. Under the Practice Direction, either party to a construction action may serve a Mediation Notice that should identify the mediation rules to be applied.

The concept of "minimum amount of participation" was introduced in this pilot scheme. Another important feature of the pilot scheme is that the party who does not wish to mediate the particular dispute needs to state the reasons why mediation is considered not appropriate. There will be considered by the Judge in determining whether a party has acted unreasonably in refusing to proceed with mediation. Unreasonable refusal to mediate may lead to an adverse cost order. The Hong Kong Judiciary has also devised a reporting questionnaire to be returned by the parties or their legal representatives to the Clerk of the Construction and Arbitration List. The report seeks to record the effectiveness of the mediation process and would preferably be returned jointly by the parties. Practice Direction 6.1 came into force on 2nd April 2009 and supersedes Practice Direction 6.3 on Construction and Arbitration List Pilot Scheme for Voluntary Mediation. Part F of the practice Direction 6.1 basically affirms mediation as a possible cost-effective means of resolving construction disputes. However the use of mediation has been promoted by imposing cost sanctions where a party unreasonably refuses to attempt mediation. Thus one of the objectives of Part F is to facilitate the Court's consideration of whether or not to impose cost sanctions in relation to a refusal to go to mediation. The articles of Part F related to cost sanction are as follows:

"*41. Where a Mediation Notice has been served, an unreasonable refusal or failure to attempt mediation may expose a party to an adverse costs order.*

42. Where a party:

(1) *has engaged in mediation up to the minimum level of expected participation agreed by the parties beforehand or as determined by the Court; or*
(2) *has a reasonable explanation for non-participation,*
he should not suffer any adverse costs order.

43. What constitutes an adverse costs order will be a matter in the Court's discretion after taking into account all relevant circumstances.

44. In determining whether a party has acted unreasonably in refusing mediation, the Court will not take account of or inquire into:

(1) *what happened during the mediation;*
(2) *why the mediation failed; or*
(3) *whether any failure in the course of mediation may be ascribed to unreasonable conduct by any party."*

Under Item 41, unreasonable refusal or failure to attempt mediation may expose a party to an adverse cost order:

17.5.1 Refusal to Mediate

In *Dunnett v. Railtrack Plc*[1] the defendant's refusal to mediate had caused an adverse cost order. This case highlights that parties who ignore the chance of resolving the dispute by ADR may have to face uncomfortable costs consequences. It is clear that litigants have a duty to consider seriously the possibility of ADR procedures being utilised for the purpose of resolving their claim or particular issues within it when encouraged by the court to do so. The question thus arises is what factors are to be considered in assessing whether a refusal to mediate is unreasonable. If a party can show good reason for refusing to mediate, that is to refuse reasonably, then it should not be penalised. In *Dunnett v. Railtrack Plc.* Lord Justice Brooke stated that the discharge of the parties' duty to help the court in active case management depends on the circumstances, including the conduct of all the parties and subject to the test of reasonableness. In *Halsey v. Milton Keynes General NHS Trust,*[2] it was held that the burden in an application to deprive a successful litigant of costs for refusal to mediate was on the unsuccessful party to show why the general rule should not be followed. Such a departure was not justified unless the unsuccessful party could show that the successful party had acted unreasonably in refusing to agree to ADR. There should not be a presumption in favor of mediation. In deciding whether the refusal was unreasonable the court would have regard to a number of factors including: (a) The nature of the

---

[1] *Dunnett v Railtrack Plc [2002]* EWCA Civ 303.

[2] *Halsey v. Milton Keynes General NHS Trust* [2004] EWCA Civ 576.

dispute; (b) the merits of the case; (c) whether other settlement methods had been attempted; (d) whether the costs of mediation would be disproportionately high; (e) delay in suggesting ADR; (f) whether the mediation had a reasonable prospect of success.

Whilst the Court was clear that this list of factors is not exhaustive, it does indicate that serious consideration is needed in deciding if a refusal to mediate is unreasonable, especially if the court encourages its use and cost benefits are recognisable in view of the circumstances. It remains good law that any decision to deny a successful party its costs is an exception to the general rule that the successful party gets its costs. It is anticipated that with Practice Direction 6.1 came into force on 2nd April 2009, further cases involving the interpretation of 'refusal to mediate' will increase.

17.5.2 Failure to Attempt Mediation

The concept of minimum participation lies in the heart of "*failure to attempt mediation*". Under Practice Direction 6.1, what constitutes minimum participation should be agreed between the parties in dispute. Item 34 states that where the Applicant and Respondent differ as a sufficient attempt at mediation, the judge may (either when having a stay application or at any other time) specify the applicable level of expected participation.

One central feature of mediation is its voluntary nature. Like any endeavor that needs the cooperation of participating parties, its effective use depends on their mutual effort. The requirement of mediating in 'good faith' as a means of enhancing the mutual effort is often included in contracts in Hong Kong. It is now quite common to have a contractual provision stating that the parties agree to mediate in good faith to resolve disputes.

17.6 Chapter Summary

Managing dispute has become one of the key management functions of construction managers. Amicably resolving construction dispute reduce conflict level and thereby indirectly improve productivity. The orthodox approaches to settle disputes, like arbitration and litigation, have failed to live up with the industry's expectation. Alternative dispute resolution techniques have been identified by many countries as alternatives. Through the analysis of the various standard forms of construction contract commonly used in Hong Kong, it can be observed that no real attempt has been made within the private construction sector to promote the use of alternative dispute resolution. Moreover, the main driving force on the adoption of ADR come from the public sector, in particular the Architectural Services Department, in pioneering the use of mediation and the dispute resolution advisor system. Mediation is now an integral part of the standard dispute resolution provision for all government projects, including building, civil engineering and M&E installation works. The DRAd system is primarily used in more complex projects like hospitals and renovation works where a greater degree of changes is anticipated. The Government of Hong Kong Special Administrative Region has made a policy decision to make Hong Kong as a regional hub for arbitration and mediation services. A review of five Common Law jurisdictions having similar construction industry structure reveals that Hong Kong has preferred the use of voluntary mediation, instead of statutory adjudication as in the case of the other four jurisdictions, to improve the efficiency of construction dispute litigation. In this regard, voluntary mediation has been introduced in the Hong Kong civil procedures rules as part of the recently launched Civil Justice Reform. To give effect to voluntary use of mediation, adverse cost order is used to discourage "*unreasonable refusal to mediate*" or "*failing to attempt to mediate*". New Practice Direction in these regards came into force on 2nd April 2009. In principle, the new measures sounds sensible, but their actual impact and effectiveness are still being tested in the Court.

Acknowledgments Part of the content of the chapter has been published in Volume 2(3) of the ASCE Journal of Legal Affairs and Dispute Resolution in Engineering and Construction and Volume 16(6) of the International Journal of Project Management and is used with the permission from ASCE and Elsevier respectively.

References

Architectural Services Department (ASD) (1996). *The use of dispute resolution advisor system in building projects* (p. 2).

Bryan, K. A., & Philips, F. P. (2007). *The emerging role of ADR within the business sector*. New York: International Institute for Conflict Prevention and Resolution Inc.

Chau, K. W. (2007). Insight into resolving construction disputes by mediation/adjudication in Hong Kong. *Journal of Professional issues in engineering Education and Practice, ASCE, 133*(2), 143–147.

Cheung, S. O., & Yeung, Y. W. (1998). The effectiveness of the dispute resolution advisor system: A critical appraisal. *International Journal of Project Management, 16*(6), 367–374.

Construction and Arbitration List High Court (2009). http://www.judiciary.gov.hk/en/publications/pilotscheme_calhc.pdf. Accessed 2 March 2009.

Construction Industry Review Committee (CIRC) (Ed.) (January 2001). *Construct for Excellence, Report of the Construction Industry Review Committee*. Hong Kong: Hong Kong SAR Government.

Gaitskell, R. (2007). International statutory adjudication: Its development and impact. *Construction Management and Economics, 25*(7), 777–784.

Hong Kong Government. (1989). Administrative Guidelines for Mediation in Construction Disputes, Lands and Works Branch Technical Circular No. 8/89.

Hong Kong Institute of Architects (HKIA), the Hong Kong Institute of Construction Managers (HKICM) and the Hong Kong Institute of Surveyors (HKIS). (2005). The agreement and schedule of conditions of building contract for use in the Hong Kong special administrative region (Standard form of building contract private edition—with quantities 2005 edition).

Hong Kong Institute of Architects, the Hong Kong Institute of Construction Managers and the Hong Kong Institute of Surveyors. (2006). The agreement and schedule of conditions of building contract for use in the Hong Kong special administrative region (Standard form of building contract private edition—without quantities 2006 edition).

Joint Contracts Tribunal (JCT) (2005a). The JCT 2005 standard building contract.

Joint Contracts Tribunal (JCT) (2005b). The JCT 2005 standard design and build contract.

Jones, D. (2006). *Adjudication: Evolutionary step or revolutionary idea?* In Problem solving: Rules, roles and regulations. Canadian College of Construction Lawyers 9th Annual Conference, Fox Harbour, Nova Scotia.

Leung, H. F. (2007). The role of the judiciary and statutes in relation to mediation. *Asian Dispute Review, October,* 128–131.

The Government of Hong Kong. (1992). *General conditions of contract for the airport core programme civil engineering works.* 1992 edition.

The Government of Hong Kong. (1999a). *The government of Hong Kong special administrative region general conditions of contract for civil engineering works.* 1999 edition.

The Government of Hong Kong. (1999b). *The government of Hong Kong special administrative region general conditions of contract for building works.* 1999 edition.

The Government of Hong Kong. (1999c). *The government of Hong Kong special administrative region general conditions of contract for design and build contracts.* 1999 edition.

Tsin, C. (1997). Dispute resolution advisor system in Hong Kong—Design and development. *Arbitration, 63*(2), 67–80.

Wall, C. (1993). The dispute resolution advisor in the construction industry. *Building Research and Information, 21*(2), 122–127.

Chapter 18
The Interrelationships Among Sources, Tactics and Outcomes in Construction Dispute Mediation

Sai On Cheung and Tak Wing Yiu

Abstract Construction is of long-duration, high value and dispute-prone. As such, dispute is a regular feature in construction and consumes resources that would otherwise be used in a more productive manner. The use of mediation has been regarded as a flexible, cost-effective, and non-threatening way to dispute resolution. Reported studies on construction mediation have been instrumental in bringing out key success factors and the advantages over other adversarial resolution processes. Moreover, mediation is a form of assisted negotiation; hence the skill of a mediator shall have pivotal effect on the mediation outcome. In fact, the appropriate use of tactics by a mediator shall have deciding effect on the chance of success. This study examines the inter-relationships among dispute sources, mediator tactics and mediation outcome. The study is broadly divided into two parts. The first part deals with the development of taxonomies of construction dispute sources, mediator tactics and outcomes, employing the technique of Principal Component of Factor Analysis (PCFA). A total of eight, nine and four taxonomies for dispute sources, mediator tactics and outcomes were identified, respectively. In the second part, moderated multiple regressions (MMR) was performed on these to detect the contingent effect of tactics on outcomes respective to the dispute sources. By examining the changes in the R^2 values of the moderated regression equations, the effectiveness of tactics in bringing about the desired outcomes were investigated. The tactics of Trust Building and Reality Test were found to be the most versatile tactics. These reaffirm the industry belief that

S. O. Cheung (✉)
Construction Dispute Resolution Research Unit, Department of Civil and Architectural Engineering, City University of Hong Kong, Hong Kong, People's Republic of China
e-mail: saion.cheung@cityu.edu.hk

T. W. Yiu
Department of Civil and Environmental Engineering, The University of Auckland, Auckland, New Zealand
e-mail: k.yiu@auckland.ac.nz

S. O. Cheung (ed.), *Construction Dispute Research*,
DOI: 10.1007/978-3-319-04429-3_18,

disputes are in fact problems that can be solved if pragmatic and sensible approaches are taken instead of the entrenched confrontational attitude. More significantly, the exercise of such approaches can only be possible in a trusting environment.

18.1 Introduction

Mediation has been regarded as a flexible, cost-effective, and non-threatening dispute resolution method (Fraley 1990; Santilli 1988). It is also identified as a suitable way to resolve construction dispute because mediated settlements are made privately, a highly desirable feature for disputants who do not wish their dispute to be publicised (Chau 1990, 1991; HKIAC 2001). The use of mediation is largely voluntary. Disputants often wish to keep their business relationships and reputations intact; therefore they are very likely to settle (Bateson 1997; Kaplan et al. 1991). It is reported that about 80 % of mediation cases resulted in binding agreements (Fraley 1990; Santilli 1988). It is also widely regarded as an effective means to resolve disputes (HKIAC 2001; HKMC 2001; Lowry 2001). In Hong Kong, mediation has now become an integral part of the dispute settlement provisions in many building and engineering contracts.

The increase in the use of mediation attracted a surge of studies focusing on the comparison between mediation and other adversarial processes, the identification of successful factors, process design and the associated cultural issues (Bateson 1997; Cheung 1999; Cheung and Yeung 1998; Fenn and Gameson 1992; Hill 1998; Hollands 1992; Marcus and Marcus 1987). The results obtained from these studies have deepened the understanding of construction mediation. Nonetheless, these studies were mainly anecdotal. Empirical research to support their claims will therefore be useful. Such a view is shared by Henderson (1996) who suggested that thorough and empirically based researches in this area would be beneficial to the furtherance of construction mediation. As mediation is a form of assisted negotiation, the mediator plays a pivotal role in facilitating a settlement. The tactics used by them have immense impact on the mediation outcome. A number of researches have been conducted to identify the types of tactic commonly used by mediators in labour and divorce disputes (Carnevale 1986a; Carnevale et al. 1989; Carnevale and Pegnetter 1985; Carnevale and Pruitt 1992; Lim and Carnevale 1990). In addition, since not all tactics are of equal strength, nor do they derive similar effects, choice of tactic should take into account of the source and the desired outcome. This is described as the contingent use of tactics. The contingent approach advocates that the tactics used by mediators should be adaptive to the context of the dispute. Matters to be taken into account include the nature of the dispute and the attitude of the disputants (Carnevale and Pegnetter 1985; Kochan and Jick 1978; Lim and Carnevale 1990; Pruitt 1981; Shapiro et al. 1985; Stevens 1963). For example, in the study conducted by Carnevale and Pegnetter

(1985), a total of 32 labour mediators were invited to rate 24 sources of dispute and 37 mediator tactics which were employed in cases they had been working on prior to the study. Correlation analyses were used to identify the contingent use of tactics by the mediators. Similar researches were also conducted on divorce mediations (Lim and Carnevale 1990) and publication union mediations (Carnevale et al. 1989). In principle, a mediator would use the most appropriate tactic to achieve the desired outcome. However, this task is not easy and requires considerable experience (Lim and Carnevale 1990). Typically, mediators employ their cognitive schemas to assess the interrelationship among the three dimensions of mediation: dispute sources, tactics and outcomes (Taylor and Crocker 1981). It is also reported that mediators tend to adopt a goal-oriented and/or strategic approach to align their tactics with the desired outcomes (Carnevale 1986b; Kerr 1954; Kressel 1972; Kressel and Pruitt 1985, 1989; Locke 1991; Rifkin et al. 1991). The schema will be more complete if dispute sources can be added into the analysis (Carnevale et al. 1989; Carnevale and Henry 1989b; Lim and Carnevale 1990). The primary objective of this research therefore is to study the tactics used by construction mediators, with a focus on the inter-relationships among dispute sources, mediator tactics and mediation outcomes.

18.2 The Research Framework

The overall framework for the study of mediator tactics is given in Fig. 18.1.

The identification of dispute sources, mediator tactics and mediation outcomes were first carried out through a literature review. In view of the relatively large number of items in each of the three dimensions, it is necessary to categorise them into factor groups described as taxonomies. Moderated regression analyses were then conducted to examine the contingent use of tactics.

The study can be broadly divided into two parts: Part (I) deals with the development of taxonomies for dispute sources, mediator tactics and outcomes; and Part (II) is the study of the contingent use of tactics. In essence, the first part of the study seeks to answer the following three questions:

(i) What are the basic dispute sources in the construction industry?
(ii) What are the generic types of tactics commonly used by construction mediators?
(iii) What are the typical outcomes of construction mediation?

18.3 The Survey

In order to achieve the above-said objectives, data were collected using a questionnaire which was designed with the three mediation dimensions in mind: (1) Construction dispute sources; (2) Mediator tactics and (3) Mediation outcomes.

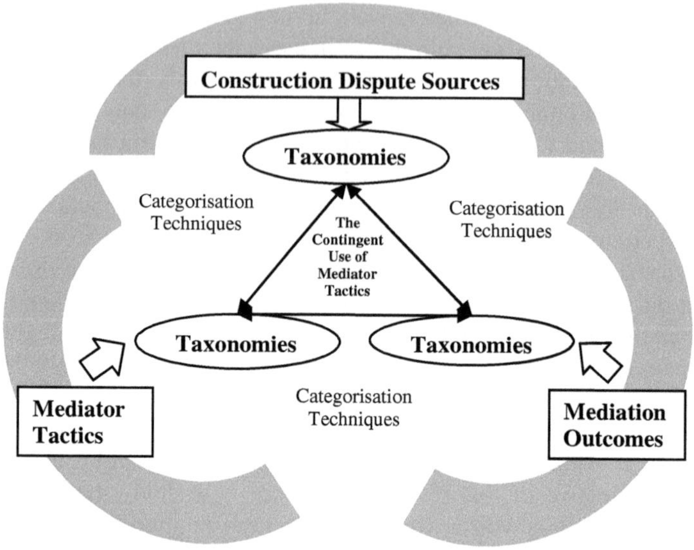

Fig. 18.1 Research framework

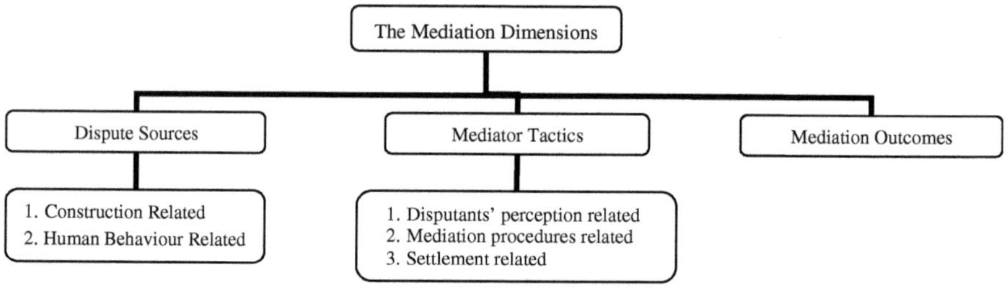

Fig. 18.2 The mediation dimensions

The first two dimensions were sub-divided into different groups primarily to reflect the nature of the dispute sources and tactics (Fig. 18.2).

As shown in Fig. 18.2, disputes in the construction industry fall mainly into two categories—Construction related and Human behaviour related. While the former is fairly self-explanatory, the impact of behaviour-based disputes should not be overlooked either, as disputes may manifest singularly due to the parties involved (Harmon 2003; Hibberd and Newman 1999). Similarly, mediator tactics can be divided into (1) Disputants' perception related; (2) Mediation procedure related; and (3) Settlement related. These groups are largely consistent with those given by Karim and Pegnetter (1983) in their overview of the institutional and behavioural studies of mediation. Based on the developed structure as shown in Fig. 18.2, literature reviews were then carried out, after which a total of thirty-three dispute sources were identified. Twenty four of them relate to the subject matter of the

Table 18.1 List of construction dispute sources

Dispute sources*
Construction related
1. Argument on acceleration costs
2. The assessment of liquidated and ascertained damages against Main Contractor
3. Clients fails to pay for variation claims
4. Late giving of possession from Client
5. Client takes over the site and denies assess to Main Contractor
6. Errors/substantial changes in Bills of Quantities
7. Argument on the prolongations costs
8. Architect/Engineer dissatisfies the work progress of Main Contractor
9. Argument on the measurement and valuation of contracted work
10. Late instructions from Architect or Engineer
11. Main Contractor fails to proceed in a competent manner
12. Delay interim payment from Client
13. Late release of retention monies to Main Contractor
14. Argument on the time extension costs claimed by sub-contractor
15. Changes of scope due to extra work
16. Inadequate site and/or soil investigation report
17. Delay works due to utility services organisation
18. Non-payment to sub-contractor by Main Contractor
19. Main Contractor ceases work on site
20. Argument on the time extension costs claimed by Main Contractor
21. Main Contractor denies assess of the site for the sub-contractor
22. Sub-contractor works delay due to Main Contractor
23. Consequences on opening for inspection
24. Sub-contractor ceases work on site
Human behaviour related
1. Negotiators lacked experience
2. Too many issues brought to table
3. Both parties not prepared for negotiations
4. Both parties want to control over proceedings
5. Both parties are not interested to settle
6. Parties have unrealistic expectations
7. No leadership within the project teams
8. No trust between the parties
9. Felt no trust on mediator

* Dispute sources were rated on a scale from (1) least significant to (5) most significant

dispute, whilst the remaining nine relate to human behaviour. The reviews also reveal thirty-two commonly used mediator tactics, among which eight are related to the disputant's perception, thirteen are related to mediation procedures, and eleven are related to settlement. Finally, a total of sixteen mediation outcomes are also identified. Lists of dispute sources, mediator tactics and mediation outcomes are showed in Tables 18.1, 18.2 and 18.3 respectively.

Table 18.2 List of mediator tactics

Mediator tactics[a]
Disputants' perception related
1. Educate the parties about the bargaining or impasse process
2. Encourage the parties themselves to verbalise their willingness to respectfully listen to each other's grievances
3. Encourage the parties to meet each other's needs
4. Help the parties to "save face"
5. Remind the parties that their position was unrealistic
6. Suggest particular settlement for parties to consider
7. Try to change the expectation of parties
8. Encourage the parties to apologise, and regret for harm suffered by another in the past
Mediation procedure related
1. Use humour to lighten the atmosphere
2. Keep in rapport with the parties
3. Argue one party's case to the other
4. Reduce the feeling of hostility towards each party
5. Focus on the impasses issues during caucuses session
6. Attempt to speak their language
7. Control the bargaining structure and timing
8. Formulate clear goals before or during the process
9. Call for frequent caucuses during mediation
10. Avoid taking sides on important issues in joint sessions
11. Assure each party that the other was being honesty
12. Keep the negotiations focused on the issues only
13. Express pleasure or displeasure at negotiation progress
Settlement related
1. Settle simple issue first
2. Help the parties to devise a framework for negotiations
3. Make compromise suggestions to the parties
4. Suggest the parties to review their needs
5. Mention the costs of disagreement
6. Simplify the agenda by eliminating or combining issues
7. Discuss other settlements in comparable cases
8. Help the parties to establish priorities among the issues
9. Suggest some tradeoffs among issues
10. Press the parties to make concessions
11. Make the parties to aware the destructiveness of the conflict

[a] Mediator tactics were rated on a scale from (1) least useful to (5) most useful

18.3.1 Data Collection

The data for this study has to be case specific. Thus, respondents were asked to select one of their most recently completed mediation cases when answering the questionnaires. The questionnaire has four sections. The first section requires the

Table 18.3 List of mediation outcomes

Mediation outcomes[a]
1. Agreement perceived to be devised from the parties
2. The parties gained satisfaction on the mediation as a tool of dispute resolution
3. Overall success
4. I felt the parties trust the mediator
5. The underlying core conflict of the dispute was resolved
6. A mutually beneficial settlement was reached
7. The needs and goals of mediator satisfied
8. The number of issues was reduced
9. Nothing ambiguously stated
10.The settlement was reached in reasonable time
11. The inter-party relations improved
12. I acquired a reputation for the effectiveness in setting the dispute
13. Both parties felt no future problems expected
14. Both parties learned to communicate
15. I improved self-esteem after the settlement of the dispute
16. I improved my cultural sensitivity

[a] Mediation outcomes were rated on a scale from (1) not achieved to (5) highly achieved

respondents to provide their background information and the particulars of the mediated cases such as the project nature, contract sum and parties involved. The next three sections were designed to address the three mediation dimensions. The respondents were asked to rate the degree of significance of the dispute source in relation to the mediated case on a Likert scale of 1 (least significant)–5 (most significant). Next, the respondents were asked to rate, on a similar scale, the degree of usefulness of the mediator tactics used, with 1 being the least useful and 5 the most useful. In the final section of the questionnaire, which concerns the mediation outcomes, respondents were asked to give a rating for their achievement levels. Again, this was done on a Likert scale of 1 (not achieved)–5 (highly achieved). The lists of items included in the questionnaire are shown in Tables 18.1, 18.2, 18.3.

Before sending out the data collection questionnaire, a list of prospective respondents was first prepared. The accredited mediators maintained by the Hong Kong International Arbitration Center (HKIAC) were the targeted respondents. The HKIAC is the leading organisation for the provision of arbitration and mediation services in Hong Kong. Furthermore, to ensure relevancy of responses, only those mediators with construction background were sent the questionnaire. A total of 85 construction mediators were identified. They were then contacted to solicit their agreement to participate in this study. At the end, 32 of them agreed to participate.

18.3.2 Input Data

The 32 accredited mediators who agreed to participate all returned their questionnaires. The respondents all hold senior positions and are well-respected within the industry. Each of them has a minimum of 5 years experience in construction mediation. In terms of the project nature, 50 % of the mediated disputes were civil, 35 % were building-related, and the remaining 15 % concerned building services and maintenance.

18.4 The Taxonomies

Principal Component Factor Analysis (PCFA) can be used to explore the structure of the interrelationships between data, and to define a set of common underlying constructs, thus enabling the development of taxonomies. Separate dimensions of the structure can firstly be identified. The interpretation of variables can then be accomplished by summarising the data according to the constructs (Hair et al. 1995). The data obtained for dispute sources (2 groups), mediator tactics (3 groups) and mediation outcomes (1 group) were each subjected to a PCFA to develop their respective taxonomies. As such, six PCFA were performed.

Before performing a PCFA, the suitability of the data was first evaluated by examining the Kaiser-Meyer-Olkin (KMO) measure of sampling adequacy. The KMO values for the six PCFAs are all within the range of 0.577–0.739 and above the threshold requirement of 0.5 (Cheung et al. 2000; Cheung and Yeung 1998; Holt 1997). To shortlist factors, the eigenvalue-greater-than-1 principle, which is the commonly used criterion, was applied. Factors having an eigenvalue greater than 1 were considered significant, and those with an eigenvalue below 1 were discarded. In order to simplify the factor structures and to obtain a more meaningful factor solution, rotation of the factor matrices was performed to reduce the ambiguities that often accompany initial unrotated factor solutions. A Varimax rotation was applied. A rule of thumb suggested by Comrey and Lee (1992) is that a factor loading value of 0.71 shall be considered a good demarcation for the inclusion of variables in factors. Accordingly, variables with loading less than 0.71 were discarded. The final factor matrices are given in Tables 18.4, 18.5, 18.6, 18.7, 18.8, and 18.9.

18.4.1 Interpretation of the Factors and Developing Taxonomies

As discussed above, dispute sources are arranged into two categories: construction related and human behaviour related. The factor matrix for construction related dispute sources, after a Varimax rotation, is shown in Table 18.4.

Table 18.4 Factor matrix for the dispute sources (construction related) after VARIMAX rotation

Dispute sources (construction related)	Factor						h^2
	1	2	3	4	5	6	
Factor 1: variation							
Client fails to pay for variation claims	**0.944**	0.014	−0.009	0.027	−0.147	0.025	0.914
Argument on the measurement and valuation of contracted work	**0.872**	0.131	0.019	−0.068	−0.062	0.114	0.799
Errors/substantial changes in Bills of Quantities	**0.813**	0.134	−0.228	0.140	0.072	0.147	0.776
Factor 2: incompetence of works							
Changes of scope due to extra work	0.173	**0.838**	0.130	−0.085	0.009	0.292	0.842
Late instructions from Architect or Engineer	0.093	**0.754**	0.250	0.093	0.142	−0.027	0.670
Factor 3: cost of delay							
Argument on the time extension costs claimed by Main Contractor	0.133	0.270	**0.861**	−0.151	−0.102	−0.026	0.865
Argument on prolongations costs	0.002	0.320	**0.797**	0.247	−0.194	0.066	0.841
Factor 4: sub-contractor related							
Sub-contractor work delay due to Main Contractor	0.123	0.010	0.094	**0.873**	0.143	0.178	0.839
Argument on the time extension costs claimed by Sub-contractor	−0.040	0.091	0.325	**0.831**	−0.578	−0.092	0.819
Non-payment to sub-contractor by Main Contractor	0.238	0.299	−0.305	**0.756**	0.063	0.065	0.818
Factor 5: cease works							
Main Contractor ceases work on the site	0.029	0.134	−0.070	0.038	**0.886**	−0.046	0.831
Sub-contractor ceases work on the site	0.008	0.263	−0.049	0.015	**0.836**	0.245	0.813
Factor 6: site availability							
Late giving of possession from Client	0.310	0.048	0.145	0.227	0.180	**0.769**	0.795
Percent of Variance	30.228	16.493	11.082	9.752	5.844	5.438	
Eigenvalue	5.743	3.134	2.106	1.853	1.110	1.033	

Table 18.5 Factor matrix for the dispute sources (human behaviour related) after VARIMAX rotation

Dispute source (human behaviour related)	Factor		h^2
	1	2	
Factor 1: parties internal problems			
Negotiators lacked experience	**0.825**	0.156	0.704
Both parties not prepared for negotiations	**0.789**	−0.177	0.654
Felt no trust on mediator	**0.788**	0.094	0.630
No trust between the parties	**0.718**	0.289	0.598
No leadership within the project teams	**0.715**	0.265	0.581
Factor 2: process overload			
Too many issues brought to table	0.161	**0.793**	0.655
Percent of Variance	43.265	14.283	
Eigenvalue	3.894	1.285	

Table 18.6 Factor matrix for the disputants' perceptions related mediator tactics after VARIMAX rotation

Disputants' perceptions related mediator tactics	Factor		h^2
	1	2	
Factor 1: encourage for self-improvement			
Encourage the parties to meet each other's needs	**0.860**	0.040	0.742
Encourage the parties to apologise, and regret for harm suffered by another in the past	**0.760**	−0.260	0.646
Encourage the parties themselves to verbalise their willingness to respectfully listen to each other's grievances	**0.750**	0.217	0.610
Help the parties to "save face"	**0.739**	0.301	0.637
Educate the parties about the bargaining or impasse process	**0.732**	0.195	0.574
Factor 2: reality test			
Try to change the expectation of parties	−0.220	**0.794**	0.679
Percent of Variance	41.186	16.690	
Eigenvalue	3.295	1.335	

18.4.2 Taxonomies of Construction Related Dispute Sources

18.4.2.1 Construction Related

Six factors were extracted for constriction related dispute sources. The three dispute sources extracted for Factor 1 were "Client fails to pay for variation claims", "Argument on the measurement and valuation of contracted work" and "Errors/ substantial changes in Bills of Quantities". Factor 1 addresses the most common sources of dispute. As these three disputes are somehow related to variation, Factor 1 is described as Variation. Variation claims are administered by the variation clauses in contracts; arguments arising from variation are common in the industry (Kumaraswamy and Chan 1998). From a statistical point of view, this factor

Table 18.7 Factor matrix for the mediation procedures related mediator tactics after VARIMAX rotation

Mediation procedures related mediator tactics	Factor				h^2
	1	2	3	4	
Factor 1: process control					
Control the bargaining structure and timing	**0.839**	0.207	−0.033	−0.047	0.750
Express pleasure or displeasure at negotiation progress	**0.798**	−0.105	−0.336	−0.127	0.776
Use humour to lighten the atmosphere	**0.720**	−0.047	0.041	0.396	0.680
Factor 2: caucuses					
Call for frequent caucuses during mediation	−0.103	**0.851**	0.086	0.122	0.758
Factor 3: trust building					
Avoid taking sides on important issues in joint sessions	0.064	−0.016	**0.926**	0.051	0.865
Keep in rapport with the parties	0.028	0.249	**0.840**	−0.184	0.802
Factor 4: analysing					
Argue one party's case to the other	0.099	0.062	−0.091	**0.928**	0.883
Percent of Variance	35.378	16.213	9.698	8.421	
Eigenvalue	4.599	2.108	12.61	1.095	

Table 18.8 Factor matrix for the settlement related mediator tactics after VARIMAX rotation

Settlement related mediator tactics	Factor			h^2
	1	2	3	
Factor 1: ice breaking				
Settle simple issue first	**0.755**	−0.195	−0.110	0.62
Help the parties to establish priorities among issues	**0.748**	0.250	0.091	0.63
Factor 2: seeking progress				
Make compromise suggestions to the parties	−0.144	**0.875**	0.234	0.841
Suggest some tradeoffs among issues	0.342	**0.815**	0.032	0.782
Press the parties to make concessions	−0.148	**0.716**	0.181	0.567
Factor 3: pressing settlement				
Mention the costs of disagreement	0.010	0.224	**0.742**	0.601
Make the parties aware of the destructiveness of the conflict	0.064	0.137	**0.737**	0.566
Percent of Variance	29.042	19.764	10.563	
Eigenvalue	3.195	2.174	1.162	

explained about 30 % of variance, which is the most significant source of dispute being mediated. Factor 2 consists of two dispute sources. They are "Changes of scope due to extra work" and "Late instructions from Architect or Engineer". These are collectively described as incompetence of works. It may be the employers' change of mind, or omissions on the part of the design team, that causes the extra work. Factor 3 includes "Argument on the time extension costs claimed by Main Contractor" and "Argument on prolongations costs". It is

Table 18.9 Factor matrix for the mediation outcomes after VARIMAX rotation

Mediation outcomes	Factor				h^2
	1	2	3	4	
Factor 1: win–win settlement					
A mutually beneficial settlement was reached	**0.800**	0.043	−0.018	0.488	0.880
Agreement perceived to be devised from the parties	**0.752**	0.075	0.197	0.147	0.631
The parties felt satisfaction on the mediation as a tool of dispute resolution	**0.711**	0.386	0.141	0.186	0.709
Factor 2: progress					
I felt the parties trust the mediator	0.043	**0.863**	0.320	0.078	0.850
The underlying core conflict of the dispute was resolved	0.480	**0.760**	0.035	0.082	0.816
The number of issues was reduced	−0.037	**0.737**	0.093	0.285	0.634
Factor 3: improvement					
I improved my cultural sensitivity	0.300	0.171	**0.831**	0.070	0.814
Both parties learned to communicate	0.197	0.063	**0.806**	0.302	0.784
I improved self-esteem after the settlement of the dispute	−0.047	0.249	**0.804**	0.225	0.762
Factor 4: time advantage					
The settlement was reached in reasonable time	0.308	0.255	0.073	**0.713**	0.674
Percent of Variance	43.771	11.775	10.283	6.806	
Eigenvalue	7.003	1.884	1.645	1.089	

apparent that these disputes are related to the delay of a project. In fact, the cost of delay is central to most construction disputes (Al-Khalil and Al-Ghafly 1999; Kaming et al. 1997; Kumaraswamy and Chan 1998; McDonald 1984). "Sub-contractor work delay due to Main Contractor", "Argument on the time extension costs claimed by sub-contractor" and "Non-payment to sub-contractor by Main Contractor" were the disputes extracted for Factor 4. Sub-contracting of construction works is a long-established practice in the industry. Multi-layered sub-contracting generally increases contractual complexity, and has thus become one of the main sources of dispute in the construction industry. Factor 5 consists of the disputes: "Main Contractor ceases work on site" and "Sub-contractor ceases work on site". Both disputes are related to the cease of contract works during the construction process. Factor 6 is composed of a single dispute source: "Late giving of possession from Client", which obviously addresses the disputes over site availability. In sum, the six factors extracted can be described as follows:

Factor 1: Variation
Factor 2: Incompetence of works
Factor 3: Cost of delay
Factor 4: Sub-contractor related
Factor 5: Cease works
Factor 6: Site availability

18.4.2.2 Taxonomies of Human Behaviour Related Dispute Sources

Human behaviour is another major contributor to construction disputes. Examples of behavioural problems include the ambiguity of the role of the architect, a lack of interpersonal skills between the parties, and a lack of responsiveness to changes (Langford et al. 1992). The result obtained in this factor analysis shows four dispute sources under Factor 1. They are "Negotiators lacked experience", "Both parties not prepared for negotiations", "Felt no trust on mediator", "No trust between the parties" and "No leadership within the project teams". These sources address the inter-group conflicts within project teams, which generally prohibit them from solving problems themselves. Factor 2 consists of "too many issues brought to table" and is described as Process Overload. In light of the above findings, the two factors for human behaviour related disputes can be described as follows:

Factor 1: Parties' internal problems
Factor 2: Process overload

The factor matrix after a Varimax rotation is presented in Table 18.5.

18.4.3 Taxonomies of Mediator Tactics

The full list of tactics is given in Table 18.1. Factor analyses were conducted on each of the tactic groups to develop the taxonomies.

18.4.3.1 Disputants' Perceptions Related Mediator Tactics

The factor matrix after a Varimax rotation is presented in Table 18.6. Five tactics were extracted for Factor 1. They were "Encourage the parties to meet each other's needs", "Encourage the parties to apologise, and regret for harm suffered by another in the past", "Encourage the parties themselves to verbalise their willingness to respectfully listen to each other's grievances", "Help the parties to save face", and "Educate the parties about the bargaining or impasse process". These tactics concern the encouragement of the parties to self-improvement. A single tactic, "Try to change the expectation of parties", was extracted for Factor 2. This tactic is commonly described as Reality Testing as it helps the disputants to examine whether their expectations are realistic. The two factors extracted can be described as follows:

Factor 1: Encourage for self-improvement
Factor 2: Reality test

18.4.3.2 Mediation Procedures Related Mediator Tactics

The factor matrix after a Varimax rotation is shown in Table 18.7. Factor 1 consists of three tactics: "Control the bargaining structure and timing", "Express pleasure or displeasure at negotiation progress" and "Use humour to lighten the atmosphere". These tactics seek to control the process during mediation and are collectively described as Process Control. "Call for frequent caucuses during mediation" was the only tactic extracted for Factor 2. Factor 3 is composed of two tactics: "Avoid taking sides on important issues in joint sessions" and "Keep in rapport with the parties". These tactics are used by mediators in order to build trust. The tactic, "Argue one party's case to the other", was extracted for Factor 4. This tactic is often used by mediators to analyse the dispute for the disputants during the mediation process. In short, the four factors extracted can be summarised as follows:

Factor 1: Process control
Factor 2: Caucuses
Factor 3: Trust building
Factor 4: Analysing

18.4.3.3 Settlement Related Mediator Tactics

As indicated in Table 18.8, the two tactics extracted for Factor 1 were "Settle simple issue first" and "Help the parties to establish priorities among issues". Both tactics are aimed to help the disputants break the inertia; hence "Ice Breaking" is used to describe this taxonomy. Factor 2 consists of three tactics: "Make compromise suggestions to the parties", "Suggest some tradeoffs among issues", and "Press the parties to make concessions". These address the ways through which mediators seek progress during the mediation process. Finally, "Mention the costs of disagreement" and "Make the parties aware of the destructiveness of the conflict" were extracted for Factor 3. Mediators use these tactics to press for a settlement. The four factors extracted are described as follows:

Factor 1: Ice breaking
Factor 2: Seeking progress
Factor 3: Pressing settlement

18.4.4 Taxonomies of Mediation Outcomes

Sixteen outcomes were used in this study. The factor matrix for these outcomes is presented in Table 18.9. Factor 1 consists of the outcomes "A mutually beneficial settlement was reached", "Agreement perceived to be devised from the parties", and "The parties gained satisfaction on the mediation as a tool of dispute resolution". It can be said that these are Win–Win settlements. Factor 2 consists of three outcomes: "I felt the parties trust the mediator", "The underlying core conflict of the dispute was resolved" and "The number of issues was reduced". These outcomes suggest that progress towards settlement has been achieved. "I improved my cultural sensitivity", "Both parties learned to communicate", and "I improved self-esteem after the settlement of the dispute" were the outcomes extracted for Factor 3. These outcomes suggest some form of improvement has been attained through the mediation. Factor 4 contains a single outcome: "The settlement was reached in reasonable time". It is obvious that this outcome describes the time advantage offered by mediation. Therefore, Factor 4 is labeled "Time Advantage". In brief, the four factors extracted can be described as follows:

Factor 1: Win–Win settlement
Factor 2: Progress
Factor 3: Improvement
Factor 4: Time advantage

The factor matrix after a Varimax rotation is presented in Table 18.9.

18.5 Discussion

The development of taxonomies is summarised in Fig. 18.3. This figure is in fact the enhanced version of the conceptual model in Fig. 18.1.

With the developed of taxonomies, the number of variables has been reduced to a more manageable size. The second part of this study investigates the interrelationships between these three dimensions (represented by the dotted arrows in Fig. 18.3).

18.6 The Contingent Use of Mediator Tactics

Mediation is a dynamic process. The tactics used by a mediator unfold with the mediation, taking into account of the issues in dispute and the concerns of the parties with the aim of steering them towards a settlement (Bercovitch and Houston 1996). However, for any mediation, it will be difficult to reach a settlement with an ineffective mediator (Thoennes and Pearson 1985). It has been suggested that the appropriate use of mediator tactics is one of the decisive factors for mediation success (Kochan and Jick 1978). Use of mediation tactics should not be detached from contexts. This is often described as the contingent use of mediator tactics.

A wide range of tactics can be used by a mediator (Hiltrop 1989; Wall and Rude 1989). For example, Kressel and Pruitt (1985, 1989) identified three generic types of mediation tactics: reflexive, contextual, and substantive. Reflexive tactics are

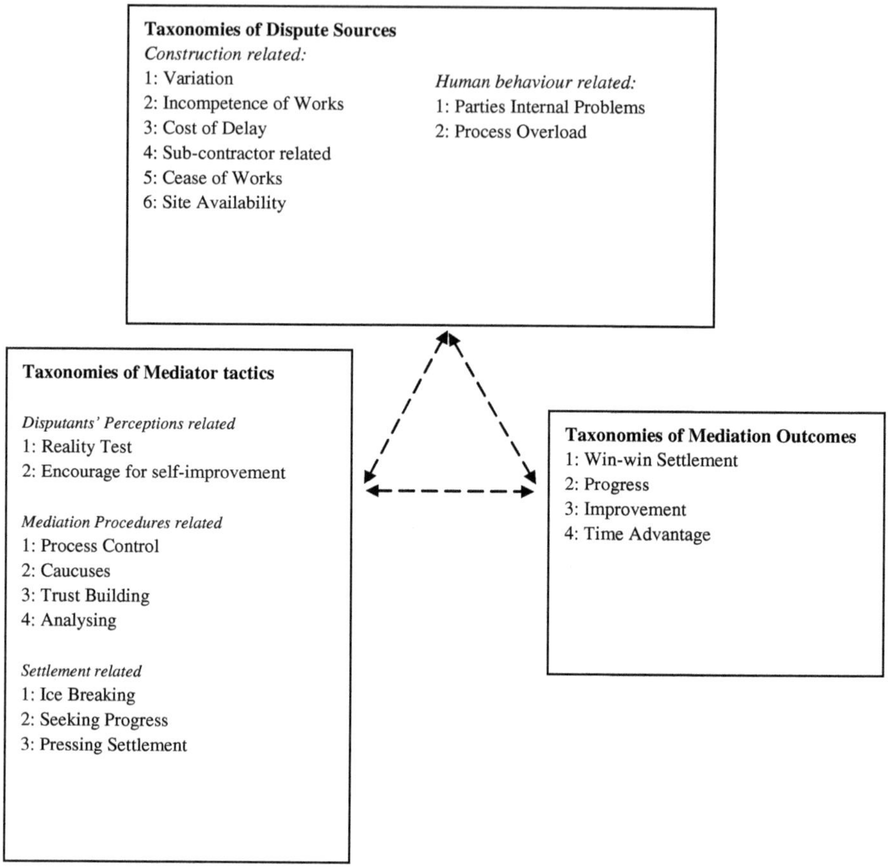

Fig. 18.3 The taxonomies of dispute sources, mediator tactics and mediation outcomes

designed to orient mediators to the dispute, and to create a foundation for their future activities. Substantive tactics deal directly with the issues in a dispute such as making suggestions for settlement. Contextual tactics involve facilitating the dispute resolution process so that the parties themselves will be able to discover an acceptable solution. Lim and Carnevale (1990) had taken this conceptualisation further and suggested four substantive tactics available to mediators: integration, pressing, compensation, and inaction. Furthermore, Wall et al. (2001), focusing on the outcomes of the tactics, suggested another set of generic mediation tactics: disputant oriented, disputant–disputant relationship, and disputants-third-party relationship. Disputant oriented tactics include, among others, information gathering, pressing, and compensation. Setting agenda, providing integrative solution are examples of the disputant–disputant relationship type of tactics. As for the disputants-third party relationship type of tactics, using a third party and making the dispute public were cited as examples. Other studies do not attempt to theorise; nevertheless, they do suggest that there are two generic groups of mediator tactics:

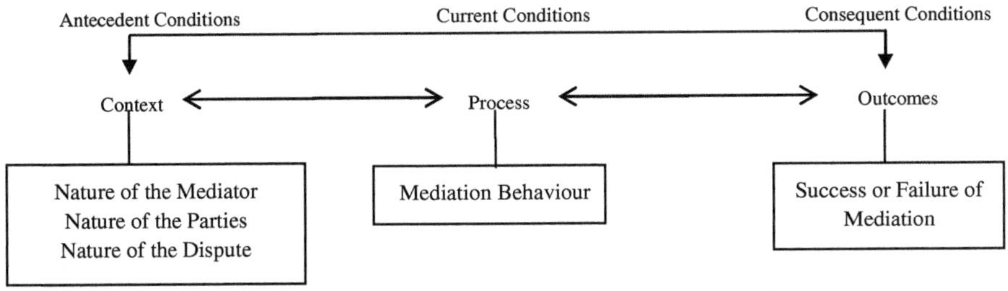

Fig. 18.4 A contingency model of mediation (Bercovitch and Houston 1996)

(1) tactics used in all mediation systemically (Boulle 2001; Maggiolo 1972), and (2) tactics used in specific situations (Carnevale and Pegnetter 1985; Lim and Carnevale 1990; Robbins and Dennenberg 1976; Steven 1963). The former is described as general, systemic, primary, universal, or non-contingent; whilst the latter is labeled as situation-and-person-specific, reactive, or contingent.

Non-contingent mediator tactics refer to those that can be universally applied. These include: (1) explaining the process to the parties, (2) providing structure and control, (3) building trust and confidence of the parties, (4) gaining an accurate understanding of the issues and the underlying impediments to a settlement, and (5) assessing the underlying attitudes of the parties (Boulle 2001; Keltner 1965; Kochan and Jick 1978; Kressel 1972; Simkin 1971). Among these, it is suggested that building trust and confidence of the parties should be the first strategic proposition, and that tactics designed to achieve this objective will be useful in all mediations. Once a certain level of trust is established, the parties are more likely to remain committed to resolving their disputes on the negotiation table. Non-contingent mediator tactics also allow the mediator to collect information regarding the issues in dispute, as well as the underlying interests and attitudes of the parties. In this way, the mediator can have a fuller understanding of the needs of both parties, and find out any hidden agendas or other underlying impediments to a settlement (Kressel 1972; Simkin 1971).

Notwithstanding, a successful mediation must be adaptive and responsive to the contexts (Bercovitch and Houston 1993; Bercovitch and Langley 1993; Carnevale and Pegnetter 1985; Lim and Carnevale 1990). This approach specifies variables with conditional criteria, each of which may have an impact on the effectiveness of the mediation. The most recent development of such approach can be found in the field of international mediation. Bercovitch and Houston (1996) developed a contingent model for international mediations. The core parts of this approach are the clustering of context, process, and outcome variables. Context variables include the nature of the mediator, the disputing parties, as well as the dispute itself. Process variables refer to the behaviours of the mediator. The approach taken in this study recognises the inter-relationships among the context, the process and the outcome (Fig. 18.4).

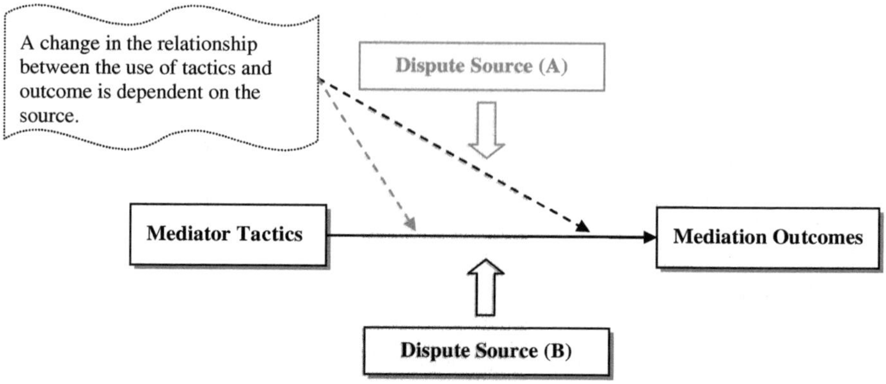

Fig. 18.5 Moderated causal relationship of mediation

The purpose of adding mediation outcomes into the analysis is to consider whether contingent mediator behaviours will improve the likelihood of favorable outcomes. Pioneer studies from this perspective can be found in the works of Hiltrop (1985, 1989) and Rubin (1980). A common assumption used in these studies is that there is only one outcome (settlement vs. non-settlement). Nevertheless, these results suggest that some mediation tactics are more likely to be associated with a particular mediation outcome in the presence of certain contextual factors. Further research work was conducted by Lim and Carnevale (1990), and their findings were similar to those of Hiltrop (1989). For example, Hiltrop (1989) found that under a low level of hostility, the use of substantive pressure was positively associated with settlement. In contrast, the use of forceful pressure was positively associated with settlement only when the level of hostility was high. Lim and Carnevale (1990) found that male mediators were more likely to use substantive/press tactics than female mediators. Furthermore, they confirmed the positive correlation between the use of pressing tactics and general settlement in highly hostile disputes. In the same study (Lim and Carnevale 1990), the effectiveness of tactics as applied to different types of dispute was also analysed. Effectiveness was defined as the degree of achievement of the desired outcome. In essence, it was an investigation of the relation between the use of a tactic and an outcome, taking into account of the source of the dispute (Lim and Carnevale 1990). Their findings show that the ability to drive a certain outcome of a tactic depends on nature of the dispute. This can be illustrated by the moderated causal relationship diagram in Fig. 18.5. In practical terms, it means that the use of mediation tactics should be contingent on the circumstances of the case.

The second part of this study builds upon the framework provided by Lim and Carnevale (1990) to examine the contingent use of tactics in construction mediations.

18.7 Methodology

The contingent use of tactics in construction mediation is examined using moderated multiple regression (MMR), which is an accepted way to assess the viability of a contingency model (Cohen et al. 2003; Lim and Carnevale 1990). To do so, mediator tactics are set as the predictors, mediation outcomes as the criteria, and dispute sources as the moderator variables. The analysis aims to answer the question whether the reported use of Mediator Tactic (T_i) interacts with the Dispute Source (D_i) in predicting the Mediation Outcome (O_i). If the interaction effect is statistically significant, it means that the expected change in mediation outcome in response to a mediator tactic will vary depending on the dispute source. However, if the interaction effect is not significant, it can then be said that the tactics used have "constant" effect on the mediation outcomes (Cohen et al. 2003; Jaccard et al. 1990). In the first part of this study, 21 taxonomies, or factors, were developed using principal component factor analyses (PCFA)—8 for dispute sources, 9 for mediator tactics, and 4 for mediation outcomes. Based on these factors, factor scales were created for use in the moderated multiple regressions (Hair et al.1995). As a result, new sets of variables are developed for each of the three mediation dimensions.

18.7.1 Moderated Multiple Regression

Regression analysis is a statistical technique which allows one to assess the relationship between dependent and independent variables. Multiple regression is an extension of bivariate regression, in which multiple independent variables (instead of just one) are combined to predict a value for a dependent variable. The result of a regression is an equation that best predicts the dependent variable given certain continuous independent variables. For instance, the equation of a multiple regression with two independent variables is shown as follows:

$$Y = a + b_1X_1 + b_2X_2 + \varepsilon \qquad (18.1)$$

where Y = dependent variable; X_1, X_2 = independent variables; a, b_1, b_2 = unknown constant; ε = random error for any given set of values for X_1, X_2

In the above regression equation, X_1 and X_2 have independent effects on the prediction of Y. If the predictive power of X_1 on Y depends on X_2, a moderation effect exists. This is also known as the interaction effect. The moderator term is a compound variable formed by multiplying X_1 by the moderator X_2. This moderation effect is included in the regression equation by introducing the moderator variable X_1X_2. The regression equation thus becomes:

$$Y = a + b_1X_1 + b_2X_2 + \varepsilon \qquad (18.2)$$

where X_1X_2 = moderator variable

The moderated regression models for the investigation of the contingent use of tactics and its effectiveness are developed in a similar manner. Firstly, the regression model for mediation outcome and mediator tactics is given by:

$$O_i = a + b_1 T_i + b_2 D_i + \epsilon \qquad (18.3)$$

where,

O_i = ith mediation outcome scale from the identified taxonomy, i = 1, 2, 3 and 4;
T_i = ith mediator tactic scale from the identified taxonomy, i = 1, 2, 3, 4, 5, 6, 7, 8 and 9;
D_i = ith dispute source scale from the identified taxonomy, i = 1, 2, 3, 4, 5, 6, 7 and 8

According to Eq. 18.3, T_i and D_i have independent effects on O_i. If T_i depends on D_i (i.e. the contingent use of mediator tactics), this equation is no longer applicable for the prediction of O_i. A moderated term is therefore added to explain this relationship. Equation 18.4 shows the moderated multiple regression model used in this study.

$$O_i = a + b_1 T_i + b_2 D_i + b_3 T_i D_i + \epsilon \qquad (18.4)$$

where $T_i D_i$ = moderator variable

18.7.2 The Procedures

As suggested by Jaccard et al. (1990) and Cohen et al. (2003), the first step of an MMR is the formation of interaction. This is achieved by establishing Eqs. 18.3 and 18.4 above. As described in pervious paragraphs, mediator tactics, mediation outcomes and dispute sources are used as the predictors, criteria and moderator variable respectively (Aiken and West 1991; Cohen et al. 2003; Darlington 1990; Jaccard et al. 1990). In this study, a total of 288 moderated multiple regression models (devised from the combination of 8 dispute source scales, 9 mediator tactic scales, and 4 mediation outcome scales) were identified. Next, the interaction effects of these models are tested. The moderated effect is said to be significant if a significant change in R^2 (i.e. ΔR^2) is produced between Eqs. 18.3 and 18.4 as a result of the inclusion of the predictor-moderator product (i.e. $T_i D_i$ term). An F-test is conducted for each model accordingly. The F-test is defined by the following equation:

$$F = \frac{(R_2^2 - R_1^2)/(k_2 - k_1)}{(1 - R_2^2)/(N - k_2 - 1)} \qquad (18.5)$$

where k_2 is the number of predictors in the expanded Eq. 18.4; k_1 is the number of predictors in the original Eq. 18.3; *N* is the total sample size; $(k_2 - k_1)$ and $(N - k_2 - 1)$

are the degrees of freedom; R_2 is the multiple R in the expanded Eq. 18.4; and R_1 is the multiple R in the original Eq. 18.3

The null hypothesis is established for which the regression coefficient for the product term (i.e. b_3) is zero. Rejection of this hypothesis is consistent with the notion that an interaction effect is present. Moreover, there are two methods for the calculation of F test statistic:

(i) By the use of F-test as shown in Eq. 18.5.
(ii) The same substantive results as that of point (i) above can also be obtained by squaring the *t* test statistic of the b_3 coefficient in Eq. 18.4 (Jaccard et al. 1990).

In this study, the second method was adopted. The critical value was obtained from the F-distribution table with different significance level ($\alpha = 0.10$, $\alpha = 0.05$ and $\alpha = 0.01$). The ΔR^2 was regarded as significant at these significance levels, and the interaction effect was significant in that particular moderated regression model. Hence, 288 numbers of the R^2 value, change in R^2 (ΔR^2) and the standardised regression coefficients of the Tactic Source x interactions (b_3) for the respective Outcome scale are obtained. Accordingly, a total of 40 (out of 288) moderated regression models were found significant (Table 18.10).

18.8 The Result

Moderated regression models that showed significant moderation effects are summarised in Table 18.1. The effectiveness of the tactics in achieving a specific outcome is contingent on the dispute source. To give an illustration; the tactic "Trust Building" is perceived to be effective in dealing with "Variation"- related disputes as far as the outcome of "improvement" is concerned. Similarly, the tactic "Encourage to Self-improvement" is effective in achieving the outcomes of "Win–win Settlement" and "Time Advantage" when the dispute source is "Variation". To save space and preserve clarity, only the results of the forty moderated regression models that showed significant moderation effects are presented in Table 18.11.

The interpretation of Table 18.11 can be explained again by way of example. Consider the first row of the table: the use of the tactic "encourage to self-improvement" in a dispute involving "variation" is contingently effective in achieving the outcomes of "win–win settlement" and "time advantage". Other scenarios can be interpreted in the same way.

Table 18.10 Summary of the 40 moderated regression models with significant interaction effect

No.	Mediation outcome (O_i)	Tactics (T_i)	Dispute source scales (D_i) that make significant contribution to the relationship of O_i and T_i
1	Win–win Settlement	Encourage to self-improvement	Variation
2			Cease of works
3		Reality test	Cost of delay
4			Cease of works
5		Process control	Cease of works
6		Caucus	Variation
7			Process overload
8		Trust building	Cost of delay
9			Process overload
10		Ice breaking	Variation
11		Seeking progress	Variation
12			Cost of delay
13			Process overload
14		Pressing settlement	Cost of delay
15	Progress	Reality test	Variation
16			Process overload
17		Trust building	Cost of delay
18			Parties' internal problems
19			Process overload
20		Analysing	Sub-contractor related
21		Ice-breaking	Variation
22			Incompetence of works
23		Seeking progress	Cost of delay
24		Pressing settlement	Cost of delay
25			Parties' internal problems
26	Improvement	Reality test	Cease of works
27		Trust building	Variation
28			Incompetence of works
29			Cost of delays
30			Process overload
31		Pressing settlement	Cost of delay
32			Process overload
33	Time advantage	Encourage to self-improvement	Variation
34			Site availability
35		Reality test	Incompetence of works
36			Cost of delay
37		Trust building	Cease of works
38			Process overload
39		Ice breaking	Incompetence of works
40		Pressing settlement	Cost of delays

18.9 Discussion

As noted from Table 18.11, it can be seen that not all tactics are perceived to have an effect on a mediation outcome in relation to the eight dispute sources. This in general therefore supports the proposition that tactics should be employed appropriate to the contexts.

Discussions regarding Table 18.11 shall be structured under the heading of the three mediation dimensions: Tactics, Sources and Outcomes. To preserve clarity, only the relatively important observations are discussed. That means, for the three mediation dimensions, (i) the two most versatile tactics; (ii) the two dispute sources that show good responses to a range of tactics and (iii) the outcome that is most responsive to tactics.

18.9.1 Tactics

From Table 18.11, it can be said that some tactics are more versatile than the other. In this connection, versatile is defined as being effective in a wide range of situations. As such, reality testing and trust building are the two most versatile tactics. This can be supported by the number of dispute displaying a moderation effect. In fact, similar observation was reported by Lim and Carnevale (1990).

According to Boulle and Nesic (2001), mediators are often referred to as 'agents of reality' in so far as their function of encouraging the parties to face the realities of their situations. The purpose of reality testing is to make the relevant party reflect more systemically and practically on a position, behaviour or attribute, and to think beyond the present situation to future consequences. Reality testing can apply to subjective factors associated with the dispute, and to objective factors which are part of the wider picture. As such reality-testing tactics can be effective to deal with the following factors:

(i) The parties' substantive, procedural and emotional interests;
(ii) Resources like time and cost;
(iii) All aspects of a proposed settlement, including its degree of specificity, its durability and its fairness to all parties; and
(iv) The possible scenarios should the dispute is not settle.

In these respects, reality test tactics are used by mediators to direct disputants to assess their positions from a more realistic perspective. In fact, disputants often over-value their own cases while under-value the opponent's assertions. This would affect the objectivity of the assessment. Reality test helps to restore objectively by pointing out the feasibility and viability of expectations. Through reminders and arm-twisting, disputants are urged to re-assess the strength and weakness, the likely cost implications of their cases. These are all useful and

Table 18.11 Summary of moderated multiple regression (MMR)

Mediator tactics	Dispute sources								Mediation outcome			
	Variation	Incompetence of works	Cost of delay	Sub-contractor related	Cease of works	Site availability	Parties internal problems	Process overload	Win–win settlement	Progress	Improvement	Time advantage
Encourage to self-improvement	*								√			√
					*				√			
						*						√
Reality test	*									√		
		*										√
			*						√			√
					*				√		√	
								*		√		
Process control					*				√			
Caucus	*								√			
								*	√			
Trust building	*										√	
		*									√	

(continued)

Table 18.11 (continued)

Mediator tactics	Dispute sources								Mediation outcome			
	Variation	Incompetence of works	Cost of delay	Sub-contractor related	Cease of works	Site availability	Parties internal problems	Process overload	Win–win settlement	Progress	Improvement	Time advantage
			*						√	√	√	
					*							√
							*			√		
								*	√	√	√	√
Analysing				*						√		
Ice breaking	*								√	√		
		*								√		√
Seeking progress	*								√			
			*						√	√		
								*	√			
Pressing settlement			*						√	√	√	√
							*			√		
								*			√	

Dispute source (*) that responds to tactic for the outcome (√).

contribute towards settlement. In Table 18.2, reality testing tactics are perceived to have contingent effect on the mediation outcome for five types of dispute source.

Boulle (2001) also points out that if the disputants trust the mediator and the process, then they are more likely to remain at the negotiating table and direct efforts to strive for a settlement. In Part (I) of this chapter, two tactics were grouped in the tactic taxonomy of Trust Building: avoid taking sides on important issues in joint sessions and keep in rapport with the parties. The first tactic features the neutrality of mediators, a quality that has been identified as fundamental for trust in the mediator and thus the mediation process to be established (Cheung and Suen 2002). The second tactic of establishing rapport with the parties seeks to impress on the disputants that the mediator can be trusted.

Trust can be described as a disputant's willingness to believe, to be open to, and to take risks with the other parties (Boulle 2001). It is not too difficult to apprehend that parties in dispute frequently distrust each other. Whilst reality testing seeks to enable a disputant to pragmatically assess his position, he needs to rely on information in his possession or provided by the other party to perform this task. Trust underpins his willingness to use that piece of information. In fact, apart from the dispute sources that are sub-contractor related or concerning site availability, the tactic of trust building would bring out improvements in at least one of the mediation outcomes. The following techniques can be used to help parties generate trust in the process in which the disputants are participating (Boulle 2001):

(i) Explaining, normalising and validating the mediation process;
(ii) By reassuring the parties, where possible, on their anxieties about the process;
(iii) By providing for equality of speaking time for the parties;
(iv) By applying the mediation guideline appropriately;
(v) By using the caucus to keep the process moving.

18.9.2 Dispute Sources and Mediation Outcomes

From Table 18.11, where a dispute is related to either variation or associated with a process overload, a range of tactics can be used to effectively achieve different mediation outcomes. In fact variation has been identified as the most critical source of dispute in construction (Cheung et al. 2000; Kumaraswamy and Chan 1998). Typically, dispute related to variation would involve some form of disagreement on position (validity). The time and cost implications are also the main sources of disagreement. Ice breaking tactics and the use of caucus help to establish communication. Reality testing techniques helps to alter positions and expectation. Encouraging to self-improve and seeking progress are often vital if a final settlement can be reached. A process will be overloaded if too many issues are brought to the table. In this connection, a skillful mediator would first establish some agreed facts so as to simplify the agenda. This would also eliminate non-productive discussions on agreed items.

Out of the four types of outcome, win–win settlement appears to be the most responsive to tactics. This result suggests that mediation is an effective mean to resolve construction disputes. Indeed, the result is not surprising as the ultimate aim of mediation is to achieve a win–win settlement. The use of mediator tactics attempt to emphasis in advance the best alternative to a negotiated agreement (BATNA) and the worst alternative to a negotiated agreement (WATNA) to the parties, so that they can decide their eventual settlement in order to avoid their disputes to involve in expensive arbitration or litigation. From Table 18.11, no moderator effect is detected for the analysing tactics as far as win–win settlement is concerned.

18.10 Chapter Summary

In construction mediation, the mediator plays a pivotal role in facilitating settlement. Whilst previous researches on construction mediation have focused on process design and critical success factors, a study of tactics used by mediators extends our understanding on how mediation works and therefore is of both academic and practical value. A study of construction mediator tactics was conducted in Hong Kong.

Due to the dynamic nature of a mediation process, mediator should employ appropriate tactics to the context in particular the dispute source. This study examines the contingent use of these tactics in relation to the dispute sources and mediation outcomes. The study is broadly divided in two parts. Part I of the study is to develop taxonomies of construction dispute sources, mediator tactics and mediation outcomes. Principal Component Factor Analyses were conducted for this purpose. As a result, 8, 9 and 4 taxonomies were identified for dispute sources, mediator tactics and outcomes respectively. Based on these results, factor scales were derived for use in Part II of the study where Moderated Multiple Regression (MMR) was used to detect the moderator effect of dispute source on the tactic-outcome. It is found that not all of the dispute sources display a moderator effect. This in general supports the proposition that certain mediator tactics are more effective in driving a certain outcome with respect to a particular dispute source. Moreover, trust building and reality test were found to be the most versatile tactics.

Acknowledgments The content of the chapter has been published in Volume 42(2) of Building and Environment and is used with the permission of Elsevier.

References

Aiken, L. S., & West, S. G. (1991). *Multiple regression: Testing and interpreting interactions.* Newbury Park, CA: Sage.

Al-Khalil, M. I., & Al-Ghafly, M. A. (1999). Important causes of delay in public utility projects in Saudi Arabia. *Construction Management and Economics, 17*, 647–655.

Bateson, D. (1997). "Mediation and adjudication in Hong Kong—are these alternative dispute resolution procedures working? *The Journal of the Chartered Institute of Arbitrators, 63*(4), 243–245.

Bercovitch, J., & Houston, A. (1996). The study of international mediation: Theoretical issues and empirical evidence. In: J. Bercovitch (Ed.), *Resolving international conflicts: the theory and practice of mediation*. Boulder, Colorado: Lynne Rienner Publishers.

Bercovitch, J., & Houston, A. (1993). Influence of mediator characteristics and behavior on the success of mediation in international relations. *International Journal of Conflict Management, 4*, 297–321.

Bercovitch, J., & Langley, J. (1993). The nature of the dispute and the effectiveness of international mediation. *Journal of Conflict Resolution, 37*, 670–691.

Boulle, L. (2001). *Mediation—skills and techniques*. Chatswood: Butterworths.

Boulle, L., & Nesic, M. (2001). *Mediation—principles, process, practice*. London: Butterworths.

Carnevale, P. J. D. (1986a). Strategic choice in mediation. *Negotiation Journal, January*, 41–56.

Carnevale, P. J. D. (1986b). Meditating disputes and decisions in organizations. In R. J. Lewicki, B. H. Sheppard, & M. H. Bazerman (Eds.), *Research on negotiation in organizations* (Vol. 1, pp. 271–283). Greenwich, CT: JAI Press.

Carnevale, P. J. D., & Henry, R. A. (1989). Determinants of mediator behavior: A test of the strategic choice model. *Journal of Applied Social Psychology, 19*(6), 481–498.

Carnevale, P. J. D., & Pegnetter, R. (1985). The selection of mediator tactics in public sector dispute: A contingency analysis. *Journal of Social Issues, 41*(2), 65–81.

Carnevale, P. J. D., & Pruitt, D. G. (1992). Negotiation and mediation. *Annual Review of Psychology, 43*, 531–582.

Carnevale, P. J. D., Lim, R. G., & Mclaughlin, M. E. (1989). Contingent mediator behavior and its effectiveness. In K. Kressel, & D. G. Pruitt (Eds.) *Mediation research—the process and effectiveness of third-party intervention*. San Francisco: Jossey-Bass Publication.

Chau, K. W. (1990). The prospect of conciliation in the construction industry in Hong Kong. *Hong Kong Engineers, July*, 24–28.

Chau, K. W. (1991). Update to conciliation in resolving Hong Kong construction disputes. *Hong Kong Engineers, March*, 34–35.

Cheung, S. O., & Suen, C. H. (2002). The contribution of the third party neutral towards amicable construction dispute resolution. *The International Construction Law Review, 19*(1), 79–86.

Cheung, S. O., Tam, C. M., Ndekugri, I., & Harris, F. C. (2000). Factor affecting clients' project dispute resolution satisfaction in Hong Kong. *Construction Management and Economics, 18*, 281–294.

Cheung, S. O. (1999). Critical factors affecting the use of alternative dispute resolution processes in construction. *The International Journal of Project Management, 17*(3), 189–194.

Cheung, S. O., & Yeung, Y. W. (1998). The effectiveness of the dispute resolution advisor system: A critical appraisal. *The International Journal of Project Management, 16*(6), 367–374.

Cohen, J., Cohen, P., West, S. G., & Aiken, L. S. (2003). *Applied multiple regression/correlation analysis for the behavioral sciences"*. Mahwah, NJ: L. Erlbaum Associate.

Comrey, A. L., & Lee, H. B. (1992). *A first course in factor analysis* (2nd ed.). Hillsdale, NL: Erlbaum.

Darlington, R. B. (1990). *Regression and linear models*. New York: McGraw-Hill.

Fenn, P., & Gameson, R. (1992). Construction conflict management. *Proceedings of the First International Construction Management Conference*. The University of Manchester, Institute of Science and Technology, London: E & FN Spon 25–27 September 1992.

Fraley, A. (1990). Mediation and the professional. *Chartered Builder, March/April*, 7–8.

Hair, A., Tatham, R. L., & Black, W. C. (1995). *Multivariate data analysis* (5th ed.). Englewood Cliffs, NJ: Prentice Hall.

Harmon, K. M. (2003). Conflict between owner and contractors: proposed intervention process. *Journal of Management in Engineering, 19*(3), 121–125.

Henderson, D. A. (1996). Mediation success: An empirical analysis. *Ohio State Journal on Dispute Resolution, 11*(1), 105–147.

Hibberd, P. R., & Newman, P. (1999). *ADR and adjudication in construction disputes*. Malden, MA: Blackwell Science.

Hill, R. (1998). The theoretical basis of mediation and other forms of ADR: Why they work. *Arbitration International, 14*(2), 173–184.

Hiltrop, J. M. (1985). Mediator behavior and the settlement of collective bargaining disputes in Britain. *Journal of Social Issues, 41*(2), 83–99.

Hiltrop, J. M. (1989). Factors associated with successful labor mediation. In K. Kressel & D. G. Pruitt (Eds.), *Mediation research: The process and effectiveness of third party intervention* (pp. 241–262). San Francisco: Jossey-Bass.

Holland, D. E. (1992). Alternative dispute resolution in construction. *Arbitration, February*, 57–59.

Holt, G. (1997). Construction research questionnaire and attitude measurement: Relative index or mean. *Journal of Construction Procurement, 3*(2), 88–94.

Hong Kong International Arbitration Centre (HKIAC) (2001). Services and Facilities (http://www.hkiac.org)

Hong Kong Mediation Centre (2001). http://www.mediationcentre.com.hk

Jaccard, J., Turrisi, R., & Choi, K. W. (1990). *Interaction effects in multiple regression*. Newbury Park: Sage.

Kaming, P. F., Olomolaiye, P. O., Holt, G. D., & Harris, F. C. (1997). Factors influencing construction time and cost overruns on high-rise projects in Indonesia. *Construction Management and Economics, 15*, 83–94.

Kaplan, N., Spruce, J., & Cheng, T. Y. W. (1991). *Hong Kong arbitration cases and materials*. Hong Kong: Butterworths.

Karim, A., & Pegnetter, R. (1983). Mediator strategies and qualities and mediation effectiveness. *Industrial Relations, 22*, 105–114.

Keltner, J. W. (1965). Communication and the labor-management mediation process: Some aspects and hypotheses. *Journal of Communication, 15*(2), 64–80.

Kerr, C. (1954). Industrial conflict and its mediation. *American Journal of Sociology, 60*, 230–245.

Kochan, T. A., & Jick, T. (1978). The public sector mediation process. *Journal of Conflict Resolution, 22*, 209–240.

Kressel, K. (1972). *Labor mediation: An exploratory survey*. Albany, NY: Association of Labor Mediation Agencies.

Kressel, K., & Pruitt, D. G. (1985). Themes in the mediation of social conflict. *Journal of Social Issues, 41*, 179–198.

Kressel, K., & Pruitt, D. G. (1989). A research perspective on the mediation of social conflict. In K. Kressel & D. G. Pruitt (Eds.), *Mediation research: The process and effectiveness of third party intervention* (pp. 394–435). San Francisco: Jossey-Bass.

Kumaraswamy, M. M., & Chan, D. W. M. (1998). Contributors to construction delays. *Construction Management and Economics, 16*, 17–29.

Langford, D. A., Kennedy, P., & Somerville, J. (1992). Contingency management of conflict: Analysis of contract interfaces. *Proceedings of the First International Construction Conflict Management and Resolution Conference*. University of Manchester Institute of Science and Technology, U.K., pp. 151–161.

Lim, R. G., & Carnevale, P. J. D. (1990). Contingencies in the mediation of disputes. *Journal of Personality and Social Psychology, 58*(2), 259–272.

Locke, E. A. (1991). The motivation sequence, the motivation hub, and the motivation core. *Organizational Behavior and Human Decision Processes, 50*, 288–299.

Lowry, L. R. (2001). Mediation: Fulfilling its promise for effective dispute resolution (Part 1). *Asian Dispute Review, June*, 24–26.

Maggiolo, W. (1972). *Techniques of mediation in labor disputes*. Dobbs Ferry, NY: Oceana Publications.

Marcus, G., & Marcus, P. (1987). Fact-based mediation for the construction industry. *The Arbitration Journal, 42*(3), 6–14.
McDonald, P. R. (1984). Construction claims costing for owners and contractors. *Construction Management and Economics, 2*, 1–12.
Pruitt, D. G. (1981). *Negotiation behavior.* New York: Academic Press.
Rifkin, J., Millen, J., & Cobb, S. (1991). Toward a new discourse for mediation: A critique of neutrality. *Mediation Quarterly, 9*, 151–164.
Robbins, E., & Dennenberg, T. S. (1976). *A guide for labour mediators*. Hawaii: University of Hawaii Press.
Rubin, J. Z. (1980). Experimental research on third-party intervention in conflict: Toward some generalizations. *Psychological Bulletin, 87*, 379–391.
Santilli, C. (1988). Mediation keeps you out of court. *Concrete Construction, February*, 190–192.
Shapiro, D., Drieghe, R., & Brett, J. (1985). Mediator behavior and the outcome of mediation. *Journal of Social Issues, 41*(2), 101–114.
Simkin, W. (1971). *Mediation and the dynamics of collective bargaining*. Washington, DC: BNA.
Stevens, C. (1963). *Strategy and collective bargaining negotiation.* New York: McGraw-Hill.
Taylor, S. E., & Crocker, J. (1981). Schematic bases of social information processing. In E. T. Higgens, C. P. Herman, & M. P. Zanna (Eds.), *Social cognition: The Ontario symposium* (Vol. 1, pp. 89–133). Hillsdale, NJ: Erlbaum.
Thoennes, N. A., & Pearson, J. (1985). Predicting outcomes in divorce mediation: The influence of people and process. *Journal of Social Issues, 41*(2), 115–126.
Wall, J. A., & Rude, D. E. (1989). Judicial mediation of settlement negotiations. In K. Kressel & D. G. Pruitt (Eds.), *Mediation research.* San Francisco, CA: Jossey-Bass.
Wall, J. A., Stark, J. B., & Standifer, R. L. (2001). Mediation—a current review and theory development. *Journal of Conflict Resolution, 45*(3), 370–391.

Chapter 19
The Efficacy of Trust-Building Tactics in Construction Dispute Mediation

Tak Wing Yiu

Abstract Distrust hinders disputing parties and mediators from achieving mediation success. Mediators therefore often use trust-building tactics to generate some degree of trust in themselves and in the mediation process. This chapter reports a study that identified the trust-building tactics used by construction mediators and examined the efficacy of these tactics with respect to their outcomes. Three study stages were designed. With reference to the mediation model of Sloan (1998), trust-building tactics and outcomes were first identified in Stage I. Next, the data were collected from accredited mediators with a questionnaire survey in Stage II. The collected data were then validated via reliability assessments in Stage III. With the use of multiple regression analyses, the efficacy of the trust-building tactics was examined by relating these tactics to their outcomes. The findings of this study suggest that the trust-building tactics used in Step 4 (i.e., explore interests) of Sloan's mediation model (1998) are influential in developing trust among disputing parties and that they can also act as a time-saving tool in the mediation process. Furthermore, it was found that mediators can earn trust by adopting the trust-building tactics used in Step 3 (i.e., issues and trust) of Sloan's model (1998). These tactics can also serve to improve the relationships between the disputing parties. The results show that the trust-building tactics used in the final step (i.e., solutions) of Sloan's model (1998) seem to have low efficacy in developing trust among disputing parties.

T. W. Yiu (✉)
Department of Civil and Environmental Engineering, The University of Auckland, Auckland, New Zealand
e-mail: k.yiu@auckland.ac.nz

S. O. Cheung (ed.), *Construction Dispute Research*,
DOI: 10.1007/978-3-319-04429-3_19,

19.1 Introduction

Mediation has become a common means of resolving disputes in construction for its flexible, cost-effective, and non-confrontational approach. Mediation also allows disputes to be settled voluntarily and privately without a loss of business relationships or damage to reputations (Chau 1992, 2007; Cheeks 2003; Hon 2006). Such settlements can therefore be made more rapidly and are more adaptable to the specific needs of the disputing parties in the mediation process (Cheung and Yiu 2007; Striengnitz 2006; Susskind and Ozawa 1983; Yiu and Cheung 2007). However, mediation involves a series of negotiation processes with no guarantee of success.

The actions taken by mediators are critical to mediation success. Integrity, reliability, and competence are the most important attributes of mediators that affect the level of trust that disputing parties have in them (Boulle 2001; Settle 1998). During the process of mediation, the major task of the mediator is to encourage the disputing parties, such as clients, contractors, and sub-contractors, into rethinking and modifying their positions (Kolb 1985; Madden 2001). However, the degree of influence that mediators have depends considerably on whether the disputing parties trust them (Kolb 1985). Therefore, mediators need to realise the importance of using trust-building tactics in the course of the mediation process to address any long-held and deep-seated concerns among the disputing parties (Blackstock 2001). If trust can be built, then mediators need to monitor the level of that trust (and its influence) on the disputing parties, and thus credible relationships can be established with them (Carnevale 1986; Carnevale et al. 1989). Such relationships can serve as a lubricant to avoid unnecessary hurdles in the mediation process (Boulle 2001; Torres 1991). For example, the early establishment of trust can ease subsequent inquiries into the course of mediation (Boulle 2001; Torres 1991). Time-savings can also be achieved when the disputing parties are willing to disclose confidential information or hidden agendas in the mediation process. If the disputing parties trust the mediator, then they are more likely to remain at the negotiating table, to remain committed to the mediation process, to believe in achieving successful mediation outcomes (Boulle 2001), and to behave co-operatively (Cheung and Yiu 2007). Mistrust, in contrast, discourages mediation success. The level of trust among disputing parties varies with attitude (Govier 1997). Mediators should apply trust-building tactics to generate some degree of trust in themselves and in the mediation process (Boulle 2001) to ensure that the disputing parties work toward a win-win settlement. Hence, the objectives of the study reported in this chapter were to identify the trust-building tactics used by construction mediators and to examine the efficacy of these tactics with respect to their outcomes. To achieve these objectives, three study stages were involved. In Stage I, a literature review was conducted to identify commonly used trust-building mediator tactics and mediation outcomes. Using these, a questionnaire survey was designed in Stage II to collect data from accredited mediators. A series of data analyses were then performed in Stage III to examine the efficacy of the trust-building tactics that were identified.

Table 19.1 Brief introduction to Sloan's mediation model (1998)

Step	Description
1. Preparation	*The aim of this step is to orient the disputing parties to the mediation process and to encourage genuine interest. Expressing genuine interest can increase awareness among the disputing parties.*
2. Introduction	*The aim of this step is to establish and maintain a collaborative tone for negotiation. The mediator can help the disputing parties to develop ground rules for the negotiation processes.*
3. Issues and trust	*The aim of this step is to identify what the disputing parties have come to the mediation to resolve. The mediator has to encourage the disputing parties to listen to, recognise, and understand each other.*
4. Explore interests	*The aim of this step is to help the disputing parties explore what interests are important to them and why. The mediator should assist them by asking questions and using interrogative and reflective skills.*
5. Solutions	*The aim of this step is to help the disputing parties generate a creative, tailored, and durable settlement.*

19.2 Stage I: Identification of Trust-Building Tactics and Mediation Outcomes

19.2.1 Trust-Building Tactics

Theoretically, the mediation process flows from one phase to the next: for example, from the mediator's opening statement to the disputing parties' opening statements and then to a joint session, private sessions, joint negotiations, and the final closure phase. As mediation proceeds, the mediators work to build and monitor the trust levels of the disputing parties. To do so, they employ different types of trust-building tactics in each phase of the process. Sloan's mediation model (1998), which conceptualises the mediation process, provides an important framework for the identification of trust-building tactics in construction mediation. According to this model, which is shown in Table 19.1, the mediation process has five steps: (1) Introduction (2) Preparation (3) Issues and Trust (4) Exploration of Interests, and (5) Solutions.

Based on the steps in Sloan's mediation model (1998), trust-building tactics were identified from the literature (Bercovitch and Derouen 2004; Boulle 2001; Cheung and Yiu 2007; Kolb 1985; Latz 2001; Moore 1996; Salem 2003). As can be seen from Table 19.2, a list of 18 trust-building tactics was compiled.

19.2.2 Mediation Outcomes

In this study, the outcomes refer to the results that were obtained from the trust-building tactics used in construction mediation. The generic types of outcomes, which were identified from the literature review and are listed in Table 19.3, can

Table 19.2 List of trust-building tactics

Trust-building tactics	Steps involved in the mediation model of Sloan (1998)
T1. Try to observe and understand how the parties interact and communicate with and treat each other	Step 1: Preparation
T2. Try to be effective and show respect/concern for the disputing parties, even when they do not trust you initially	
T3. Try to use humor to lighten the atmosphere	
T4. Express very clearly what you can and cannot do. Try to focus on designing a negotiation process	Step 2: Introduction
T5. Try to comfort the parties first by solving minor issues	
T6. Try to develop ground rules for the mediation process	
T7. Set realistic targets for the disputing parties	
T8. To allow the parties to express themselves freely, you have to know how to listen and when to keep silent during the mediation process	Step 3: Issues and Trust
T9. Implement caucusing to understand and explore the parties' concerns and their bottom lines	
T10. When a dilemma occurs, encourage the disputing parties to ask for help, thus acknowledging their need for assistance from the other disputing parties	
T11. Try to be patient and understand the feelings of the disputing parties	Step 4: Explore Interests
T12. Share your personal details and experience of mediation with the disputing parties	
T13. Try to simplify the agenda, develop a framework, and prioritise the issues	
T14. Be well-prepared for the issues that the disputing parties want to clear up	
T15. Try to encourage the disputing parties to make incremental agreements in which success can be measured along the way	Step 5: Solutions
T16. To assist the effective resolution of the dispute, learn not only about the immediate issue, but also about its background/history	
T17. Gain insight into how the disputing parties react when you make suggestions	
T18. Keep explicit promises and do not lie to the disputing parties	

be reduced to: (1) trust-building between the disputing parties (2) trust-building between the disputing parties and the mediator, and (3) the negative (or positive) implications of using these trust-building tactics (Bercovitch and Derouen 2004; Boulle 2001; Butler 1991; Cheung and Yiu 2007; Kolb 1985; Latz 2001; Lui et al. 2006; Moore 1996; Salem 2003).

Table 19.3 List of mediation outcomes

Mediation outcomes
Trust-building (party-party)
O5. The disputing parties voluntarily resolved the identified issues by themselves
O7. Interactions between the disputing parties were facilitated
O13. The disputing parties became less defensive and more willing to share information
Trust-building (mediator-parties)
O8. Your reputation was enhanced
O11. The parties were willing to rely on you and accepted risk and vulnerability
O6. In private sessions (caucusing), the disputing parties were willing to share confidential information that was crucial in reaching a mutually acceptable solution
Failure to build trust
O1. Mistrust appeared due to the different cultural, racial, and historical background of the disputing parties
O2. It was difficult to persuade the disputing parties to disclose confidential information
O3. It was difficult to built trust, and there was a lack of frank communication between you and the disputing parties
O4. The disputing parties avoided face-to-face conversations by sending their representatives to act on their behalf
O9. Mistrust led to positional persistence and failure of the negotiated settlement
Improvement of relationships
O12. A win-win settlement was achieved, and the relationships among the disputing parties were improved
Deadlock
O10. Mistrust delayed the settlement or resulted in no settlement
Time-savings
O14. The mediation process was shortened.

19.3 Stage II: Data Collection

To accomplish the research objectives, a four-part questionnaire survey was performed to collect case-specific data from construction mediators. The first part required the respondents to provide personal information, such as their sex, their age, and the number of years of experience they have had in construction mediation. Next, the respondents were asked to provide particulars about their mediated cases, such as the project nature, contract sum, and the parties involved. The last two parts of the survey were designed to identify the trust-building mediation tactics used and their respective outcomes, based on the items in Tables 19.1 and 19.2. Seven-point Likert scales were used to measure the degrees of usefulness (1: least useful; 7: most useful) and agreeableness (1: least agreeable; 7: most agreeable) on each trust-building tactic and mediation outcome, respectively. The respondents targeted in this study are accredited mediators from the General Panel of the Hong Kong International Arbitration Centre (HKIAC), which is composed of 226 accredited

Table 19.4 Profile of respondents (by amount of experience)

	Years of experience (No.)	Percentage
No experience: newly accredited mediators	8	26.7
Less than 5 years	6	20
5–10 years	1	3.3
11–15 years	1	3.3
16–20 years	3	10
More than 20 years	11	36.7
Total	30	100.0

Table 19.5 Types of reported mediation cases

	Frequency (No.)	Percentage
Building	5	22.8
Civil	11	50
Building services	2	9.1
Maintenance	2	9.1
Others	1	4.5
Building & civil	1	4.5
Total	22[a]	100.0

[a] 8 respondents did not provide this information

mediators from 30 professions. Ninety-six of these mediators who specialise in building, construction, and engineering were selected as the target respondents, and their participation was requested. If they agreed to participate in the questionnaire survey, then questionnaires were sent to them by post, fax, or e-mail. The respondents, particularly newly accredited mediators, made a number of enquiries regarding the completion of the questionnaire. Many of them wanted to know about the possibility of completing the questionnaires by referring to their experience with mediating simulated cases in lieu of reporting real mediated cases. To ensure the relevance of the responses, this suggestion was rejected. Thirty of the 96 targeted respondents completed the questionnaire survey, for a response rate of 31 %. This sample size is comparable to those in previous studies of construction mediation (Cheung and Yiu 2007; Yiu and Cheung 2007). All of the respondents are accredited mediators with a construction background, such as construction lawyers, quantity surveyors, and engineers, who actively participate in mediating construction disputes in Hong Kong. They all hold senior positions and are well-respected by the industry. For example, 36.7 % of the respondents have more than 20 years of experience in construction mediation. Fifty percent of the reported mediation cases concerned civil projects. The profiles of the respondents by experience and type of reported mediation cases are shown in Tables 19.4 and 19.5, respectively. As for the scale of the reported cases, 31.8 % of them had contract sums of less than HK$50 million. Most of the disputes (36.4 %) had arisen between the employer and the main contractor. Tables 19.6 and 19.7 show the reported mediation cases by value and the parties involved, respectively.

Table 19.6 Reported mediation cases by value (in HK$)

	Frequency (No.)	Percentage
<50 million	7	31.8
50–200 million	6	27.3
200–500 million	6	27.3
Above 500 million	3	13.6
Total	22[a]	100.0

[a] 8 respondents did not provide this information

Table 19.7 Parties involved in reported mediation cases

	Frequency (No.)	Percentage
Employer	6	27.3
Main contractor	2	9.1
Nominated subcontractor	1	4.5
Employer and main contractor	8	36.4
Main contractor and domestic subcontractor	4	18.2
Main contractor and others	1	4.5
Total	22[a]	100.0

[a] 8 respondents did not provide this information

19.4 Stage III: Efficacy of the Trust-Building Tactics

A series of data analyses were performed on the collected data using the Statistical Package for Social Sciences (SPSS). Descriptive statistics were first performed to identify (1) the most commonly used trust-building tactics in construction mediation and (2) the most frequent outcomes achieved using them. Next, reliability assessments were performed for the Cronbach's alpha coefficients and item-total correlations to test the internal consistency of the responses (Streiner and Norman 1997). Finally, multiple regression analysis (MRA) was employed to examine the efficacy of the trust-building tactics by relating them to the outcomes. The independent contribution of each trust-building tactic to the prediction of outcomes was investigated by the MRA regression coefficients.

19.4.1 Descriptive Statistics

The descriptive statistics, such as the minimum, maximum, and mean scores for each trust-building tactic and its outcomes are shown in Table 19.8.

In general, these descriptive statistics show that the trust-building tactics of implementing caucuses (T9), giving the disputing parties the opportunity to

Table 19.8 Descriptive statistics of trust-building tactics and outcomes

Variables	Minimum	Maximum	Mean	Standard deviation	Item-total correlations[a]	Cronbach's α coefficients[b]
Trust-building tactics						
T1.	3	7	5.33	0.92	0.47	0.89
T2.	2	7	4.67	1.35	0.70	0.88
T3.	2	7	5.83	1.26	0.21[c]	0.89
T4.	2	7	4.77	1.38	0.47	0.89
T5.	1	7	4.77	1.52	0.35	0.89
T6.	2	7	4.83	1.29	0.55	0.88
T7.	3	7	5.10	1.21	0.47	0.89
T8.	4	7	5.60	0.86	0.52	0.89
T9.	4	7	5.84	1.11	0.53	0.88
T10.	3	7	5.57	1.17	0.67	0.88
T11.	3	7	5.60	1.22	0.74	0.88
T12.	1	6	3.33	1.54	0.66	0.88
T13.	1	7	4.30	1.66	0.59	0.88
T14.	2	7	5.60	1.45	0.64	0.88
T15.	3	7	5.47	1.07	0.19[c]	0.89
T16.	2	7	4.90	1.32	0.52	0.88
T17.	2	7	4.63	1.27	0.61	0.88
T18.	2	7	5.60	1.45	0.60	0.88
Outcomes						
O1.	1	7	4.30	1.69	0.51	0.78
O2.	2	7	4.77	1.31	0.28[c]	0.80
O3.	3	7	5.40	0.97	0.24	0.80
O4.	2	6	4.33	1.42	0.28	0.80
O5.	1	7	4.17	1.95	0.71	0.76
O6.	2	7	5.27	1.20	0.42	0.79
O7.	3	7	5.20	0.93	0.48	0.79
O8.	2	6	4.90	1.24	0.36	0.80
O9.	2	7	5.37	1.89	0.20[c]	0.81
O10.	1	7	5.50	1.41	0.13[c]	0.82
O11.	2	6	5.00	1.08	0.58	0.78
O12.	4	7	5.63	1.00	0.60	0.78
O13.	2	7	5.23	0.90	0.57	0.79
O14.	2	7	4.60	1.52	0.73	0.76

[a] Threshold of item-total correlations: 0.30 (Streiner and Norman 1997)
[b] Threshold of Cronbach's α coefficients: 0.70 (Cronbach 1951)
[c] Item failed to meet the threshold for item-total correlations

express themselves freely (T8), being patient and understanding the feelings of the disputing parties (T11), being well-prepared for the issues (T14), and keeping explicit promises (T18) are the most common means of trust-building. Furthermore, the mediators also reported that the improvement of relationships (O12) is the outcome most frequently achieved when using trust-building tactics.

19.4.2 Reliability Assessments

The internal consistency of the responses to trust-building tactics and outcomes was assessed by the Cronbach's alpha coefficients and item-total correlations. Cronbach's alpha provides an estimate of reliability in most situations, as the major source of measurement error is the sampling of content (Cronbach 1951). In addition, reliability that is based on internal consistency considers the sources of errors that are based on the "sampling" of the situational factors that accompany the administration of items (Nunnally and Bernstein 1994). Item-total correlation measures the relationship between an item and the total score of the set of items within the scale (Robinson et al. 1991). This correlation represents not only the relationships among the items, but also the internal consistency of the model. A low corrected item-total correlation value indicates that an item is inconsistent with the other items and is not measuring what the rest of the test is trying to measure (Ferketich 1991). The results of these two assessments are also reported in Table 19.8. As per the rule of thumb suggested by Nunnally and Bernstein (1994), a Cronbach's alpha of 0.70 and an item-total correlation of 0.3 or above are the threshold values for these assessments. In this study, the Cronbach's alpha coefficients for both the trust-building tactics and their outcomes exceeded the threshold value of 0.70, which indicates that the scale items are homogeneous (Bowling 1997). These rules were adopted in previous studies by Ferketich (1991), Robinson et al. (1991), and Knapp and Brown (1995). The item-total correlations for the trust-building tactics ranged between 0.19 and 0.74, and a range of between 0.20 and 0.73 was achieved for the outcomes. Two of the trust-building tactics, T3 and T15, and three of the outcomes, O2, O9, and O10, failed to meet the threshold of 0.30 (Streiner and Norman 1997) and were discarded from further analysis.

19.4.3 Multiple Regression Analysis

MRA, which is a statistical technique used to analyse the relationship between a single dependent variable and several independent variables (Hair et al. 1995), was employed to relate the use of the trust-building tactics to their outcomes. The efficacy of the trust-building tactics could thus be examined via the independent contribution of each tactic (i.e., the independent variable) to the prediction of the outcomes (i.e., the dependent variable). To achieve this, composite scales were calculated for each sub-group of trust-building tactics and outcomes, as shown in Tables 19.1 and 19.2. These composite scales were obtained by averaging the scores given by the respondents for each sub-group to form new sets of variables for the performance of MRA. Five composite scales of trust-building tactics were developed, each representing one of the tactics used in each step of Sloan's mediation model (1998). Likewise, five factor scales of mediation outcomes were also developed, representing the five sub-groups of outcomes that were defined in

Table 19.3. For each MRA, the dependent variable was one of the five composite scales for outcomes, and the independent variables were the composite scales of the trust-building tactics. As previously discussed, two trust-building tactics, T3 and T15, and three outcomes, O2, O9, and O10, failed to achieve the threshold values of the reliability assessments and were thus excluded from the calculation of the respective composite scales.

In light of the above, a total of five MRAs were performed. Equation 19.1 shows the MRA models.

$$\mathrm{O}_i = \mathrm{a} + \mathrm{b}_1\mathrm{T}_1 + \mathrm{b}_2\mathrm{T}_2 + \mathrm{b}_3\mathrm{T}_3 + \mathrm{b}_4\mathrm{T}_4 + \mathrm{b}_5\mathrm{T}_5, \tag{19.1}$$

where O is the dependent variable (the composite scales of outcomes); T is the independent variable (the composite scales of trust-building tactics); and $i = 1, 2,$ 3, 4, and 5.

The results obtained from the five MRAs are shown in Table 19.9. The R^2 values for the five regression models, which represent the combined effect of the entire variant in prediction, range from 0.393 to 0.522. Comparable results were reported by Oetzel (1998), Gross and Guerrero (2000), Sharland (2001) and Cheung et al. (2006). For ease of discussion, the relative contributions of the trust-building tactics to the outcomes can be compared via the normalised regression coefficients (Cheung et al. 2006): the higher the normalised regression coefficient, the greater its contribution to the prediction of the outcome.

During the mediation process, the mediator works to build and maintain the trust of the disputing parties. However, the levels of trust may improve, deteriorate, or remain the same during the course of mediation (Boulle 2001). As can be seen from Table 19.9, the highest normalised regression coefficient (i.e., 0.379) of Model 3 indicates that the trust-building tactics used in Step 5 (i.e., Solutions) may not improve the level of trust among the disputing parties. Thus, the trust-building tactics used in the last step of Sloan's model (1998) appear to have low efficacy in developing trust among disputing parties. This means that if the level of distrust is high, then the disputing parties are defensive, which makes it difficult for the mediation process to reach a joint decision. Perhaps, as supported by the MRA results obtained from Models 1 and 2, it would be more pragmatic to apply these tactics at an earlier stage of the mediation process.

Model 1 shows that the trust-building tactics used in Step 4 of the Sloan's model (1998) (i.e., Explore Interests) are influential in developing trust among disputing parties. This can be seen from Table 19.9 in which the normalised regression coefficient of these trust-building tactics is 0.437, which is the highest among the independent variables. According to Sloan (1998), the aim of exploring interests is to help the disputing parties to identify what interests are important to them and why. The trust-building tactics used in this step, such as understanding the feelings of the disputing parties, sharing experiences, and simplifying/prioritising agendas or issues, may encourage the disputing parties to demonstrate genuine concern about the interests and enlarge the range of settlement alternatives. More importantly, the use of these tactics can encourage the disputing parties to understand

Table 19.9 Overall results of multiple regression analysis

Model	Dependent variables (Outcomes)	Independent variables (Trust-building tactics used in the five steps of Sloan's mediation model) (1998)	R^2	Standardised regression coefficients (Beta)	Normalised regression coefficients
1	Trust-building (party-party)		0.471		
		Preparation (Step 1)		0.124	0.087
		Introduction (Step 2)		−0.304	0.214
		Issues and trust (Step 3)		0.221	0.156
		Explore interests (Step 4)		0.620	0.437
		Solutions (Step 5)		−0.151	0.106
2	Trust-building (mediator-party)		0.522		
		Preparation (Step 1)		0.049	0.057
		Introduction (Step 2)		0.018	0.021
		Issues and trust (Step 3)		0.371	0.428
		Explore interests (Step 4)		0.285	0.328
		Solutions (Step 5)		0.144	0.166
3	Failure to build trust		0.474		
		Preparation (Step 1)		0.021	0.025
		Introduction (Step 2)		0.211	0.243
		Issues and trust (Step 3)		0.083	0.096
		Explore interests (Step 4)		0.223	0.257
		Solutions (Step 5)		0.329	0.379
4	Improvement of relationship		0.486		
		Preparation (Step 1)		0.388	0.287
		Introduction (Step 2)		−0.106	0.078
		Issues and trust (Step 3)		0.607	0.449
		Explore interests (Step 4)		−0.091	0.068
		Solutions (Step 5)		−0.160	0.118

(continued)

Table 19.9 (continued)

Model	Dependent variables (Outcomes)	Independent variables (Trust-building tactics used in the five steps of Sloan's mediation model) (1998)	R^2	Standardised regression coefficients (Beta)	Normalised regression coefficients
5	Time-savings		0.393		
		Preparation (Step 1)		0.151	0.128
		Introduction (Step 2)		−0.146	0.124
		Issues and trust (Step 3)		0.027	0.023
		Explore interests (Step 4)		0.662	0.561
		Solutions (Step 5)		−0.193	0.164

each other (Blackstock 2001) and can thus facilitate communication so that the interests of each can be explored more effectively. Model 5 also suggests that the trust-building tactics used in this step can be a significant predictor of time-savings in the mediation process.

Step 3 of Sloan's model (1998) (i.e., Issues and Trust) suggests that the mediator should encourage the disputing parties to listen to, recognise, and understand the issues of the dispute, and, most importantly, he or she should attempt to restore trust in the long run and deal with the problems of the dispute in the short run. This is thus the step that can generate trust in the mediator. The result obtained from Model 2 is consistent with these propositions: the normalised regression coefficient of the trust-building tactics used in Step 3 is the highest—0.428. This result is also supported by Boulle (2001) and Sloan (1998), who noted that such trust-building tactics as providing equal speaking time and separate meetings for the disputing parties can generate trust in the mediator, which means that these parties may be able to take risks with him or her that they would not take with each other (Boulle 2001). As mediation is a form of facilitated negotiation, the role of the mediator is critical. If the mediator can be trusted, then the disputing parties are willing to engage with him or her openly and disclose important and confidential information. This helps the mediation process to proceed effectively and eventually improves the relationships between the disputing parties. Model 4 thus supports the notion that the trust-building tactics used in this Step are also significant and contribute to the improvement of the relationship between the disputing parties.

19.5 Chapter Summary

The appropriate use of trust-building tactics can have an immense impact on mediation outcomes. The aim of the study reported in this chapter was to identify the trust-building tactics used in construction mediation and to examine the efficacy of these tactics with respect to their outcomes. The research design was based on the mediation model of Sloan (1998), and, with the use of MRA, the key findings of this chapter can be concluded as follows. (1) The trust-building tactics used in Step 4 (i.e., Explore Interests) of Sloan's mediation model (1998) are influential in developing trust among disputing parties and can also act as a time-saving tool in the mediation process. (2) Mediators can earn trust by adopting the trust-building tactics used in Step 3 (i.e., Issues and Trust) of the Sloan's model (1998) and can also improve the relationship between the disputing parties. (3) The trust-building tactics used in the last step (i.e., Solutions) of Sloan's model (1998) appear to have low efficacy in developing trust among disputing parties.

Acknowledgments Special thanks to Wai Ying Lai for collecting data for the study. The content of this chapter has been published in Volume 135(8) of the Journal of Construction Engineering and Management is used with the permission from ASCE.

References

Bercovitch, J., & Derouen, K. (2004). Mediation in internationalized ethnic conflicts: Assessing the determinants of a successful process. *Armed Forces and Society, 30*(2), 147–170.

Blackstock, M. D. (2001). Where is the trust? Using trust-based mediation for first nations dispute resolution. *Conflict Resolution Quarterly, 19*(1), 9–30.

Boulle, L. (2001). *Mediation skills and techniques*. Sydney: Butterworths.

Bowling, A. (1997). *Research methods in health: Investigating health and health services*. Maidenhead: Open University Press.

Butler, J. K. (1991). Toward understanding and measuring conditions of trust: Evolution of a conditions of trust inventory. *Journal of Management, 17*, 643–663.

Carnevale, P. J. D. (1986). Strategic choice in mediation. *Negotiation Journal, January*, 41–56.

Carnevale, P. J. D., Lim, R. G., & McLaughlin, M. E. (1989). Contingent mediator behavior and its effectiveness. In K. Krussel, & D. G. Pruitt (Eds.), *Mediation research—the process and effectiveness of third-party intervention*. San Francisco: Jossey-Bass Publication.

Chau, K. W. (2007). Insight into resolving construction disputes by mediation/adjudication in Hong Kong. *Journal of Professional Issues in Engineering Education and Practice, ASCE, 133*(2), 143–147.

Chau, K. W. (1992). Resolving construction disputes by mediation: Hong Kong experience. *Journal of Management in Engineering, ASCE, 8*(4), 384–392.

Cheeks, J. R. (2003). Multistep dispute resolution in design and construction industry. *Journal of Professional Issues in Engineering Education and Practice, 129*(2), 84–91.

Cheung, S. O., & Yiu, T. W. (2007). A study of construction mediator tactics—part I: Taxonomies of dispute sources, mediator tactics and mediation outcomes. *Building and Environment, 42*(2), 752–761.

Cheung, S. O., Yiu, T. W., & Yeung, S. F. (2006). A study of styles and outcomes in construction dispute negotiation. *Construction Engineering and Management, 132*(8), 805–814.

Cronbach, L. J. (1951). Coefficient α and the internal structure of tests. *Psychometrika, 16*, 297–334.

Ferketich, S. (1991). Focus on psychometrics. Aspects of item analysis. *Research in Nursing & Health, 14*(2), 165–168.

Govier, T. (1997). *Social trust and human communities*. Kingston, Ont: McGill- Queen's University Press.

Gross, M. A., & Guerrero, L. K. (2000). Managing conflict appropriately and effectively: An application of the competence model to Rahim's organizational conflict styles. *The International Journal of Conflict Management, 11*(3), 200–226.

Hair, J. F., Anderson, R. E., Tatham, R. L., & Black, W. C. (1995). *Multivariate data analysis* (6th ed.). Upper Saddle River, NJ: Prentice Hall.

Hon, C. Y. (2006). The common problems and suggested solutions of mediation in the construction industry in Hong Kong. *Surveyors Times, The Hong Kong Institute of Surveyors, 4*, 36–46.

Knapp, T. R., & Brown, J. K. (1995). Focus on psychometrics. Ten measurement commandments that often should be broken. *Research in Nursing & Health, 18*(6), 465–469.

Kolb, D. M. (1985). To be a mediator: Expressive tactics in mediation. *Journal of Social Issues, 41*(2), 11–26.

Latz, M. E. (2001). Give and take: In mediation, it is important to negotiate effectively with the mediator. *ABA Journal, 87*, 66.

Lui, S. S., Hang, Y. N., & Hon, A. H. Y. (2006). Coercive strategy in interfirm cooperation: Mediating roles of interpersonal and interorganizational trust. *Journal of Business Research, 59*, 466–474.

Madden, J. P. (2001). Recipe for success in construction mediation. *Dispute Resolution Journal, 56*(2), 16.

Moore, C. W. (1996). *The mediation process: Practical strategies for resolving conflict*. San Francisco, CA: Jossey-Bass Publishers.
Nunnally, J. C., & Bernstein, I. H. (1994). *Psychometric theory*. New York: McGraw-Hill.
Oetzel, J. G. (1998). The effects of self-control and ethnicity on self-reported conflict styles. *Communication Reports, 11*(2), 133–143.
Robinson, J. P., Shaver, P. R., & Wrightsman, L. S. (1991). Criteria for scale selection and evaluation. In J. P. Robinson, P. R. Shaver, & L. S. Wrightsman (Eds.), *Measures of personality and social psychological attitudes*. San Diego, CA: Academic Press.
Salem, R. (2003). *Trust in mediation*. Retrieved from http://www.beyondintractability.org/essay/trust_mediation/
Settle, J. (1998). *The element of trust in mediation: Practice pointers drawn from theory*. Retrieved from http://www.convenor.com/madison/trust.htm
Sharland, A. (2001). The negotiation process as a predictor of relationship outcomes in international buyer-supplier arrangements. *Industrial Marketing Management, 30*(7), 551–559.
Sloan, G. (1998). Turning conflict into consensus: Mediation theory. In C. Darling (ed.) *Process and skills*. Continuing Legal Education Society: Vancouver, B.C.
Streiner, D., & Norman, G. (1997). *Health measurement scales: A practical guide to their development and use* (2nd ed.). Oxford: Oxford Medical Publications.
Striengnitz, M. (2006). Conflicts over coastal protection in a National Park: Mediation and negotiated law making. *Land Use Policy, 23*, 26–33.
Susskind, L., & Ozawa, C. (1983). Mediated negotiation in the public sector: Mediator accountability and the public interest problem. *American Behavioral Scientist, 27*(2), 255–279.
Torres, R. (1991). Improving the quality of internal evaluation: The evaluator as consultant-mediator. *Evaluation and Program Planning, 14*, 189–198.
Yiu, T. W., & Cheung, S. O. (2007). A study of construction mediator tactics—part II: The contingent use of tactics. *Building and Environment, 42*(2), 762–769.

Chapter 20
Logrolling "Win–Win" Settlement in Construction Dispute Mediation

Yingying Qu and Sai On Cheung

Abstract Reaching "win–win" settlement is the desired outcome of mediation. Logrolling is a strategy for achieving "win–win" trade-off. In this study, a logrolling strategy in mediation is proposed through which parties can improve the joint value by bargaining exchange and get convergence along the efficient frontier. A multi-objective decision making (MODM) model is employed to propose the efficient frontier and assist parties to engender "win–win" settlement. To operationalise the logrolling strategy, a web-based logrolling system is developed to assist parties to achieve "win–win" settlement in a user-friendly environment. The system includes 3 processes: reality test, preference identification and logrolling. Reality test is proposed to test parties' concession rate. Preference identification assists parties to identify their utility value of the bargaining alternatives. Logrolling is to provide user-friendly strategies for parties to make efficient trade-off that involves (1) when to concede (2) on which issue (3) for which party and (4) how much should be conceded. Finally a mock mediation experiment was conducted to examine whether the logrolling system can assist parties to achieve "win–win" settlement, where the system simulates a *Mediator* in action. The results are evaluated by comparing the difference between the mediator's expected logrolling outcomes and the subjects' actual logrolling outcomes. The logrolling-difference degree (L-DD) is used to measure this difference. It is found that the average of L-DD in bargaining range and reaching agreement are 11.43 % and 8.46 % respectively, which indicates that the logrolling system is having good potential in assisting parties to achieve "win–win" settlement.

Y. Qu (✉) · S. O. Cheung
Construction Dispute Resolution Research Unit, Department of Civil and Architectural Engineering, City University of Hong Kong, Hong Kong, People's Republic of China
e-mail: yingyingqu2@gmail.com

S. O. Cheung
e-mail: saion.cheung@cityu.edu.hk

S. O. Cheung (ed.), *Construction Dispute Research*,
DOI: 10.1007/978-3-319-04429-3_20,

20.1 Introduction

Conflict and dispute regularly feature in construction projects. Mediation has been identified as an effective means to resolve dispute due to its flexibility, cost-effectiveness and non-threatening process. From the information of the Hong Kong International Arbitration Center (HKIAC), out of the arbitrations handled by the HKIAC, the construction industry provided 24 % of the disputes in 2012. Chau (2007) reported that 82 % of all disputes got settled either by mediation or through negotiation at the mediation stage. The Chief Executive of the Government of the Hong Kong Special Administrative Region, in his 2007–2008 policy address announced the vision of developing Hong Kong as a regional centre for mediation service.

Reaching “win–win” settlement is the desired outcome of mediation. A “win–win” settlement can be seen as one that encourages parties to uphold their contracts when one party achieve its profits and the other party would still be better off (White 2009). However parties negotiating face to face often have difficulty in identifying and realising “win–win” settlement (Neale and Bazerman 1991; Pruitt 1981; Sebenius 1992).

In the last two decades, negotiation support system (NSS) and e-negotiation system (ENS) have been widely applied in conflict resolution. Negotiation support system (NSS) is developed on stand-alone computers or local network to implement decision making models (Lim and Benbasat 1993). They can help users to understand and formalise the objectives and preferences, and can help users to understand the problem structure and search for solution. Examples of such systems include MEDIATOR (Jarke and Jelassi 1987), RAINS (Bronisz et al. 1988), HIPRE (Hämäläinen and Pöyhönen 1996), RAMONA (Teich et al. 1995). E-negotiation systems (ENS) refer to those web-based systems that are equipped with decision making analysis, communication and coordination functions (Bichle et al. 2003; Insua et al. 2003). They can provide the reactions of the counterparts and the construction of arguments and counter-arguments. They can help to set up virtual laboratories and collect data from people around the world by user-friendly interfaces. They can facilitate the parties to communicate, store and access bargaining information. They also help the parties in achieving an agreement, by offering potential compromises and proposing concessions which may lead to a settlement. The role of these systems is thus similar to that of a mediator who communicates the parties’ true interests and preferences. There are some examples of ENS application: Inspire (Kersten and Noronha 1999), Web-HIPRE systems (Hämäläinen et al. 2001), Kasbah (Maes et al. 1999), WebNS (Yuan et al. 2003), Negoisst (Schoop and Quix 2001), MeMo (Weigand et al. 2003), Negotiator Assistant (Druckman et al. 2004).

Logrolling is a strategy for achieving integrative trade-off, by which each party concedes on low priority issues in exchange for concessions on issues of higher priority to them (Neale and Bazerman 1991; Pruitt and Rubin 1986; Lax and Sebenius 1986). However there is no literature either on application of logrolling strategy in mediation or on computer-simulated mediator in facilitating "win–win" settlement. This study fills this gap by developing a web-based logrolling system to assist parties to achieve "win–win" settlement in mediation. In this study, a logrolling strategy is proposed through which parties can improve the joint value by bargaining exchange and get convergence along the efficient frontier. Based on that, a web-based logrolling system is developed to assist parties to achieve "win–win" settlement in a user-friendly environment. The logrolling system can help parties to make efficient trade-off, by suggesting (1) when to concede (2) on which issue (3) for which party and (4) how much should be conceded.

The remainder of the chapter is organised as follows. Review of literature on logrolling is firstly outlined. The conceptual model of logrolling strategy in mediation is then presented. To operationalise the logrolling strategy, a web-based logrolling system is developed. A mock mediation experiment was reported to examine whether the logrolling system can assist parties to achieve the "win–win" settlement. In the experiment, the logrolling system simulates a *Mediator* in action.

20.2 Logrolling in Negotiation

The strategy of logrolling is closely related to the concept of efficient frontier. In economics, "frontier" is where alternative is worse than what they could achieve (Mas-Colell et al. 1995). The logrolling process is described as procedures that generate jointly improving outcomes from non-Pareto optimal alternative towards a Pareto optimal one.

Kersten (2001) compared several logrolling models on negotiation. The logrolling solution paths in Fig. 20.1a, c, d, e, f are similar that each subsequent offer gives higher utility than the previous one. Fig. 20.1a can be viewed as an example of a single negotiated text (SNT) process. The SNT is one of the earliest logrolling models (Raiffa 1982). SNT is a tentative negotiation proposal that is to be examined and improved by all parties. The method produced a series of SNTs, which are jointly improving and ends when all parties accept one SNT as their final agreement. In Fig. 20.1a, an initial offer *a* is made that yields very low utility by both parties. Each party then proposed a better offer *b*, *c*, *d*, until the efficient offer *e*. Figure 20.1c refers to a negotiation in which the parties know the sets of offers. Knowledge of the sets of offers allows the parties to verify the efficiency of each offer. The difference between two processes in Fig. 20.1a and d is whether the parties expand the utility set. In Fig. 20.1a the parties do not modify the utility set *U*. In Fig. 20.1d, the parties propose new offers that are outside of the utility set they considered earlier. In Fig. 20.1e both integrative and distributive offers are

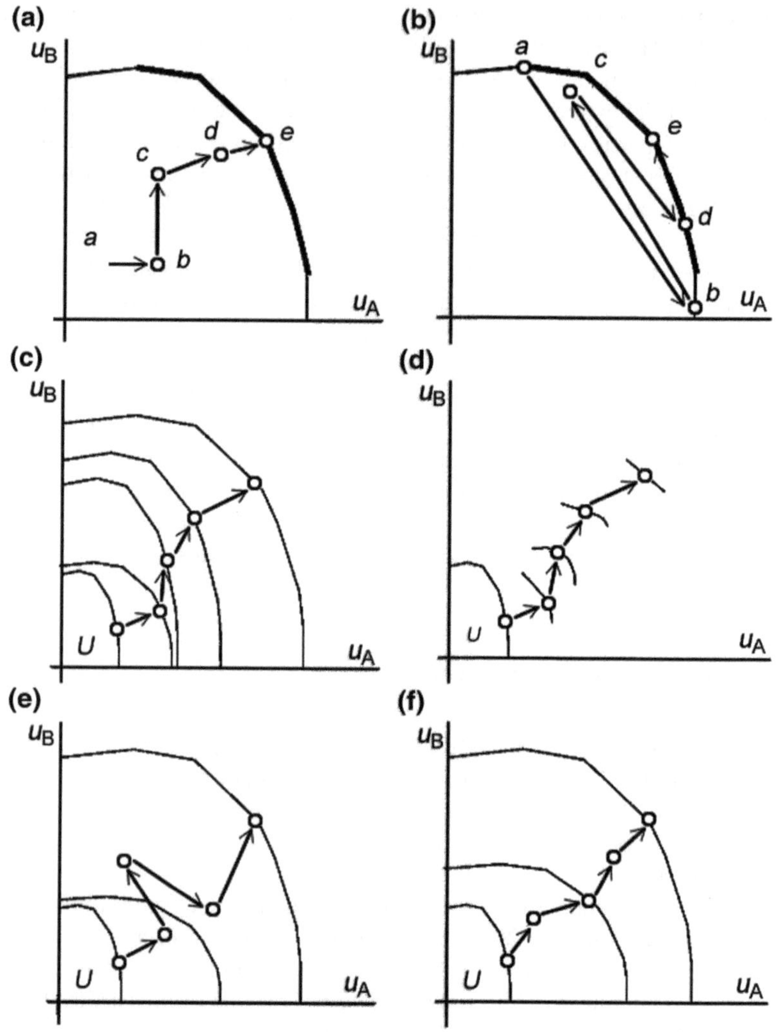

Fig. 20.1 Logrolling solution path comparison (adopted from Kersten 2001)

made, while in Fig. 20.1f there are only integrative offers. The process in Fig. 20.1b is a simple example of positional bargaining. The parties begin to make offers at the worst position which yields high utility value for their opponents. In Fig. 20.1b party A makes offer *a* to which party B replies with counteroffer *b*. Then party A proposes *c* and party B replies with offer *d*. In response to the counteroffers, each party makes a small concession and the process continues till both parties achieve a compromise *e*.

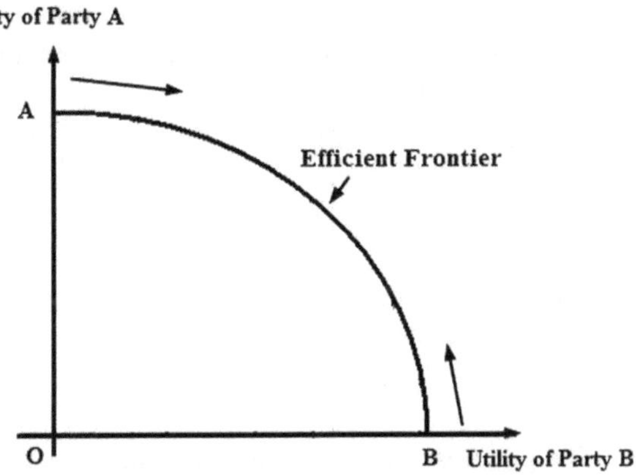

Fig. 20.2 Logrolling strategy

20.3 Application of Logrolling Strategy in Mediation

A logrolling process is proposed through which parties could improve joint value by bargaining exchange and get convergence along the efficient frontier. The related conceptual model of logrolling in mediation is shown in Fig. 20.2, which specifies an optimal logrolling solution path for negotiators. The parties are to begin with their most preferred position. Party A begins with point A and Party B begins with point B. Both parties move along the efficient frontier towards pareto-optimal solution.

To achieve it, a multi-objective decision making model is proposed to approximate efficient frontier and assist parties to engender "win–win" mediation settlement. Multi-objective decision making (MODM) approach accompanied with multi-attribute utility theory (MAUT) and applied in a multi-criteria decision making (MCDM) setting has been widely applied to generate options and identify potential agreements in dispute resolution.

There are several decision making criteria for judging the pareto-optimal outcome: maxi-min equity solution, Nash solution and utilitarian solution. Maxi-min equity solution is an approach that seeks to balance the difference between two parties. It has been suggested that one party is not only motivated by self-interest but also "a strong aversion to disadvantage themselves" (Nowak et al. 2000). It seems that the aversion to disadvantage (or "envy principle") affected the animal species as well. Brosnan and de Waal (2003) reported that high percentages of capuchin monkeys rejected the opportunity to trade rocks for cucumber slices when they saw other monkeys receiving grapes, either in exchange for their rocks or without being required to exchange anything. Nash solution is the famous principle for solving "efficiency" in non zero-sum two-person bargaining game. Raiffa (1985) and Lax and Sebenius (1986) pointed out that a lot of disputes are settled with "the value left on the table", since disputants focus on the pie to be shared, but fail to realise "this small pie" can be enlarged. Nash solution is measured as maximisation of the product of the two parties' utilities when the status quo point is normalised to zero (Raiffa et al. 2002). Utilitarian solution is to maximise the sum of the two parties' utilities (Thompson 1990).

The proposed MODM model:

With i as disputing party, j is defined as issue $j = 1,\ldots, J$; U_{ijn} is defined as the preferred utility of party i on issue j's each bargaining alternative n; w_{ij} is defined as the weighting value on issue j preferred by party i where $\sum_{j=1}^{J} w_{ij} = 1$; define M_{ijn} as utility value considering the weight of current issue j of party i. Therefore, M_{ijn} can be calculated as follows:

$$M_{ijn} = U_{ijn} \times w_{ij} \tag{20.1}$$

The gain/loss rate is taken to be one party's gain in terms of the other party's loss. For example, the utility gain of *Contractor* in terms of utility loss of *Client* between alternative $n + 1$ and n, on issue j can be calculated as follows:

$$Rate_{Contractor} = \frac{\left|M_{Contractor,\, j,\, (n+1)} - M_{Contractor,\, j,\, n}\right|}{\left|M_{Client,\, j,\, (n+1)} - M_{Client,\, j,\, n}\right|} \tag{20.2}$$

Both parties are proposed to concede on minimum loss in exchange for maximum gain from the other party for every bargaining round. Therefore, the *Benefit* is defined as evaluation of joint value on each issue. Following the above example, the related formula is as follows:

$$Benefit = M_{Client,\, j,\, n} + Rate_{Contractor} \times M_{Contractor,\, j,\, n} \tag{20.3}$$

The *benefit* can be used to measure and improve the efficiency of the logrolling process. And we can get the efficient logrolling proposals as the maximum *benefit*. In other words, when *Max.Benefit* is satisfied, the efficient points on the frontier can be calculated, with a vector $[M_{Client,\, j,\, n},\ M_{Contractor,\, j,\, n}]$, $j = 1,\ldots, J$, thus the efficient frontier is simulated and the utility value of efficient point on the frontier is $[\sum_j M_{Client,\, j,\, n}, \sum_j M_{Contractor,\, j,\, n}]$.

The optimal agreement is generated based on the following three criteria:

Utilitarian solution: $Max. \sum_i \sum_j M_{ij}$

Nash solution: $Max. \prod_i \sum_j M_{ij}$

Maxi-min equity solution: $Max. \left\{ \min \frac{\sum_i M_{ij}}{\sum_i \sum_j M_{ij}} \right\}$

To simulate the conceptual model, data from the case reported by Cheung et al. (2004) is used. This is a two-party, three-issue case. "This construction dispute begins with the date of completion, which was 1 Jan 2001 in the contract. But due to the delay of sub-contractor and late Architect Instruction, the completion date shifted from 1 Jan 2001 to 1 Mar 2001. The issues are *Extension of Time (EOT)*, which Main Contractor argued for 60 days but Architect only granted 40 days,

Acceleration Cost (AccCost) which was estimated as $30,000 by Project Manager, as well as *Lost and Expenses (L/E)*." From their work, the issues, bargaining alternatives and two parties' input data are listed in Table 20.1.

The points that satisfy the condition $M_{Client,j,n} + Rate_{Contractor} \times M_{Contractor,j,n}$ are listed in Table 20.2. The points constitute the efficient frontier as shown in Fig. 20.3. In Table 20.2 the optimal solutions can be generated according to the 3 criteria as aforestated. From utilitarian solution and Nash solution, the points F and G are the optimal choices, which are highlighted in Fig. 20.3. To achieve Maximin equity solution, the point G is selected finally.

The multi-objective decision making model is not only a generator of optimal solution, but also specifies a logrolling solution path for negotiators to achieve "win–win" settlement. In consideration of self-interest motivation, the proposed mediation begins with the parties' most preferred positions. In this case, *client* begins with point A and *contractor* with point L respectively. In each of the next scenario, parties are suggested to move in a direction that makes efficient trade-off. For example from A to B, *client* is suggested to increase the budget for L/E from 6000 to 6100, with no change on the other two issues. Meanwhile, the *contractor* is persuaded to cut down EOT arguments from L to K in Table 20.2. In this way moving along the efficient frontier, both parties are suggested to make concessions till convergence.

20.4 Logrolling System for Construction Dispute Mediation

To operationalise the logrolling strategy, a web-based system is developed. The system includes 3 processes: reality test, preference identification and logrolling. Reality test and preference identification are designed for logrolling information collection. Logrolling process is to provide user-friendly strategies for parties to make efficient trade-off addressing (1) when to concede, (2) on which issue, (3) for which party and (4) how much should be conceded.

20.4.1 Reality Test

Reality test is proposed to test parties' concession rate and assist disputing parties to get ready for achieving "win–win" settlement. Some negotiation support systems (Korhonen et al. 1986; Rangaswamy and Shell 1997) follow the assumption advocated by Keeney and Raiffa (1991) that "all inventing and creating of issues and potential settlements have occurred," and the parties are ready to negotiate over the identified options. However this assumption has significant limitation in practice. When engaged in negotiation, parties may hesitate to make concessions,

Table 20.1 Input data for *Contractor/Client* M_{ijz}

Issue	EOT	EOT	EOT	EOT	EOT	EOT	L/E	L/E	L/E	L/E	L/E	L/E
Bargaining alternative	35	36	37	38	39	40	6000	6100	6200	6300	6400	6500
Contractor	0	25	40	50	58	60	0	10	18	24	28	30
Client	30	28	24	18	10	0	40	38	33	25	15	0
Issue	AccCost	AccCost	AccCost	AccCost								
Bargaining alternative	10000	11000	12000	13000								
Contractor	0	5	8	10								
Client	30	25	15	0								

(Adopted from Cheung et al. 2004)

Table 20.2 Set of points on the efficient frontier

	EOT	L/E	AccCost	Contractor utility	Client utility	Utilitarian solution	Nash solution	Maxi-min equity solution	
								Contractor (%)	Client (%)
A	36	6000	10000	20	98.4	118.4	1968	16.89	83.11
B	36	6100	10000	35	96	131	3360	26.72	73.28
C	37	6100	10000	50	92	142	4600	35.21	64.79
D	38	6100	10000	60	86	146	5160	41.09	58.91
E	38	6200	10000	68	81	149	5508	45.64	54.36
F	38	6200	11000	73	76	149	5548	48.99	51.01
G	39	6200	10000	76	73	149	5548	51	49
H	39	6200	11000	81	68	149	5508	54.36	45.64
I	39	6300	11000	87	60	147	5220	59.18	40.82
J	39	6400	11000	91	50	141	4550	64.54	35.46
K	39	6400	12000	94	40	134	3760	70.15	29.85
L	40	6400	12000	96	30	126	2880	76.19	23.81

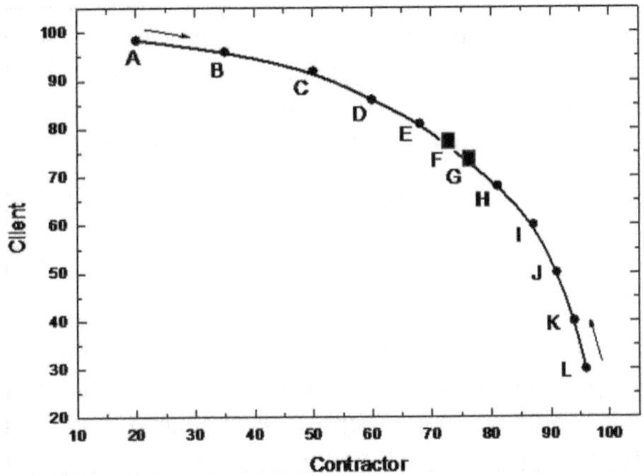

Fig. 20.3 Efficient frontier

because they are not sure whether the other party will make the concession or not. It is also quite common for parties to over-value their own cases while under-value the opponent's assertions (Neale and Bazerman 1991). As a result, the disputants always walk away at very early stage. According to Boulle and Nesic (2001), mediators are 'agents of reality' in so far as their function of encouraging the parties to consider the realities of the dispute.

If two parties reach agreement, they definitely have options that are mutually acceptable. Otherwise parties still need to adjust their concession rate to close the gap.

Table 20.3 Symbol definition

Symbol	Definition
$CR_{Contractor}$	Concession rate of *contractor*
CR_{Client}	Concession rate of *client*
$V_{Contractor}$	*Contractor*'s optimistic proposal
V_{Client}	*Client*'s optimistic proposal
$BL_{Contractor}$	*Contractor*'s bottom line
BL_{Client}	*Client*'s bottom line
Min.Rate	Parties' mutual concession rate in mediation
n	Bargaining alternatives on one issue
N	Total number of bargaining alternatives
VPre	Parties' offer after ($N-1$)th *concession*
VExp	Parties' final offer after *N*th concession
POW	Exponentiation

Principle 1: Party A and Party B reach agreement *if and only if* Party A's final offer is better than Party B's bottom line *AND* Party B's final offer is better than Party A's bottom line.

In this respect, $CR_{Contractor}$ and CR_{Client} are defined as the concession rate of *Contractor* and *Client* respectively; define $V_{Contractor}$ and V_{Client} as parties' optimistic proposal; define $BL_{Contractor}$ and BL_{Client} as Party's bottom line. The definitions of the symbols used are summarised in Table 20.3.

Take an example, if Client argued money for remedy and Contractor tried to reduce the payment. The bottom line for party i can be calculated as follows:

$$BL_{Contractor} = V_{Contractor} \times (1 + CR_{Contractor}) \tag{20.4}$$

$$BL_{Client} = V_{Client} \times (1 - CR_{Client}) \tag{20.5}$$

Minimum concession rate Min.Rate is defined as the parties' mutual concession rate in mediation. Therefore, Min.Rate = Minimum (CR_i). Define n ($n \geq 1$) as bargaining alternatives on one issue and the total number of alternatives is N. Each bargaining alternative represents one bargaining round. Suppose parties concede at the same rate on each round, in this example, the Principle 1 can be formalised as follows:

$$\begin{aligned} BL_{Contractor} \times (1 + MinRate)^N \geq BL_{Client} AND \\ BL_{Contractor} \times (1 - MinRate)^N \leq BL_{Contractor} \end{aligned} \tag{20.6}$$

Based on Principle 1, concession rate can be evaluated with given bargaining alternatives n. The algorithm for reality test is presented as Algorithm 1.

Algorithm 1: Reality Test

1. **Input** $CR_{Contractor}$, which is *Contractor's* concession rate;
2. **Input** CR_{Client}, which is *Client's* concession rate;
3. **Input** V_{Client}, which is *Client's* optimistic proposal
4. **Input** $V_{Contractor}$, which is *Contractor's* $V_{Contractor}$ optimistic proposal
5. $Min.Rate$ = Minimum ($CR_{Contractor}$, CR_{Client})
6. Calculate $BL_{Contractor} = V_{Contractor} \times (1 + CR_{Contractor})$
7. Calculate $BL_{Client} = V_{Client} \times (1 - CR_{Client})$
8. Define $VPre$ as parties' offer after $(N-1)^{th}$ concession
Define $VExp$ as parties' final offer after N^{th} concession
9. $V\mathrm{Pr}e_{Contractor} = BL_{Contractor} \times POW((1 + Min.Rate), N - 1)$
10. $V\mathrm{Pr}e_{Client} = BL_{Client} * POW((1 - Min.Rate), N - 1)$
11. $VExp_{Contractor} = BL_{Contractor} \times POW((1 + Min.Rate), N)$
12. $VExp_{Client} = BL_{Client} \times POW((1 - Min.Rate), N)$
13. **If** $((VPre_{Contractor} \geq BL_{Client}) AND\ (VPre_{Client} \leq BL_{Contractor}))$ **Then**
14. $Suggestion$ = The advice of decreasing concession rate to both *contractor* and *client*;
15. **Else If** $((VExp_{Contractor} \geq BL_{Client}) AND\ (VExp_{Client} \leq BL_{Contractor}))$ **Then**
16. $Suggestion$ = The successful of bargaining in reality test to both *contractor* and *client*;
17. **Else**
18. $Suggestion$ = The advice of increasing concession rate to both *contractor* and *client*;
19. **End If**
20. **Output** *Suggestion*;

The related flow chart of reality test is shown in Fig. 20.4. The contractor and client are interacting via a message center with mediator. The test of concession rate in each round will return to both contractor and client. If the expected final offer doesn't reach the other party's bottom line, the mediator (system) will suggest another round until success.

To further illustrate the operation of reality test and the user interface, a series of screenshots from the system are provided in Fig. 20.5. Firstly, the *contractor* and *client* input the concession rate via different computers. Usually the concession rate begins with 3–5 % in mediation. After *contractor* and *client* submitted their concession rates, *mediator* can automatically receive a message, which contains the parties' concession rates. After that, system will provide a suggestion for *mediator* automatically whether the *contractor* and *client* cannot reach any agreement in this round. Then the 'Message Center' will send suggestion messages from the *mediator* to both the *contractor* and *client*. The parties receive the suggestion and work on the next round. If the *contractor* and *client*'s expected final offer reaches each other's bottom line, *mediator* will send a success message to them.

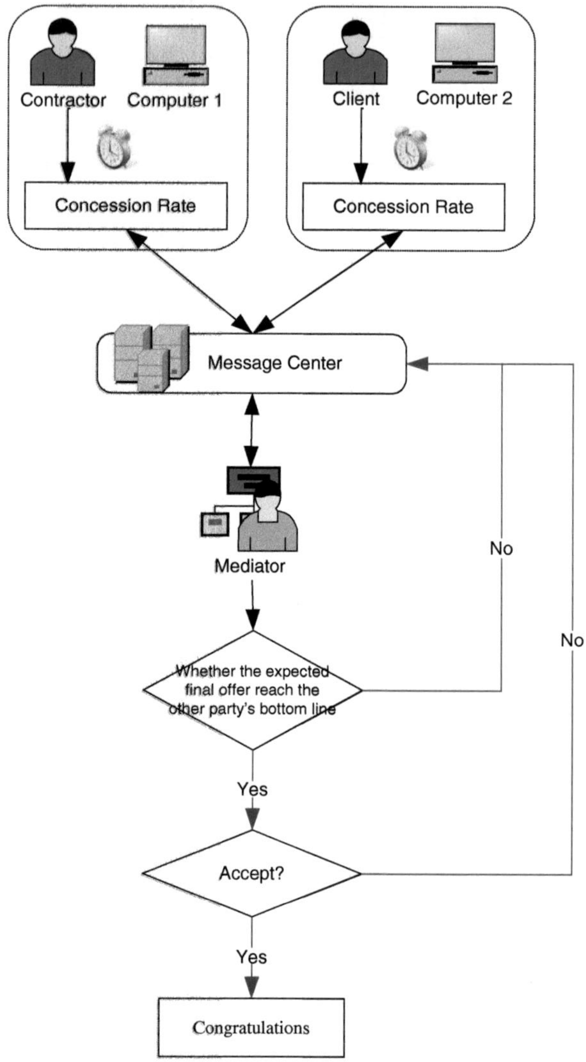

Fig. 20.4 Flow chart of reality test

20.4.2 Preference Identification

Preference identification is used to assist parties to identify the preference of each bargaining alternative. Non-linear utility distribution is applied widely. One notable example is the "law of diminishing marginal utility", which states that an individual consumes or acquires more of a good, the marginal utility of additional amounts of those good decreases (Northcraft et al. 1998). Some negotiation researches also outlined how non-linear preference functions dramatically alter the dynamics of negotiation exchanges (Northcraft et al. 1995, 1998; Pennings and Smidts 2003). However, there is no negotiation support system with non-linear

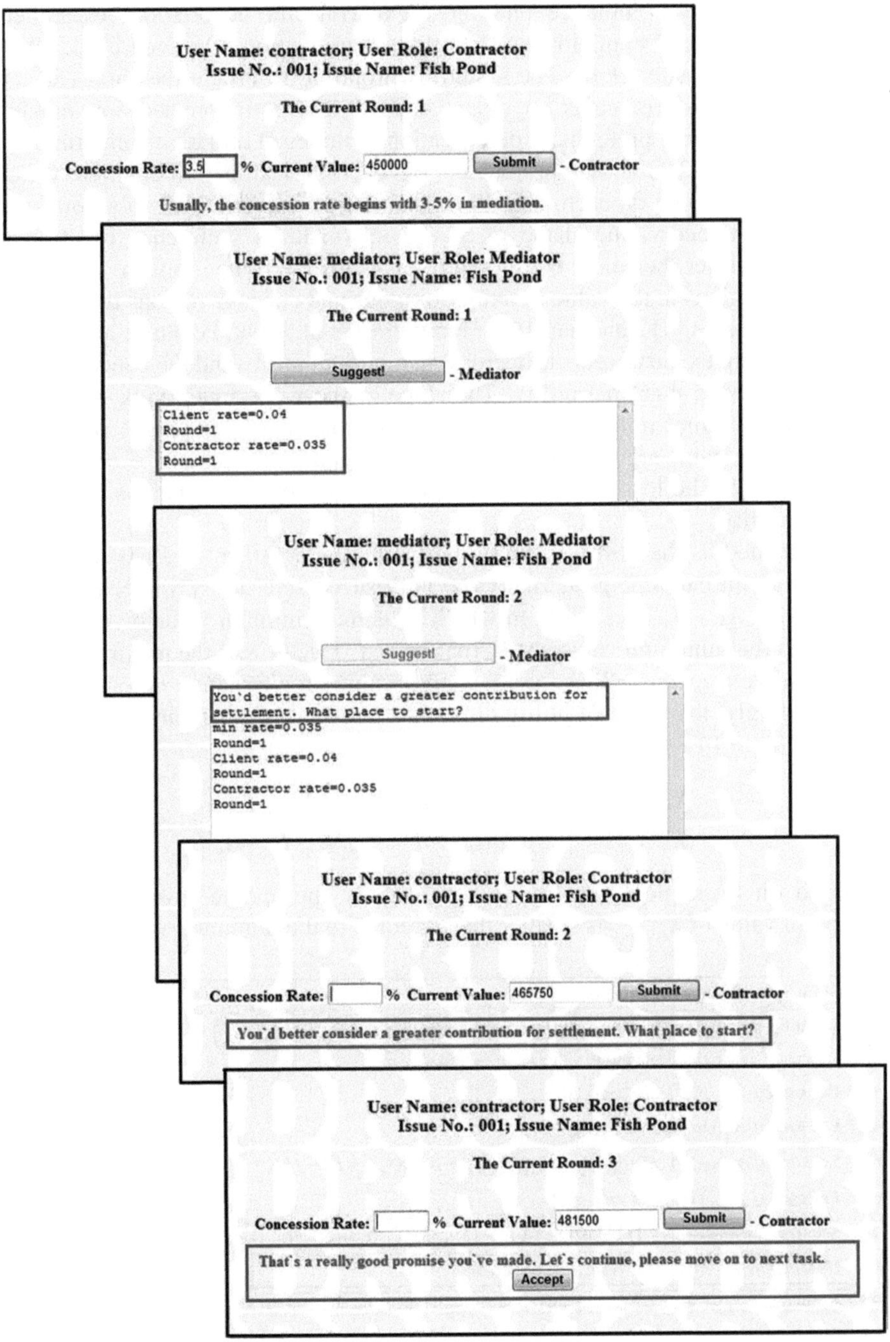

Fig. 20.5 User interface of reality test

utility distribution in practice and most experimental negotiation researchers imposed on subjects' explicitly linear utility identification (De Dreu et al. 1994; Thompson and Hastie 1990). The system is proposed to bridge the gap.

Negotiators' preference is an essential part in negotiation decision making process. The parties' preference identification includes (1) relative weightings on each issue and (2) relative preference on each bargaining alternative which involves utility range on each bargaining alternative and utility distribution.

In prospect theory, the shape of a decision maker's preference function is assumed to differ between the domain of gains and the domain of losses. "...convex regions in the value function for gains and concave regions in the value function for losses" (Kahneman 1979; Northcraft et al. 1998; Pennings and Smidts 2003). The convex curve reflects decreasing marginal utility and the concave curve reflects increasing marginal utility. The slope of the value function is steeper for losses than for gains, reflecting the fact that decisions involving prospective losses or prospective gains are distinguished by how much is at stake.

Principle 2: The loss-framed utility distribution should be taken as increasing marginal utility.

t_n is defined as the utility value on bargaining alternative. n; X_n is defined as utility range on bargaining alternative n, with successive utility value, thus $X_n = [t_n, t_n + R], X_{n+1} = t_{n+1}, t_{n+1} + R]$ in which, t_n is the minimum value on alternative n; t_{n+1} is the minimum value on alternative $n + 1$; t_n + R is the maximum value on alternative n t_{n+1} + R is the maximum value on alternative $n + 1$.

Accordingly, the range of utility difference between two bargaining alternative $[n, n + 1]$ *and* $[n - 1, n]$ can be calculated as follows:

$$\begin{aligned} X_{n,\,n+1} &= [|t_n - (t_{n+1} + R)|, |(t_n + R) - t_{n+1}|] \\ X_{n-1,n} &= [|t_{n-1} - (t_n + R)|, |(t_{n-1} + R) - t_n|] \end{aligned} \tag{20.7}$$

Based on Principle 2, the utility range X_n is appropriate *if and only if* the maximum value of $X_{n-1,\,n}$ is smaller than or equal to the minimum value of $X_{n,\,n+1}$.

Algorithm 2: Preference Identification

Input the start and end number as *start* and *end*, respectively.
Define t_n as the candidate utility value on the alternative n;
For each ((X_{n-1}) **in** [*start*, *end* - *R*])
For each ((X_n) **in** [*start*, *end* - *R*])
Set *Min.difference*(*n*-1, *n*) =Minimum $(|t_{n-1} - (t_n + R)|, |(t_{n-1} + R) - t_n|)$;
For each ((X_{n+1}) **in** [*start*, *end* - *R*])
Set *Max.difference* (*n*, *n*+1) =Maximum $(|t_n - (t_{n+1} + R)|, |(t_n + R) - t_{n+1}|)$;
If (*Min.difference*(*n*-1, *n*) >=*Max.difference*(*n*, *n*+1)) **Then**
Record (X_{n-1}), (X_n), (X_{n+1}) as appropriate utility range and add into list *L*;
End If
End For
End For
End For
Output *L*;

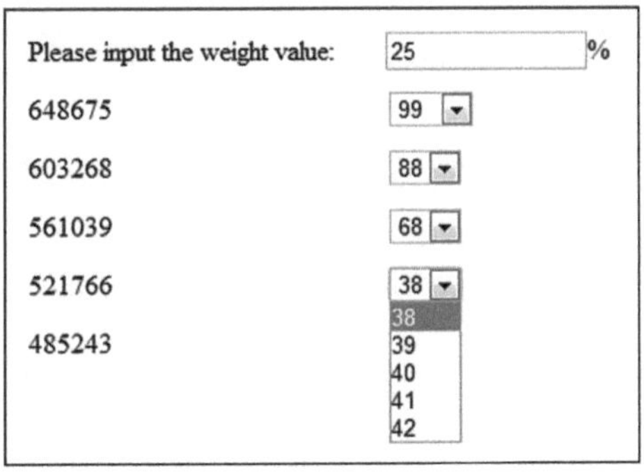

Fig. 20.6 User interface of preference identification

In this research, *start* = 1, *end* = 100, *R* = 5. The utility ranges are calculated by Algorithm 2, which should satisfy Principle 2. It is found that there are many complying results. In this implementation, the ranges of [96–100], [85–89], [66–70], [38–42], and [1–5] were selected for its large coverage over the scale [1–100].

An example of user interface of utility range [38–42] on bargaining alternative "521766" is shown in Fig. 20.6. The utility ranges are arranged in descending order. Here the users are required to identify most preferred utility value from the range on the corresponding bargaining alternative.

Contractor and *Client* come to "Preference Identification". In this part, they are required to submit survey regarding (1) the relative preferences among the issues, which are the weightings among all the issues; (2) the relative preferences among the bargaining alternatives. The complete user interface for *Contractor* is shown in Fig. 20.7.

20.4.3 Logrolling

The flow chart of Logrolling is shown in Fig. 20.8. After *contractor* and *client's* submission of preference information, *Mediator* analyses the data and calculates the optimal proposals, which are then returned to *contractor* and *client* as *Mediator's* suggestion round by round. *Contractor* and *client* need to confirm whether accept or reject.

In logrolling process, the mediation system will provide user-friendly strategies for parties to make efficient trade-off, which involves (1) when to concede; (2) on which issue; (3) for which party; and (4) how much should be conceded. Parties need to confirm whether they accept or reject. Fig. 20.9 shows the suggestion for *Contractor* on the first round.

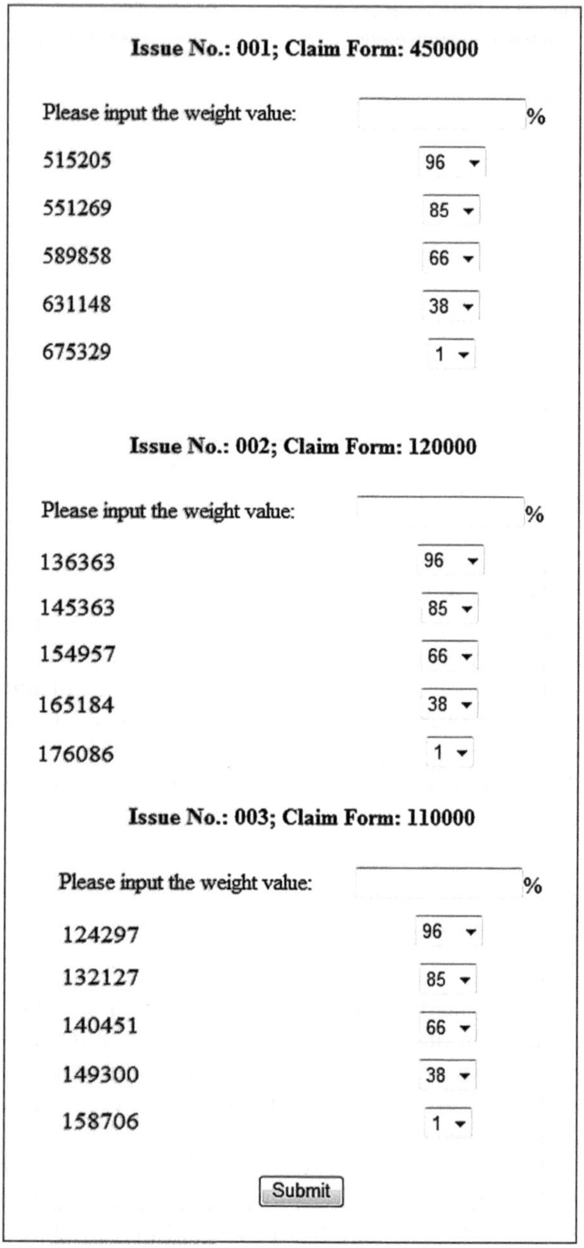

Fig. 20.7 Complete user interface of preference identification

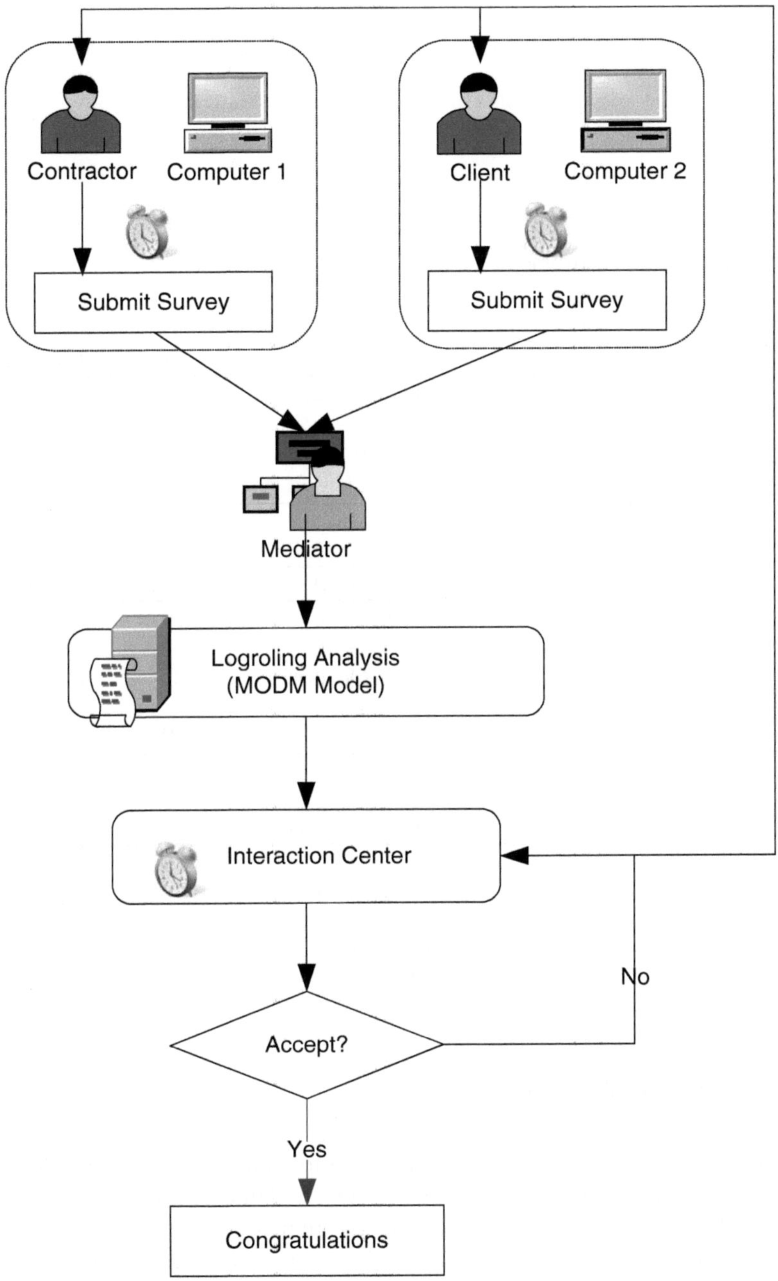

Fig. 20.8 Flow chart of logrolling

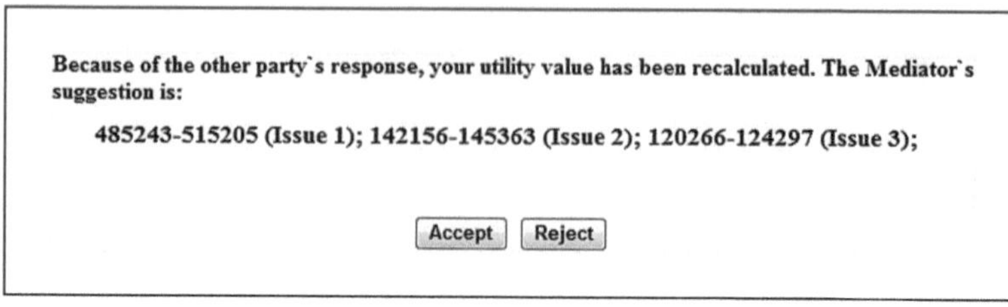

Fig. 20.9 Strategy for contractor on the first round

Table 20.4 Working experience

	Year of experience	No.	Percentage (%)
a.	Below 10 years	50	83.3
b.	10 years	4	6.7
c.	Above 20 years	6	10
	Total	60	100

Table 20.5 Nature of works

	Nature of Works	No.	Percentage (%)
a.	Building	50	83.3
b.	Civil	8	13.3
c.	A and A/Maintenance	2	50
			3.3
	Total	60	100

Table 20.6 Education level

	Education Level	No.	Percentage (%)
a.	Degree	44	73.3
b.	Associate degree	12	3.3
c.	Master	2	20
d.	High diploma	2	3.3
	Total	60	100

20.5 Use of the System

A mock mediation experiment was conducted to examine whether the logrolling system can assist parties to achieve "win–win" settlement, where the system is to serve as a *Mediator* in action. The participants are randomly selected and invited to participate in the experiment. Totally 60 construction practitioners (30 pairs) participated in the experiment. The detailed information in participants' year of experience, nature of work and education level is shown in Tables 20.4, 20.5 and 20.6.

Before the experiment, the subjects were required to fill in the Personal Information Form and were given a hypothetical case *Peter and Brothers Gardening and Landscaping Ltd V ABC Property Management Ltd* and the experiment manual on the use of mediation system. The case is from openly available materials for Arbitration training of the CIArb and it had been adopted and modified to suit Hong Kong context.

The dispute occurred between the Client ABC property Management Limited and the Contractor Peter and Bothers Gardening and Landscaping Limited. Villa Rocha is a private luxury estate in Hong Kong, completed in June 2006. On 16th April 2008 Client engaged a Contractor. The contract price agreed was 2,500,000 Hong Kong dollars. The Client paid a deposit of 1,250,000 Hong Kong dollars and the Contractor started work. On 2nd June 2008, the Client contended the work of the contractor was of poor quality and workmanship. The Client therefore refused to pay the balance and claimed 1,130,000 dollars for remedy. On 7th July 2008, the Contractor issued a Mediation Notice. There are 3 dispute issues, fish pond, turf area and glass house.

20.5.1 Fish Pond

The defects of the fish pond are the location and the leakage. From the *Client*'s statement the *Contractor* transposed the diagram. The fish pond is marked on the right, but was constructed on the left. However from the *Contractor's* statement the wrong location is due to the ambiguous sketch. And they pointed out that the left side is a perfect location for fish since that side is shadier. As for the leakage, the *Client* emphasised the terrible situation that the pond was empty in the next morning even if it was fully filled with water the night before. But the *Contractor* argued that the leakage was really minor. They also failed to agree on the type of remedial work. The *Client* claimed HK$750,000 (claim form) for moving the pond to the correct position and replacing the lining. The *Contractor* responded HK$450,000 (response form) for the 'extra' work.

20.5.2 Turf Area

From the *Client's* statement, the turf was bared and had not been properly laid. A compensation of HK$200,000 was demanded by the *Client*. While the *Contractor* contends that the damage was caused by the *Client's* car, a request of HK$120,000 for re-doing the turfing was raised. The *Client* asked the *Contractor* to come around and see what had happened. But the *Contractor* refused to come unless the *Client* agreed to pay the bill in full.

Table 20.7 Case summaries

	Client		Contractor	
	Defects	Damages	Defences	Offer
Fish Pond	(1) Wrong location	HK$750,000	(1) Wrong location was because of the ambiguous sketch.	HK$450,000
	(2) Leakage		(2) Leakage was minor.	
Turf area	Bared	HK$200,000	Patches resulted from unplanned use of car parking	HK$120,000
Glass House	(1) Leakage	HK$180,000	(1) Leakage was because of the use of inappropriate adhesive chosen by the *Client*.	HK$110,000
	(2) Cloudy pane		(2) Cloudy pane was caused by the *Client*'s incorrect cleaning.	
			(3) Addition erection work cost 3 more man days.	
Total		HK$1,130,000		HK$680,000

20.5.3 Glass House

The *Client* claimed HK$180, 000 for the leakage and cloudy pane of the newly constructed glass house. The *Contractor* contended that the cloudy pane was caused by the *Client*'s incorrect cleaning, and the leakage was because of the poor quality of adhesive selected by the *Client*. Another *Contractor's* defense point is the additional installation work. From the *contractor's* statement the prices in the agreement of 16th April 2008 were estimates and subject to change if any of the work proved to be more or less difficult than anticipated. The complicated installation work cost more man days than anticipated. Based on these, the *Contractor* only offered HK$110,000 for *Client*'s damages (Table 20.7).

In the experiment, the dyad was randomly assigned to the roles of Client or Contractor, and each dyad was told not to speak with each other face to face in the experiment. The subjects are required to generate bargaining range and reach agreement using the logrolling system.

The logrolling system can assist subjects to achieve "win–win" settlement, which means that the difference between mediator's expected logrolling outcomes and the subjects' actual logrolling outcomes is not significant. The outcomes suggested by computer-simulated mediator are taken as the mediator's expected outcomes. The outcomes generated by subjects are taken to be the subjects' actual outcomes. The logrolling-difference degree (L-DD) is defined as difference between the two and to be used for the measure. The smaller the L-DD, the closer are the actual outcomes to the efficient frontier.

The four steps for calculating logrolling-difference degree (L-DD) are described as follows:

Step 1: Collect the data of expected logrolling outcomes and actual logrolling outcomes. The data are then used to plot curves on bargaining range and reaching agreement respectively (see Figs. 20.10 and 20.11).

Set x_1 as the value on the curve of expected logrolling outcomes

Set y_1 as the value on the curve of expected logrolling outcomes where $x = x_1$

Set as the value on the curve of actual logrolling outcomes

Set y_2 as the value on the curve of actual logrolling outcomes where $x = x_2$

Step 2: Calculate the corresponding mapping points on two curves respectively. Since the points on two curves may be both different in dimensions x and y, this mapping step is to unify the value in dimensions x, so that the data on two curves can be comparable for comparison in dimension y.

Set y_1's mapping value on the curve of actual logrolling outcomes where $x = x_1$

Set y_2's mapping value on the curve of expected logrolling outcomes where $x = x_2$

Step 3: Calculate L-DD based on points in two curves in dimension y. The L-DD on each pair of points is calculated as follows:

$$L - DD_{\text{pointi}} = \frac{|y_2 - y_1|}{Maximum(y_1, y_2)} \tag{20.8}$$

Step 4: Evaluate the average L-DD.

$$L - DD = \frac{1}{n} \times \sum_{i=1}^{n} L - DD_{\text{pointi}} \tag{20.9}$$

Taking group 1 as an example, the logrolling outcome on bargaining range and reaching agreement are shown in Figs. 20.10 and 20.11.

The L-DD between *Subjects*' actual logrolling outcomes and *Mediator*'s expected logrolling outcomes of group 1 in bargaining range task is 5.75 %, and 3.08 % in reaching agreement task. It is found that the *subjects*' actual logrolling curve is fitted with the *Mediator*'s expected curve (shown as Fig. 20.11). It reflects that the logrolling system can help user to make rational decision. Repeating the same procedures as for bargaining range, the L-DD of reaching agreement was obtained. It was found that this L-DD is smaller than task of bargaining range, since subjects can achieve more efficient trade-off in the logrolling process, assisted with the computer-simulated mediator.

The number of bargaining rounds and logrolling-difference degree (L-DD) between mediator's expected logrolling outcomes and subjects' actual logrolling outcomes for 30 groups are calculated in Table 20.8. It is found that the average of

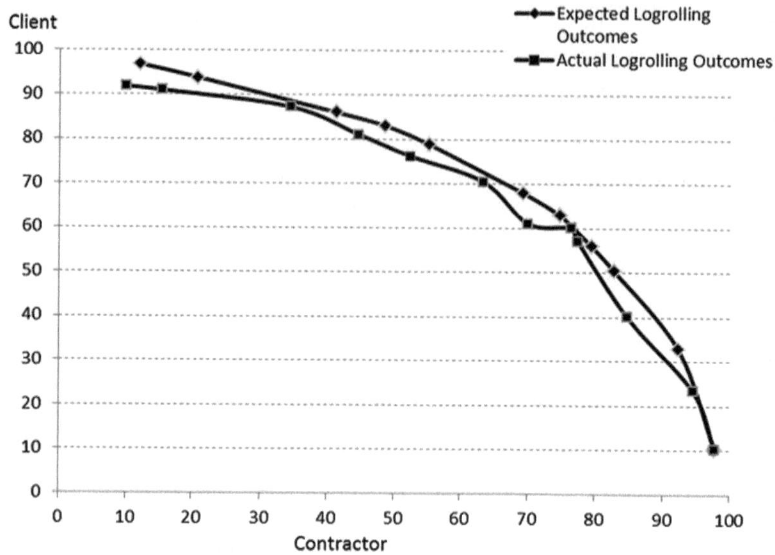

Fig. 20.10 Logrolling result in bargaining range

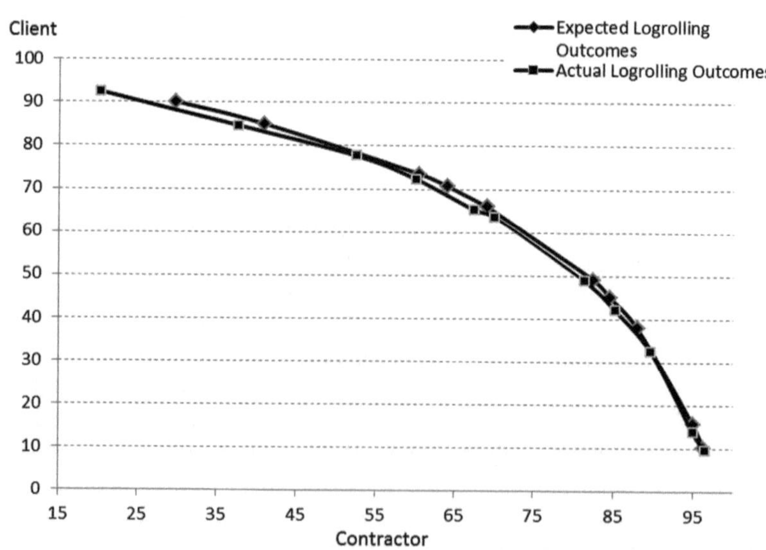

Fig. 20.11 Logrolling result in reaching agreement

Table 20.8 Experiment results

Group No.	Bargaining range task		Reaching agreement task	
	bargaining rounds	Logrolling-difference degree (%)	Bargaining rounds	Logrolling-difference degree (%)
1	7	5.75	7	3.08
2	7	5.79	7	11.03
3	6	10.97	6	11.94
4	6	12.49	6	11.19
5	7	9.31	6	15.41
6	6	20.95	6	10.87
7	7	20.93	6	14.68
8	7	9.45	7	2.76
9	6	12.86	7	5.86
10	6	11.43	6	4.86
11	6	8.05	7	5.69
12	6	16.59	7	7.88
13	7	4.29	6	1.70
14	7	30.23	7	16.32
15	7	8.52	7	13.09
16	7	10.87	7	9.85
17	7	19.79	7	7.60
18	6	10.35	6	7.21
19	6	8.90	6	9.82
20	6	2.01	6	5.57
21	6	12.56	6	3.67
22	6	2.71	6	10.48
23	7	16.49	7	9.70
24	7	5.00	7	18.34
25	6	3.32	6	10.31
26	6	17.83	6	1.31
27	6	21.54	6	8.77
28	6	5.12	7	3.36
29	6	3.69	6	5.86
30	7	15.15	6	5.65
Mean		**11.43**		**8.46**
SD		**6.80**		**4.44**

L-DD in bargaining range and reaching agreement are 11.43 % and 8.46 % respectively, which indicates that the logrolling system is having good potential in assisting parties to achieve "win–win" settlement. Figures. 20.12 and 20.13 show the fluctuation of the L-DD among the 30 groups in bargaining range and reaching agreement respectively. The fluctuation in bargaining range task is obviously greater. It is also found that 57 % subjects fulfilled the experiment within 6 bargaining rounds, while 43 % subjects needed 7 bargaining rounds to complete. The experiment test provides support for the hypothesis that the logrolling system is having good potential in assisting parties to achieve "win–win" settlement.

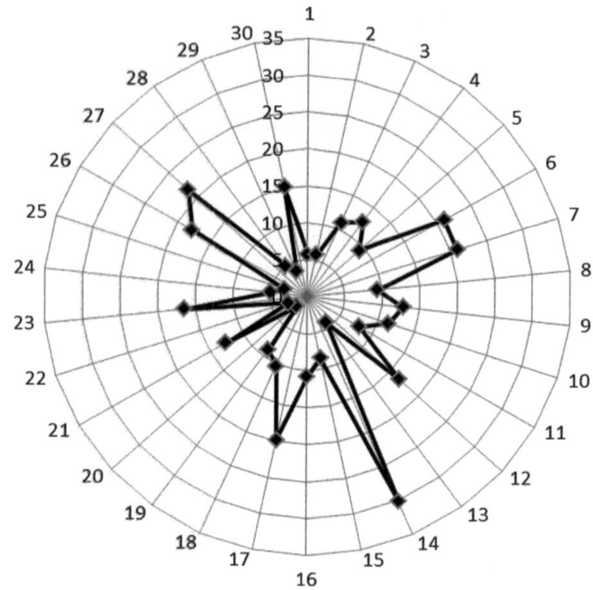

Fig. 20.12 Fluctuation of logrolling-difference degree in bargaining range

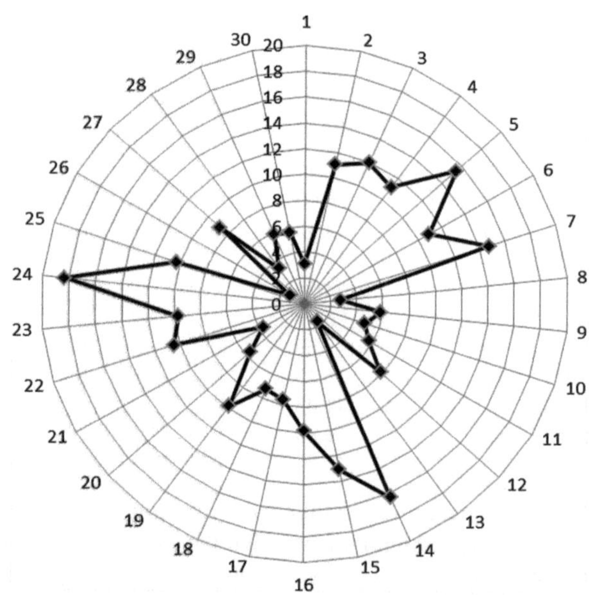

Fig. 20.13 Fluctuation of logrolling-difference degree in reaching agreement

20.6 Discussion

In the last two decades, negotiation support system (NSS) and e-negotiation system (ENS) have been widely applied in conflict resolution. For example, they can help users to understand and formalise the objectives and preferences (Bronisz et al. 1988; Hämäläinen and Pöyhönen 1996; Jarke and Jelassi 1987; Teich et al. 1995). They can facilitate the parties to communicate, store and access bargaining information (Hämäläinen et al. 2001; Kersten and Noronha 1999; Maes et al. 1999; Schoop and Quix 2001; Weigand et al. 2003; Yuan et al. 2003). The logrolling system also put a great deal of effort in its practical exploration in dispute resolution. Reaching "win–win" settlement is the desired outcome of mediation. A "win–win" settlement can be seen as one that encourages parties to uphold their contracts when one party achieve its profits and the other party would still be better off (White 2009). However parties negotiating face to face often have difficulty in identifying and realising "win–win" settlement (Neale and Bazerman 1991; Pruitt 1981; Sebenius 1992). Logrolling is a solution for achieving integrative trade-off, by which each party concedes on low priority issues in exchange for concessions on issues of higher priority to them (Lax and Sebenius 1986; Neale and Bazerman 1991; Pruitt and Rubin 1986). The model of logrolling in negotiation is closely related to the concept of efficient frontier. In economics, "frontier" is where alternative is worse than what they could achieve (Mas-Colell et al. 1995). Thus in this study, a logrolling strategy is proposed through which parties can improve the joint value by bargaining exchange and get convergence along the efficient frontier. A multi-objective decision making (MODM) model is employed to simulate the efficiency frontier and assist parties to engender "win–win" settlement. The model is test and applied in a two-party, three-issue case reported by Cheung et al. (2004) in a project management environment. It is found that the MODM model is not only a generator of optimal solution, but also specifies a logrolling solution path for negotiators to achieve "win–win" settlement. To operationalise the logrolling strategy, a web-based logrolling system has been developed to assist parties to achieve "win–win" settlement in a user-friendly environment. The logrolling system can help parties to make efficient trade-off, by suggesting (1) when to concede (2) on which issue (3) for which party and (4) how much should be conceded. Finally, a mock mediation experiment was conducted to examine the performance of the logrolling system, where the system is to serve as a Mediator in action. The experiment test provides support for the hypothesis that the logrolling system is having good potential in assisting parties to achieve "win–win" settlement. The mediator can also use the suggestion of the 'computer-simulated mediator' to help negotiators to facilitate the "win–win" settlement in the actual mediation process. Furthermore, the logrolling system can also be used for mediation training. By completion of the training, the trainee will be able to demonstrate: (a) they have acquired knowledge on the theoretical framework of logrolling process in mediation and two critical components, (b) they can analyse and explain logrolling model and logrolling strategy, and (c) they have the ability of putting the theories they learnt into practice on making use of the logrolling system to generate an optimal bargaining range and reach a "win–win" agreement.

There are several limitations in the use of the proposed logrolling system. In this study the logrolling system follows the logrolling assumption that high-priority issues of one party are of low-priority to the other party and vice versa. However bargaining situations can be more complex than this. Secondly, the proposed system did not consider the difference in users' risk preference, since the utility value of the MODM model largely depends on the people's risk preference which involving risk aversion (loss aversion), risk seeking, risk neutral, or combination of the above. Incorporating with risk attitude can strengthen and improve the performance of logrolling system in assisting parties to achieve "win–win" settlement. Furthermore the system in this chapter is in one-group and two-party mode. The number of parties in one group depends on the case situation, however in this mode, the experiment has to be done pair by pair, which definitely increased the time cost. As the future work, the system can be extended in n-group and n-party mode that groups of subjects can use the system simultaneously. To solve it, the database access technology which is based on table retrieval in MySQL connection currently, should be replaced by "view" technique, which is a virtual relation of data tables to improve the efficiency of data operations. For supporting more users mediating online by logrolling system at the same time and reduce data processing, internet information services (IIS) could be enhanced by technical optimisation, for example, larger data cache size and more resources allocation. Besides, this platform can also be extended and immigrated as a web service, which is a software system designed to support interoperable machine-to-machine interaction over a network. In this way, the users can use smart devices, such as mobile phone, other than just personal computer to use the logrolling system promptly and conveniently.

20.7 Chapter Summary

Reaching "win–win" settlement is the desired outcome of mediation. A "win–win" settlement can be seen as one that encourages parties to uphold their contracts when one party achieve its profits and the other party would still be better off. However this cannot be easily achieved. Logrolling is a strategy for achieving "win–win" trade-off, by which each party concedes on low priority issues in exchange for concessions on issues of higher priority to them. In this study, the logrolling strategy in mediation is proposed through which parties can improve the joint value by bargaining exchange and get convergence along the efficient frontier. A multi-objective decision making (MODM) model is employed to propose the efficient frontier and assist parties to engender "win–win" settlement. To operationalise the logrolling strategy, a web-based logrolling system is developed to assist parties to achieve "win–win" settlement in a user-friendly environment. The logrolling system can help parties to make efficient trade-off, by suggesting (1) when to concede (2) on which issue (3) for which party and (4) how much should be conceded. Finally a mock mediation experiment was conducted to examine whether the logrolling system can assist parties to achieve "win–win" settlement, where the system simulates a *Mediator* in action. The results are evaluated by comparing the difference between the mediator's expected logrolling outcomes and the subjects' actual logrolling outcomes. The logrolling-difference degree (L-DD) is used to measure this difference. It is found that the average of L-DD in bargaining range and reaching agreement are 11.43 % and 8.46 % respectively, which indicates that the logrolling system is having good potential in assisting parties to achieve "win–win" settlement.

Acknowledgments The content of this chapter has been published in Volume 24 of Automation in Construction and is used with the permission of Elsevier.

References

Bichle, M., Kersten, G., & Strecker, S. (2003). Towards a structured design of electronic negotiations. *Group Decision and Negotiation, 12*(4), 311–335.

Boulle, L., & Nesic, M. (Eds.). (2001). *Mediation—principles process and practice*. London: Butterworths.

Bronisz, P., Krus, L. & Wierzbicki, A. P. (1988). Towards interactive solutions in a bargaining problem. In A. Lewandowski & A. P. Wierzbicki (Eds.), *Aspiration based decision support systems*. Berlin: Springer.

Brosnan, S. F., & de Waal, F. B. M. (2003). Monkeys reject unequal pay. *Nature, 425*, 297–299.

Chau, K. W. (2007). Insight into resolving construction disputes by mediation/adjudication in Hong Kong. *Journal of Professional Issues in Engineering Education and Practice, 133*(2), 143–147.

Cheung, S. O., Yiu, K. T. W., & Suen, H. (2004). Construction negotiation online. *Journal of Construction Engineering and Management, 130*(6), 844–852.

De Dreu, C. K. W., Carnevale, P. J., Emans, B. J. M., & van de Vliert, E. (1994). Effects of gain-loss frames in negotiation: Loss aversion, mismatching, and frame adoption. *Organizational Behavior and Human Decision Processes, 60*(1), 90–107.

Druckman, D., Druckman, J. N., & Arai, T. (2004). e-mediation: Evaluating the impacts of an electronic mediator on negotiation behavior. *Group Decision and Negotiation, 13*(6), 481–511.

Hämäläinen, R. P., & Pöyhönen, M. (1996). On-line group decision support by preference programming in traffic planning. *Group Decision and Negotiation, 5*, 485–500.

Hämäläinen, R. P., Kettunen, E., Marttunen, M., & Ehtamo, H. (2001). Evaluating a framework for multi-stakeholder decision support in water resources management. *Group Decision and Negotiation, 10*(4), 331–353.

Insua, D. R., Holgado, J., & Moreno, R. (2003). Multicriteria e-negotiation systems for e-democracy. *Journal of Multi-Criteria Decision Analysis, 12*(3), 213–218.

Jarke, M., & Jelassi, M. T. (1987). MEDIATOR: Toward a negotiation support system. *European Journal of Operational Research, 31*(3), 314–334.

Kahneman, D. (1979). Prospect theory: An analysis of decision under risk. *Econometrica, 47*(2), 263–291.

Keeney, R. L., & Raiffa, H. (1991). Structuring and analyzing values for multiple-issue negotiations. In H. Peyton Young, H (Ed.) *Negotiation analysis*. Ann Arbor: University of Michigan Press.

Kersten, G. E. (2001). Modeling distributive and integrative negotiations—Review and revised characterization. *Group Decision and Negotiation, 10*(6), 493–514.

Kersten, G. E., & Noronha, S. J. (1999). www-based negotiation support: Design, implementation, and use. *Decision Support Systems, 25*(1), 135–154.

Korhonen, P., Moskowitz, H., Wallenius, J., & Zionts, S. (1986). An interactive approach to multiple criteria optimization with multiple decision-makers. *Naval Research Logistics Quarterly, 33*(4), 3–29.

Lax, D. A., & Sebenius, J. K. (Eds.). (1986). *The manager as negotiator: bargaining for cooperation and competitive gain*. New York: Free Press.
Lim, L. H., & Benbasat, I. (1993). A theoretical perspective of negotiation support systems. *Journal of Management Information Systems, 9*(3), 27–44.
Maes, P., Guttman, R. H., & Moukas, A. G. (1999). Agents that buy and sell: Transforming commerce as we know it. *Communications of the ACM, 42*(3), 81–91.
Mas-Colell, A., WhinstonM, D., & Green, J. R. (Eds.). (1995). *Microeconomic theory*. New York: Oxford University Press.
Neale, M. A., & Bazerman, M. H. (Eds.). (1991). *Cognition and rationality in negotiation*. New York: Free Press.
Northcraft, G. B., Brodt, S. E., & Neale, J. L. (1995). The prospect of negotiating with non-linear utilities: Why some concessions are more equal than other. *Organizational Behavior and Human Decision Processes, 63*(3), 298–310.
Northcraft, G. B., Preston, J. N., Neale, J. L., Kim, P. H., & Thomas-Hunt, M. C. (1998). Non-linear preference functions and negotiated outcomes. *Organizational Behavior and Human Decision Processes, 73*(1), 54–75.
Nowak, M. A., Page, K. M., & Sigmund, K. (2000). Fairness versus reason in the ultimatum game. *Science, 289*, 1773–1775.
Pennings, J. M. E., & Smidts, A. (2003). The shape of utility functions and organizational behavior. *Management Science, 49*(9), 1251–1263.
Pruitt, D. G. (Ed.). (1981). *Negotiation behavior*. New York: Academic Press.
Pruitt, D. G., & Rubin, J. Z. (Eds.). (1986). *Social conflict: Escalation, stalemate and settlement*. New York: Random House.
Raiffa, H. (1985). Post-settlement settlements. *Negotiation Journal, 1*(1), 9–12.
Raiffa, H. (Ed.) (1982). *The Art and Science of Negotiation*. Cambridge: Belknap Press.
Raiffa, H., Richardson, J., & Metcalfe, D. (Eds.). (2002). *Negotiation analysis: The science and art of collaborative decision making* (pp. 348–364). Cambridge: Belknap Press of Harvard University Press.
Rangaswamy, A., & Shell, G. R. (1997). Using computers to realize joint gains in negotiations: Toward an electronic bargaining table. *Management Science, 43*(8), 1143–1163.
Schoop, M., & Quix, C. (2001). Doc.com: A framework for effective negotiation support in electronic commerce. *Computer Networks, 37*(2), 153–170.
Sebenius, J. K. (1992). Negotiation analysis: A characterization and review. *Management Science, 38*(1), 18–38.
Teich, J. E., Wallenius, H., Kuula, M., & Zionts, S. (1995). A decision support approach for negotiation with an application to agricultural income policy negotiations. *European Journal of Operational Research, 81*(1), 76–87.
Thompson, L. L. (1990). Negotiation behavior and outcomes: Empirical evidence and theoretical issues. *Psychological Bulletin, 108*(3), 515–532.
Thompson, L. L., & Hastie, R. (1990). Negotiator's perceptions of the negotiation process. In B. H. Sheppard, M. H. Bazerman, & R. J. Lewicki (Eds.), *Research in negotiation in organizations*. Greenwich: JAI Press.
Weigand, H., Schoop, M., De Moor, A., & Dignum, F. (2003). B2B negotiation support: the need for a communication perspective. *Group Decision and Negotiation, 12*(1), 3–29.
White, M. D. (Ed.). (2009). *Theoretical foundations of law and economics*. Cambridge: Cambridge University Press.
Yuan, Y., Head, M., & Du, M. (2003). The effects of multimedia communication on web-based negotiation. *Group Decision and Negotiation, 12*(2), 89–109.

MIX
Papier aus verantwortungsvollen Quellen
Paper from responsible sources
FSC® C105338

If you have any concerns about our products,
you can contact us on
ProductSafety@springernature.com

In case Publisher is established outside the EU,
the EU authorized representative is:
Springer Nature Customer Service Center GmbH
Europaplatz 3, 69115 Heidelberg, Germany

Printed by Libri Plureos GmbH
in Hamburg, Germany